Algorithms for Intelligent Systems

This book series publishes research on the analysis and development of algorithms for intelligent systems with their applications to various real world problems. It covers research related to autonomous agents, multi-agent systems, behavioral modeling, reinforcement learning, game theory, mechanism design, machine learning, meta-heuristic search, optimization, planning and scheduling, artificial neural networks, evolutionary computation, swarm intelligence and other algorithms for intelligent systems.

The book series includes recent advancements, modification and applications of the artificial neural networks, evolutionary computation, swarm intelligence, artificial immune systems, fuzzy system, autonomous and multi agent systems, machine learning and other intelligent systems related areas. The material will be beneficial for the graduate students, post-graduate students as well as the researchers who want a broader view of advances in algorithms for intelligent systems. The contents will also be useful to the researchers from other fields who have no knowledge of the power of intelligent systems, e.g. the researchers in the field of bioinformatics, biochemists, mechanical and chemical engineers, economists, musicians and medical practitioners.

The series publishes monographs, edited volumes, advanced textbooks and selected proceedings.

More information about this series at http://www.springer.com/series/16171

C. Kiran Mai · A. Brahmananda Reddy ·
K. Srujan Raju
Editors

Machine Learning Technologies and Applications

Proceedings of ICACECS 2020

Springer

Editors
C. Kiran Mai
Department of Computer Science and Engineering
VNR VJIET
Hyderabad, Telangana, India

A. Brahmananda Reddy
Department of Computer Science and Engineering
VNR VJIET
Hyderabad, Telangana, India

K. Srujan Raju
Department of Computer Science Engineering
CMR Technical Campus
Hyderabad, Telangana, India

ISSN 2524-7565 ISSN 2524-7573 (electronic)
Algorithms for Intelligent Systems
ISBN 978-981-33-4045-9 ISBN 978-981-33-4046-6 (eBook)
https://doi.org/10.1007/978-981-33-4046-6

This Springer imprint is published by the registered company Springer Nature Singapore Pte Ltd.
The registered company address is: 152 Beach Road, #21-01/04 Gateway East, Singapore 189721, Singapore

Team ICACECS-2020

Patrons

Dr. D. N. Rao, President, Vignana Jyothi Society
Sri K. Harishchandra Prasad, General Secretary, Vignana Jyothi Society

Chief Patrons

Dr. C. D. Naidu, Principal, VNR VJIET
Dr. B. Chennakesava Rao, Director-Advancement, VNR VJIET
Dr. A. Subhananda Rao, Director-Research & Development, VNR VJIET

Programme Chairs

Dr. C. Kiranmai, VNR VJIET, Hyderabad, India
Dr. B. V. Kiranmayee, VNR VJIET, Hyderabad, India

Programme Co-chairs

Dr. A. Brahmananda Reddy, VNR VJIET, Hyderabad, India
Dr. Deepak Sukheja, VNR VJIET, Hyderabad, India
Dr. P. Subhash, VNR VJIET, Hyderabad, India

General Chair

Dr. Raghava Rao Mukkamala, CBDA, CBS, Denmark

Honorary Chairs

Dr. Lakhmi C. Jain, University of Technology Sydney, Australia and Founder, KES International
Dr. A. Govardhan, Rector, JNTUH, Hyderabad, India

Editorial Board

Dr. C. Kiranmai, VNR VJIET, Hyderabad, India
Dr. A. Brahmananda Reddy, VNR VJIET, Hyderabad, India
Dr. K. Srujan Raju, CMRTC, Hyderabad, India

Programme Committee

Dr. G. Ramesh Chandra, CSE, VNR VJIET
Dr. S. Nagini, CSE, VNR VJIET
Dr. P. V. Siva Kumar, CSE, VNR VJIET
Mrs. V. Baby, CSE, VNR VJIET
Dr. T. Sunil Kumar, CSE, VNR VJIET
Dr. Y. Sagar, CSE, VNR VJIET
Mr. M. Gangappa, CSE, VNR VJIET
Dr. P. Subhash, CSE, VNR VJIET

Advisory Committee

Dr. Aynur Unal, Director, Member of the Executive Team, UK
Dr. Pawan Lingras, Saint Mary's University, Canada
Dr. Raghava Rao Mukkamala, CBDA, CBS, Denmark
Dr. Rajeev Srivastava, IIT BHU, India
Dr. C. Krishna Mohan, IIT, Hyderabad, India
Dr. N. S. Choudary, IIT, Indore, India
Dr. P. Premchand, Osmania University, Hyderabad, India

Dr. Rajkamal, Former-VC, DAVV, Indore, India
Dr. R. B. V. Subramanyam, NITW, Warangal, India
Dr. O. B. V. Ramanaiah, JNTUH, Hyderabad, India
Dr. K. P. Supreethi, JNTUH, Hyderabad, India
Dr. Sujatha Banothu, Osmania University, Hyderabad, India
Dr. I. L. Narasimha Rao, Cyber Peace Foundation, New Delhi, India
Dr. Y. Padmasai, VNR VJIET, Hyderabad, India
Dr. G. Ramesh Chandra, VNR VJIET, Hyderabad, India
Dr. G. Suresh Reddy, VNR VJIET, Hyderabad, India
Dr. K. Anuradha, VNR VJIET, Hyderabad, India
Dr. Poonam Upadhyay, VNR VJIET, Hyderabad, India
Dr. R. Manjula Sri, VNR VJIET, Hyderabad, India
Dr. A. Mallika, VNR VJIET, Hyderabad, India
Dr. G. Srinivasa Gupta, VNR VJIET, Hyderabad, India
Dr. T. Srinivasa Rao, VNR VJIET, Hyderabad, India
Dr. T. Jayashree, VNR VJIET, Hyderabad, India
Dr. Sankar K. Pal, Indian Statistical Institute, Kolkata, India
Dr. Krithi Ramamritham, IIT Bombay, India
Dr. Abdul Sattar, Griffith University, Australia
Dr. N. Mangathayaru, VNR VJIET, Hyderabad, India
Anil Sukheja, Scientist "E" ISRO Ahmedabad, India

Technical Committee

Dr. Umesh Kumar Singh, Vikram University, Ujjain, India
Dr. Padmaja Joshi, CDAC, India
Dr. Pilli Emmanuel Subhakar, MNIT, Jaipur, India
Dr. M. B. Srinivas, BITS Pilani, Hyderabad, India
Dr. K. Seetha Ram Babu, Chairman CSI-Hyderabad Chapter, India
Dr. A. V. Krishna Prasad, Secretary CSI-Hyderabad Chapter, India
Dr. D. Vasumathi, JNTUH, Hyderabad, India
Dr. A. P. Siva Kumar, JNTUCEA, Andhra Pradesh, India
Dr. T. V. Rajanikanth, SNIST, Hyderabad, India
Dr. Dolly Sharma, Shiv Nadar University, UP, India
Dr. Virender Saraswat, Dr. B. R. Ambedkar University, Agra, India
Dr. T. Kishore Kumar, NIT Warangal, India
Dr. B. Vishnu Vardhan, JNTU Mantani, Peddapalli, India
Dr. N. Sandhya, VNR VJIET, Hyderabad, India
Dr. P. Neelakantan, VNR VJIET, Hyderabad, India
Dr. K. Vijaya Kumar, CMRCET, Hyderabad, India
Dr. P. Vijaya Pal Reddy, Matrusri Engineering College, Hyderabad, India
Dr. M. Raja Sekar, VNR VJIET, Hyderabad, India
Dr. G. L. Prajapat, IET, Devi Ahilya University, Indore

Dr. Parag Parandkar, REVA University, Bengaluru, India
Dr. S. Nagini, VNR VJIET, Hyderabad, India
Ms. V. Baby, VNR VJIET, Hyderabad, India
Dr. P. V. Siva Kumar, VNR VJIET, Hyderabad, India
Dr. D. Raman, VCE, Shamshabad, Hyderabad, India
Dr. T. Sunil Kumar, VNR VJIET, Hyderabad, India
Dr. Y. Sagar, VNR VJIET, Hyderabad, India
Dr. P. Subhash, VNR VJIET, Hyderabad, India
Dr. K. Srinivas, VNR VJIET, Hyderabad, India
Dr. A. Kousar Nikhath, VNR VJIET, Hyderabad, India
Dr. Niranjan Shrivastava, IMS DAVV, Indore, India
Dr. Uttam Kumar Roy, Jadavpur University, Kolkata, India
Dr. D. Srinivasa Rao, VNR VJIET, Hyderabad, India
Dr. G. Madhu, VNR VJIET, Hyderabad, India
Dr. K. Srinivas, VNR VJIET, Hyderabad, India
Dr. Chalumuru Suresh, VNR VJIET, Hyderabad, India
Dr. P. Kishore, VNR VJIET, Hyderabad, India
Dr. S. Rajendra Prasad, VNR VJIET, Hyderabad, India
Dr. C. Kiran, VNR VJIET, Hyderabad, India
Dr. S. Pranavanand, VNR VJIET, Hyderabad, India
Dr. D. Ravi Kumar, VNR VJIET, Hyderabad, India
Ms. D. N. Vasundhara, VNR VJIET, Hyderabad, India
Mr. G. S. Ramesh, VNR VJIET, Hyderabad, India
Ms. Y. Bhanusree, VNR VJIET, Hyderabad, India
Mr. G. Nagaraju, VNR VJIET, Hyderabad, India
Ms K. Jhansi Lakshmi Bai, VNR VJIET, Hyderabad, India
Mr. R. Ranthi Kumar, VNR VJIET, Hyderabad, India
Ms. L. Indira, VNR VJIET, Hyderabad, India
Ms. Priyabhatnagar, VNR VJIET, Hyderabad, India
Ms. Krithi Ohri, VNR VJIET, Hyderabad, India

Organizing Committee

Dr. C. Kiran Mai, CSE, VNR VJIET
Dr. B. V. Kiranmayee, HOD-CSE VNR VJIET
Dr. A. Brahmananda Reddy, CSE, VNR VJIET
Dr. Deepak Sukheja CSE, VNR VJIET
Dr. P. V. Siva Kumar, CSE, VNR VJIET
Dr. P. Subhash, CSE, VNR VJIET
N. V. Sailaja, CSE, VNR VJIET
D. N. Vasundara, CSE, VNR VJIET
G. S. Ramesh, CSE, VNR VJIET
P. Radhika, CSE, VNR VJIET

G. Nagaraju, CSE, VNR VJIET
R. Kranthi Kumar, CSE, VNR VJIET
Priya Bhatnagar, CSE, VNR VJIET
Kriti Ohri, CSE, VNR VJIET
V. Hareesh, CSE, VNR VJIET

Publication and Proceedings Committee

Dr. G. Ramesh Chandra, CSE, VNR VJIET
Dr. P. Subhash, CSE, VNR VJIET
Dr. K. Srinivas, CSE, VNR VJIET
Ms. Chenna Basamma, CSE, VNR VJIET

Review Committee

Dr. Mohammed Shafiul Alam Khan, Royal Holloway, University of London, UK
Dr. Hikmat Ullah Khan, COMSATS Institute of Information Technology, Wah Campus, Islamabad, India
Dr. Radhakrishnan Palanikumar, King Khalid University, Saudi Arabia
Dr. Kanwalinderjit Gagneja, Florida Polytechnic University, USA
Dr. Ravi Vadapalli, HPCC, Texas Tech University, USA
Dr. Jagdish Shivhare, Distinguished Scientist and Member-IRBA, USA
Dr. Junyuan Zeng, The University of Texas at Dallas, USA
Dr. Neeraj Mittal, The University of Texas at Dallas, USA
Dr. G. L. Prajapati, IET, Devi Ahilya University, Indore, India
Dr. Vivek Tiwari, IIIT Raipur, India
Dr. Muralidhar Kulkarni, NIT Surathkal, Mangalore, Karnataka, India
Dr. S. Raghavan, NIT Tiruchirappalli, India
Dr. T. Kishore Kumar, NIT Warangal, India
Dr. K. P. Supreethi, JNTUH, Hyderabad, India
Dr. Sujatha Banothu, Osmania University, Hyderabad, India
Dr. T. V. Rajini Kanth, Sreenidhi Institute of Science and Technology, Hyderabad, India
Dr. Amjan Shek, BVRIT, Hyderabad, India
Dr. P. Vijyapal Reddy, Matrusri Engineering College, Hyderabad, India
Dr. Aynur Unal, Palo Alto, CA, USA
Dr. Manu Pratap Singh, Dr. Bhimrao Ambedkar University, Agra, India
Dr. V. K. Saraswat, Dr. Bhimrao Ambedkar University, Agra, India
Dr. Halavath Balaji, Sreenidhi Institute of Science and Technology, Hyderabad, India
Dr. D. Stalin Alex, Guru Nanak Institute of Technology, Hyderabad, India
Dr. Nidhi Arora, SAGE University, Indore, India

Dr. N. Rajasekhar, GRIET, Hyderabad, India
Dr. Sanjeev Tokeykar, IET, Devi Ahilya Vishwavidyalaya, Indore, India
Dr. B. T. Krishna, UCEV, JNTUK, AP, India
Dr. A. Narendra Babu, LBRCE, Mylavaram, AP, India
Dr. S. P. Singh, Mahatma Gandhi Institute of Technology, Hyderabad, India
Dr. Yatendra Nath Singh, IIT Kanpur, India
Dr. Atul Negi, University of Hyderabad, India
Dr. Ganapati Panda, IIT Bhubaneswar, India
Dr. Kuldeep Kumar, Birla Institute of Technology and Science, Pilani, India
Dr. N. Rama Subramanian, NIT Tiruchirappalli, India
Dr. L. Anjaneyulu, NIT Warangal, India
Dr. P. Sreehari Rao, NIT Warangal, India
Dr. Saugata Sinha, VNIT Nagpur, India
Dr. Prabhat Kumar Sharma, VNIT Nagpur, India
Dr. Dipankar Pal, BITS Pilani, K. K. Birla Goa Campus, India
Dr. B. Ravindhar Reddy, Annamacharya Institute of Technology and Sciences, India
Dr. M. Chinna Rao, Srinivasa Institute of Engineering and Technology
Mr. Sachin Chirgaiya, SVVV, Indore, India
Dr. Jitendra Choudhary, Medicaps University, Indore, India
Dr. Dharmendra Choukse, IPS Academy, Indore, India
Dr. Delshi Howsalya, Bharat Institute of Engineering and Technology, Hyderabad,
Dr. Dhirendra Pandey, Babasaheb Bhimrao Ambedkar University, Lucknow, India
Dr. Jayesh Gangrade, IPS Academy, Indore, India
Dr. S. Govinda Rao, GRIET, Hyderabad, India
Reetu Gupta, SDBCT, Indore, India
Dr. Chanchala Joshi, Vikram University, Ujjain, India
Dr. K. Purna Chand, BVRIT Narsapur, Hyderabad, India
Dr. K. Sreekala, Mahatma Gandhi Institute of Technology, Hyderabad, India
Dr. Kamakshaiah, Geethanjali College of Engineering and Technology, Hyderabad, India
Dr. Priyesh Kanungo, DAVV, Indore
Dr. Kousar Nikhath, VNR VJIET, Hyderabad, India
Dr. B. Krishna, Vaagdevi College of Engineering, Warangal, India
Dr. Sheo Kumar, CMR Engineering College, Hyderabad, India
Dr. Korra Lakshman, NIELIT, Aurangabad, India
Dr. Srinivas Madana, Jyothishmathi Institute of Technological Sciences, Karimnagar, India
Dr. M. Mahalakshmi, CMR College of Engineering and Technology, Hyderabad, India
Dr. Bharti Malukani, Prestige Institute of Management and Research, Indore
Dr. Maya Rathore, Christianeminent College, Indore
Dr. Merugu Suresh, CMR College of Engineering and Technology, Hyderabad, India
Dr. N. Vijay, Malla Reddy College of Engineering and Technology, Hyderabad, India
Dr. Sitanshu Sekhar Sahu, BITS Mesra, Ranchi, India
Dr. B. Anuradha, S. V. University College of Engineering, Tirupati, India

Dr. M. Asha Rani, JNTUH, Hyderabad, India
Dr. D. Sreenivasa Rao, JNTUH, Hyderabad, India
Dr. G. Nagaraju, VNR VJIET, Hyderabad, India
Dr. Nagaveni, Acharya Institute of Technology, Bengaluru, India
Dr. Y. Jeevan Nagendra Kumar, GRIET, Hyderabad, India
Dr. A. Narayana Rao, NBKRIST, Nellore, India
Dr. Neelima Vontela, Jyothishmathi Institute of Technology and Science, Karimnagar, India
Dr. Bhupendra Pandya, Vikram University, Ujjain, India
Dr. Parag Parandkar, REVA University, Bengaluru, India
Dr. Subhash Parimalla, VNR VJIET, Hyderabad, India
Dr. Krishna Prasad Ponnekanti, S. V. College of Engineering, Tirupathi, India
Dr. K. Prasanna Lakshmi, GRIET, Hyderabad, India
Dr. Adiraju Prashanth Rao, Anurag University, Hyderabad, India
Dr. V. Prashanthi, GRIET, Hyderabad, India
Mohd. Qayyum, King Khalid University, Abha, Saudi Arabia
Dr. K. Prabhakar Nayak, MIT, Manipal, India
Dr. J. Ramesh, PSG College of Technology, Coimbatore, India
Dr. K. V. Raghavendra, Malla Reddy College of Engineering and Technology, Hyderabad, India
Dr. Vijaya Latha, GRIET, Hyderabad, India
Dr. K. Vijayalakshmi, Sreenidhi Institute of Science and Technology, Hyderabad, India
Dr. Rohit Raja, Sreyas Institute of Engineering and Technology, Hyderabad
Dr. E. Raju, Vaagdevi Engineering College, Warangal, India
Dr. Rakesh Kumar Tripathi, B. M. College, Indore, India
Dr. V. Raman, Vardhaman College of Engineering, Hyderabad, India
Dr. V. Ramesh, Presidency University, Bengaluru, India
Dr. Manish Sahajwani, IPS Academy, Indore, India
Dr. Seethramulu, ICFAI University, Hyderabad, India
Dr. M. Shanmukhi, Mahatma Gandhi Institute of Technology, Hyderabad, India
Dr. M. Sharadha Varalakshmi, St. Peters Engineering College, Hyderabad, India
Dr. Rashid Sheikh, Acropolis Institute of Technology and Research, Indore
Dr. Niranjan Shrivastava, DAVV, Indore, India
Dr. K. Srinivas, VNR VJIET, Hyderabad, India
Dr. Y. Suresh, Ballari Institue of Technology and Management, Ballari, India
Dr. M. Suresh Kumar, Vaagdevi College of Engineering, Warangal, India
Dr. G. Suresh Reddy, VNR VJIET, Hyderabad, India
Dr. R. S. Thakur, MNIT, Bhopal, India
Dr. Umesh Kumar Singh, Vikram University, Ujjain, India
Dr. Subhashini Valluru, MLRIT, Hyderabad, India
Dr. V. Venkateshwarlu, Vaagdevi College of Engineering, Warangal, India
Dr. M. Venugopala Chari, CBIT, Hyderabad, India
Dr. V. Akhila, GRIET, Hyderabad, India
Dr. A. Swarna Bai, Scientist E, RCI, DRDO, Hyderabad, India

Dr. Parag Parandkar, REVA University, Bengaluru, India
Dr. Niranjan Shrivastava IMS DAVV, Indore, India

Web and Publicity Committee

G. S. Ramesh, CSE, VNR VJIET
P. Radhika, CSE, VNR VJIET
R. Kranthi Kumar, CSE, VNR VJIET
S. Kranthi Kumar, CSE, VNR VJIET
N. Sandeep Chaitanya, CSE, VNR VJIET

Preface

Computer Engineering and Communication Systems are entwined more now than at any other time in history. The interplay of Information and Communication Technologies, the rise in Internet of Things (IoT) applications and smart computing, the inroads that technology has taken into personal lives through wearable devices, and so on have significant roles to play in integrating computer engineering with communication systems. International Conference on Advances in Computer Engineering and Communication Systems (ICACECS-2020) is themed majorly focusing on the application of Machine Learning concepts in smart innovations, Industry 4.0 technologies, and data analytics thereby celebrating the emerging technology trends in Computer Engineering and Communication Systems. The Conference is organized as six parallel tracks, viz., Mezzanine Technologies; Big Data and Data Analytics; Cloud, IoT, and Distributed Computing; Smart Systems; Network and Communication Systems; and Education Technology and Business Engineering.

The Primary objective is to encourage National and International communication and collaboration and promote professional interaction and lifelong learning on emerging technologies. The conference consists of keynote lectures, tutorials, workshops, and oral presentations on all aspects of advanced computing and communications. The aim of this International Conference on **Advances in Computer Engineering and Communication Systems (ICACECS)** is to present a unified platform for advanced and multidisciplinary research toward the design of smart computing and information and Communication systems.

It is organized specifically to help the academicians and professionals from IT and communication organizations/industries to derive benefits from the advances of next generation computer and communication technology. The conference provides a platform for experts and researchers from all over the world to discuss contemporary developments and novel concepts in the fields of advanced computing and communication technologies. This International Conference is being organized ONLINE by Computer Science and Engineering Department, VNR VJIET, Bachupally, Nizampet, Hyderabad, from 13 to 14 August 2020.

Quality articles are published in this volume after performing a plagiarism check and a review process by three eminent experts in the respective domain. Eminent Academicians and Researchers are delivering Keynote addresses on contemporary thrust areas. The resource pool is drawn from IITs, NITs, IIITs, IDRBT, and Universities along with Software companies like TCS, Capgemini, Microsoft, Cyber Jagrithi, etc.

A galaxy of 32 eminent personalities is chairing and acting as Conference Jury to review the presentations. The papers are classified into seven tracks which will be delivered ONLINE in 2 days using MICROSOFT TEAMS. The Pre-conference Workshops on 'Machine Learning & Artificial Intelligence', 'Block Chain Technologies', and 'Cyber Security' will be organized on GOOGLE Platform from 12 to 14 August 2020.

The field of Information technology is growing in leaps and bounds. It is worth noting that, on the one hand, the world has realized the absolute need for integrative engineering through the confluence of multiple STEM disciplines, and on the other, Industry 4.0 skills are being identified. We at VNR VJIET look at Industry 4.0 skills as "Mezzanine Technologies", inspired by the multiple functions that a mezzanine floor facilitates in an infrastructure facility. ICACECS-2020 celebrates the rise of these Mezzanine Technologies and Collaborative Innovation. As the name suggests, these technologies fit into the middle of other existing technologies to provide functional enhancement and significant empowerment.

The refereed conference proceedings of ICACECS-2020 are published in this volume. Out of 182 paper submissions from all over India, 32 papers are accepted for being published in this volume, after being reviewed scrupulously. The book volume focuses on thoroughly refereed post-conference enlargements and reviews on the progressive topics in Artificial Intelligence, Machine Learning, and Deep Learning. We assure that we have put in every effort to ensure that the participatory experience in the e-Conference will not feel like a compromise but will add value in its own stride and enable more people to participate. We hope that the Conference served its purpose for the best elucidation and progress of science, engineering, and technology for societal benefit.

Hyderabad, India
August 2020

Dr. C. Kiran Mai
Dr. A. Brahmananda Reddy
Prof. Dr. K. Srujan Raju

Acknowledgments

We would like to acknowledge the support extended by AICTE, by partially funding the event. It furthered and accelerated our thoughts to practice.

We thank all the authors for their contributions and timely response. A special thanks to our reviewers who invested their valuable time and scrupulously evaluated the submissions for the best outcomes.

We express our profound gratitude for the inspiring and informative presentations of our Keynote speakers on the frontline technologies, creating the curiosity to explore more.

The co-operation extended by the Sessions Chair is immense, elevating the presentation skills among the participants.

We would like to thank our Chief Guest and Guest of honor for making it to the e-conference, boosting our spirit to achieve more for providing technology solutions to the Society—Involve and Evolve.

Thanks to the founding members of Vignana Jyothi and their wisdom and social responsibility, we firmly believe that all technology must serve the urgent need for advancing the society, at micro- and/or macro levels.

Our sincere thanks are extended to all the Patrons, Chairpersons, members of the Editorial Board, eminent members of the Programme Committee and Advisory Committee for their guidance and support, and the enthusiastic people among the Technical Committee for their coordination and help in execution.

We would like to extend our appreciation for the amazing work done by our self-reliant and motivated team of VNR VJIET. The amazing dedication and effort of the team enabled us to reach our goal, in spite of the Pandemic.

A special acknowledgement to Ms. Sharmila Mary Panner Selvam, Project Coordinator, Books Production, Springer, for the prompt communication and Support.

Finally, we sincerely thank the Team comprising Prof. Lakshmi C. Jain, and editors of these proceedings, for the constant direction and sustenance.

About the Institute, VNR VJIET

"Vallurupalli Nageswara Rao Vignana Jyothi Institute of Engineering and Technology" (VNR VJIET), an Autonomous Institute, approved by AICTE, affiliated to JNTUH, is accredited by **NAAC with A++, a score of 3.73**, and also 7 UG and 4 PG programmes are accredited by NBA. The institute, since its inception, has charted distinct pathways to academic excellence. It is one of the most distinguished and premier institutions of higher education in the State of Telangana. Its complexity, diversity, and comprehensiveness are a fountainhead of creativity and innovation.

VNR VJIET, a modern world-class academic institute, certified as **College with Potential for Excellence** by **UGC**, has a strong inclination toward sustainable development through research and expansion of innovative technologies. In an era of global progression propelled by technology, research at academic institutes will foster economic growth and help in attaining self-reliance in technology and innovation. The Institute is committed to fundamental long-term research and innovation in leading-edge technologies and performs a diverse and expanded set of activities like

- Producing high-quality engineers with the required skills and knowledge at different levels (undergraduate, postgraduate).
- Exploring new horizons through fundamental research.
- Continuously improving the knowledge repository and domain expertise.
- Encouraging new ideas and proposals through research awards and remuneration.
- Becoming a source for innovation, addressing societal needs, and developing new products leading to revenue generation.

About the Department—Computer Science and Engineering

Computer Science and Engineering (CSE) department was started in the year 1995 and is significant for its ability of research and development activities and excellence in academics. The department is NBA Accredited and is recognized as a Research

Center by JNTUH, the affiliating University. It offers two courses at the Undergraduate level, Computer Science and Engineering (CSE) with an intake of 240 and Computer Science and Business Systems (CSBS) with 60 admitted every year. It also offers two PG Courses in Software Engineering (SE) and Computer Science Engineering (CSE) with an intake of 18 each. The state-of-the-art infrastructure of the department is commendable.

The Department is dynamic and has several publications to its credit listed in Scopus, Web of Science, SCI indexed, etc. It is involved in pioneering research in Advanced Research areas like Image Processing, Data Mining, Networks, Artificial Intelligence, Databases, and Wireless Area Network. Computer Science and Engineering Department executes several funded projects from reputed agencies like DST, DRDO, AICTE, UGC, etc. The student involvement in the research projects nurtures them in handling societal problems thus making them the technology solution providers. Involve and evolve with the changing technology keep the fraternity updated. Committed faculty, strong aptitude to learn, and the engagement in research enable to bridge the industry-academia gap and lead the department to the number one position.

Contents

About the Editors

Dr. C. Kiran Mai working as Professor in the Department of Computer Science & Engineering, VNR VJIET, has over 31 years of experience in the field of academic research and technological education. She has a multi-disciplinary approach due to varied roles taken up, including but not limited to teaching, research and administration. She was awarded as "Best teacher in Computer Science" in the year 2010, by the professional body—International Society for Technology in Education (ISTE). She also worked at various administrative positions in the institute (Principal, Dean Academics and Head of the Department) and has an extensive experience in internal administrative tasks and communication. As Head of the Department, she took the prime lead, in setting up the UG and PG laboratories and a research lab in virtual reality and real-time computing. While serving in the administrative positions as Vice-Principal and Principal, she was instrumental in designing the policies and strategies for the institute and also the administrative manuals. Dr. Kiran Mai also administered the processes and could get five departments of the institute recognized as research centers by the JNTUH, the affiliating university. The institute was twice NBA accredited and NAAC accredited with 3.71 CGPA and was also sanctioned UGC autonomous status, while she was in the role of Vice-Principal. As Dean Academics, she played a key role in curriculum revision, enhanced the learning by doing component for practical courses and introduced the concept of WIT & WIL (Why I am Teaching, What am I teaching and Why I am Learning, What am I learning). With her industrial experience, where she headed ISO 9002 division – document control, she could frame and document the processes and procedures in the institute with ease. This made the institute also ISO certified. Being a member of the Internal Quality Assurance Cell (IQAC), she administers the quality procedures in the institute and performs periodical audit of the academic and administrative processes. Under her leadership, the institute was recognized by UGC as College with Potential for Excellence, got the UG and PG courses re-accredited. She published 36 papers in various reputed national and international journals, conducted faculty development program in deep learning and intelligent systems and staff development program in data mining, with the funding from AICTE. She actively participated in the research projects and guided nearly 75 UG projects and 20 PG projects. Currently, four research scholars are working under her guidance. She co-chaired many international conferences. Her research paper on data mining for deforestation using Polyanalyst, presented at the IEEE conference held at Seoul, South Korea, in 2005, was selected as the best paper. She was on the Editorial Board for two Korean journals. Her areas of interest are network communications, data engineering and block chain technologies.

Dr. A. Brahmananda Reddy working as Associate Professor in the Department of Computer Science & Engineering at VNR Vignana Jyothi Institute of Engineering and Technology (VNR VJIET), Hyderabad. He done his Ph.D. from Jawaharlal Nehru Technological University Anantapuramu (JNTUA) Anantapuramu, in the area of Text Mining, M.Tech. – Computer Science in 2007 from JNTUCEA, Anantapuramu, and B.Tech. in 2004 in Computer Science and Engineering. He has over 13 years of experience in the field of academic research and technological education and has more than 15 research papers published in various reputed national/international conferences and journals which are listed in Scopus, Web of Science, IEEE, Inderscience and Springer Proceedings. He is a member of IEEE and a lifetime member of ISTE and CSI. Dr. Brahmananda Reddy's research interests include data mining, text mining, natural language processing, semantic web and social networks, machine learning, deep learning and image processing. He is conducting many seminars/workshops/FDPs for the benefit of students and faculty in and out by the eminent personalities from various reputed institutions and also delivered guest lectures on various topics in various academic institutions. He has guided many UG and PG projects in various mezzanine technologies. He is in charge for students' Industrial Visits and establishing MoUs between Institute and Industries to reducing the gap between academia and industry. He is a jury member for the Smart India Hackathon Grand Finale (SIH 2018, SIH 2019 and SIH 2020) Software Edition organizing by AICTE and MHRD in association with state governments and reputed industries. He performed as Session Chairs for various international conferences and Reviewer for various reputed international conferences and journals.

Dr. K. Srujan Raju is currently working as Dean Student Welfare and Heading Department of Computer Science & Engineering at CMR Technical Campus. He obtained his Doctorate in Computer Science in the area of Network Security. He has more than 20 years of experience in academics and research. His research interest areas include computer networks, information security, data mining, cognitive radio networks and image processing and other programming languages. Dr. Raju is presently working on 2 projects funded by Government of India under CSRI & NSTMIS, has also filed 7 patents and 1 copyright at Indian Patent Office, edited more than 14 books published by Springer Book Proceedings of AISC series, LAIS series and other which are indexed by Scopus also authored books in C Programming & Data Structure, Exploring to Internet, Hacking Secrets, contributed chapters in various books and published more than 30 papers in reputed peer-reviewed journals and international conferences. Dr. Raju was invited as Session Chair, Keynote Speaker, a Technical Program Committee member, Track Manager and Reviewer for many national and international conferences also appointed as Subject Expert by CEPTAM DRDO – Delhi & CDAC. He has undergone specific training conducted by Wipro Mission 10X & NITTTR, Chennai, which helped his involvement with students that is very conducive for solving their day to day problems. He has guided various student clubs for activities ranging from photography to Hackathon. He mentored more than 100 students for incubating cutting-edge solutions. He has organized many conferences, FDPs, workshops and symposiums. He has established the Centre of Excellence in IoT and Data Analytics. Dr. Raju is a member of various professional bodies and received Significant Contributor Award and Active Young Member Award from Computer Society of India and also served as a Management Committee member, State Student Coordinator & Secretary of CSI – Hyderabad Chapter.

Prediction of Anemia Disease Using Classification Methods

Sagar Yeruva, B. Pavan Gowtham, Yendluri Hari Chandana, M. Sharada Varalakshmi, and Suman Jain

1 Introduction

A Normal Blood flows through a small circular shape which carries oxygen to organs of human body parts which is circular in shape and the life span of each cell is approximately 120 days and a new blood cell is generated for every 120 days [1]. SCA is a kind of abnormal blood disease which affects hemoglobin within the RBCs. The shape of the sickle cell is disc shape which is sticky and rigid, which causes stoppage of blood flow in the human body. It is also observed that the life span of sickle cell is 10–20 days [2]. Due to the presence of sickle cell in hemoglobin, it may cause severe episodes of pain, death of tissue, and serious complications, in some cases it may lead to death [3].

In above Fig. 1, we can clearly see that the normal blood flow passing oxygen to all parts without stoppage of blood, where sickle cell shape is sticky and due to that blood flow will stop at any stage. Due to the stoppage blood flow may cause severe body pains and heart strokes, etc. Sickle cell was observed in the black population, later it has been observed in other ethnic group, which includes Middle East, Mediterranean

S. Yeruva · B. P. Gowtham (✉) · Y. H. Chandana
Department of CSE, VNR Vignana Jyothi Institute of Engineering & Technology, Hyderabad, Telangana, India
e-mail: pavangowtham2495@gmail.com

S. Yeruva
e-mail: sagar_y@vnrvjiet.in

M. S. Varalakshmi
Department of CSE, St. Peter's Engineering College, Hyderabad, Telangana, India
e-mail: sharada.mangipudi07@gmail.com

S. Jain
Thalassemia and Sickle Cell Society, Rajendra Nagar, Hyderabad, Telangana, India
e-mail: sumanjaindr@gmail.com

C. K. Mai et al. (eds.), *Machine Learning Technologies and Applications*,
Algorithms for Intelligent Systems,
https://doi.org/10.1007/978-981-33-4046-6_1

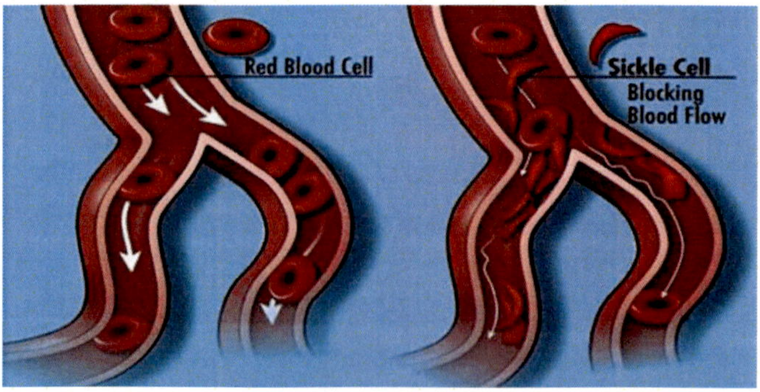

Fig. 1 Normal blood flow and sickle cell blood flow

Sea, and Central India. SCA observed more in Greece and Italy [4]. Treatment for SCA by taking early medications like Antibiotics, Blood Transfusion, Bone marrow. Most of Doctor recommends blood Transfusion for at least one–two months and patients can take antibiotics to manage complications, including chronic pain. SCD is a rare blood disorder in the human body but now it has been observed very quietly in newborn babies' hemoglobin. SCD is a crescent-shaped cells which are stiff and sticky and interact with other cells. If one organ is affected slowly it will be spreading throughout the body which may lead to death. To prevent this kind of disease, early medication is required as soon as possible to avoid serious complications (Blood Transfusion).

When a person has hemoglobin S which is caused by two abnormal genes then the person has Sickle Cell Disease.

1.1 Symptoms and Signs of Sickle Cell Anemia

The following are the symptoms and signs of sickle cell anemia from reference [5].

Episodes of pain: RBC blocks the blood flow through tiny blood vessels to chest, joints, and abdomen. Chronic pain has been observed in adults and adolescents which result in bones, ulcers, and joint damages.

Painful swelling of hands and feet: The presence of sickle cells in the blood generally causes a blockage for the blood flow that results in painful swellings in hands and feet.

Frequent infections: Organ damage (fights infection, i.e., spleen), may cause due to sickle cells in bodies. Doctors commonly suggest antibiotics and vaccinations that can prevent severe infections (pneumonia).

Delayed growth: If RBCs are getting reduced in the human body due to lack of oxygen, which may cause delayed growth.

Vision problems: RBC blocks the blood flow through tiny blood vessels due to sickle cells. These cells do not supply blood to the eyes which may lead to damage of retina. It causes blur vision problems.

1.2 Sickle Cell Complications

Heart Stroke: Heart strokes may occur if RBCs block the blood flow in brain. Symptoms of strokes include weakness, continuously speaking problems and loss of consciousness [5].

Acute chest syndrome: It causes severe chest pain, difficulty in breathing and also causes continuous fever. It is a very serious complication, which may also lead to life threatening.

Leg Ulcers: SCA (Sickle Cell Anemia) can cause open sores, in which ulcers may occur in legs.

Pulmonary Hypertension: These patients have also observed a symptom of high Blood Pressure in their lungs called as Pulmonary Hypertension and is observed in adults rather than children.

Gallstones: The breakdown of RBC may produce a substance, i.e., bilirubin. If bilirubin is having a high level in the body which leads to gallstones.

1.3 Statistical Information

1.3.1 In Indian Scenario

In [6] an Indian scenario, first observed in the Nilgiri Hills of northern Tamilnadu in 1952. The SCA observed more in the Deccan plateau of central India and north of Tamilnadu and Kerala.

1.3.2 In Worldwide Scenario

In [6] a worldwide scenario, sickle cells was firstly an unknown disease but now is been spreading over worldwide and it is observed particularly in Spanish (Western Hemisphere, The Caribbean, Central America and South America), Sub-Saharan Africa, Saudi Arabia, Mediterranean countries include(Greece, Italy, Turkey), India.

Sickle cell deaths are observed more in African-American, with children (four years of age) with SCD fell by 42% from 1999 through 2002. SCD is observed in Newborn Screening among mortality children in California, New York, and Illinois.

In the period of 1990–1994.The mortality rate is increased 1.5 per 100 in African-American children with SCD in California and Illinois in 1995. SCD is one of the major public health concerns where the average of 75,000 Hospitalization due to SCD in the US, approximately cost $475 millions.

2 Literature Review

In this section we present various approaches/methods available in the existing sources that are mainly helpful in the identification of sickle cell disease. Various scientists/researchers have made their efforts in the progress of identifying SCA in the early stage of life with good accuracy levels.

Akrimi et al. [7] has presented various image processing techniques like (1) Optimization (2) Segmentation and Mean filter are used. Mean filter is used for obtaining geometry, texture, and color features related to RBC images by using a photo imaging microscope. The SVM is used to classify whether it is normal or abnormal, high accuracy is achieved with validation measures of sensitivity (100%), specificity (0.998%), and Kappa (0.9944%).

In this paper, Elsalamony et al. [8] used geometrical shape signatures method for detecting sickle cells and elliptocytosis in blood samples. In the process of detection they have used 30 colorful microscopic images and achieved 100% accuracy by using three neural network layers for detecting anemia disease.

In [9] the observations based on shape signatures, sickle cells, Burr cells, Elliptocytosis are identified. They use 45 colorful microscopic images in 15 samples by using CHT (Circular Hough Transforms), WS (Watershed Segmentation), and Morphological methods to enhance images. Methods used for identifying the anemia disease are (1) Support Vector Machine (SVM), (2) Back-Propagation (BP), and (3) Self-Organizing Map (SOM).

Soltanzedeh et al. [10] has mentioned the experimentation on blood samples, usage of morphological methods for Elliptocytes, Discocytes and Echinocytes cells and used statistical analysis for calculating distance from each edge to mass center. By using this method they have achieved high accuracy for recognizing Elliptocytes (98.63%), Discocytes (96.7%) and Echinocytes (95.36%) respectively.

Elsalamony et al. [11] used two classification techniques like J48 and Random tree for predicting sickle cell disease which is highly affected disease in tribal zones of Gujarat and after that the author compared J48 and Random tree classification techniques in the mining process. They have used the WEKA tool for this prediction process, which is an open source tool.

3 About Dataset

As part of this project [A collaborative Research project granted by Jawaharlal Nehru Technological University, Hyderabad, TEQIP-III (funding agency) for Rs.3,00,000/- for the duration of one year (August, 2019 to July, 2020)], we had executed a Memorandum of Understanding (MoU) between our Institute, Vallurupalli Nageswara Rao Vignana Jyothi Institute of Engineering & Technology (VNRVJIET) and Thalassemia and Sickle Cell Society (TSCS) (MoU dated Oct 15, 2019). The details of TSCS can be found from https://www.tscsindia.org/ [12].

The MoU aimed with the following objectives:

I. Data Sharing and
II. Technology Transfer.

We have received a dataset of 1387 records of patients who have approached TSCS for the diagnosis purpose. These records are shared to us as part of MoU and the records are pre-processed to maintain confidentiality, anonymity that meets data privacy of patients. This data obtained from TSCS, belongs to the patients visited during August, 2017 to August, 2019.

4 Implementation

The model architecture includes three stages:

4.1 Dataset Preparation

As we mentioned above, about the dataset collected from Thalassemia and Sickle Cell Society consists 1387 patients with 13 parameters which include: (1) AGE, (2) HB—Hemoglobin, (3) HCT—Hematocrit, (4) RDW—RBC Distribution Width, (5) MCV—MeanCorpuscularVolume, (6) MCH—MeanCorpuscularHemoglobin, (7) MCHC—MeanCorpuscularHemoglobinConcentration (8) RBC—Red Blood Cell, (9) RETIC—Reticulocytes, (10) HBF—Fetal Hemoglobin, (11) HBAo, (12) HBA2, (13) Diagnosis—Diagnosis is output variable, which we need to predict based on a set of features (inputs) either it is Normal cell, Sickle cell, or Thalassemia cells.

4.2 Analyzing Dataset and Splitting Dataset

Thalassemia society has labeled each patient data record (i.e., Diagnosis) with AS, AT, BT, EA, EE, ET, NN, SC, SS, ST, TM, TT. 12 labeled Diagnosis is converted into three main groups.

(1) NN = N (Normal Cells)
(2) AS, SC, SS, ST = S (Sickle Cells)
(3) AT, BT, EA, EE, ET, TM, TT = T (Thalassemia Cells) (Fig. 2).

We use Label Encoder in python(as part of Sci-kit Learn Library) is used to convert categorical data into numbers (i.e., N-0, S-1, & T-2) for understanding predictive model easily.

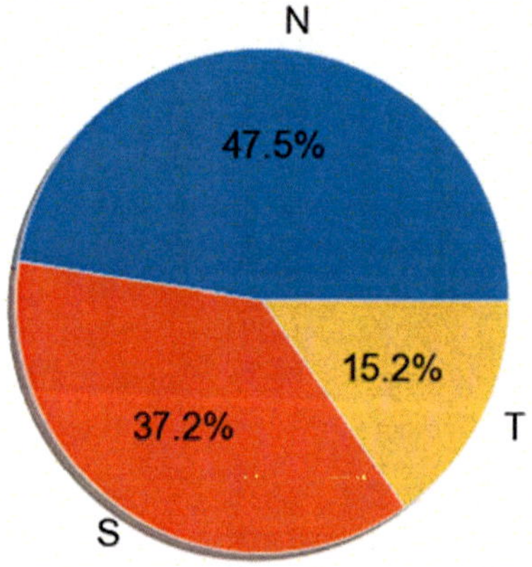

Fig. 2 Dataset collected from TSCS

The available dataset is divided into training data and testing data (80 and 20%.) The training dataset contains the target variable (i.e., output) in which the model learns from the features (input data) in order to get generalization on the data.

4.3 Model Selection [13]

Model selection is an important part as we are applying machine learning algorithms for predicting the best outcome or result.

Supervised technique is used to train on a set of input and output pairs and learn the model for correlations between inputs and outputs. Supervised learning problems are grouped into two problems:

(1) Regression analysis: when the target or output variable is real and continuous values.
(2) Classification: Problems for filtering the data which is not required.

We have a dataset with 13 independent variables (parameters or features) and one dependent variable is target variable or output variable which predicts whether patients are having either N (Normal) or S (sickle cell) or T (Thalassemia). In this project we have used various classification algorithms for the diagnosis purpose and are elaborated as follows:

We test this dataset using data mining techniques like classification method [14] for the detection of the target classes described as above.

Classifiers: Classifier is a method in which computers learn from the features(input) and predicts the target variable.

We have tested the dataset with the following classifiers which include: (I) SVM, (II) KNN, (III) Logistic Regression, (IV) Decision Tree, (V) Random Forest.

SVM classifier: SVM is a supervised learning technique which evaluates the statistics used for analysis of regression and classification. SVM algorithm provides

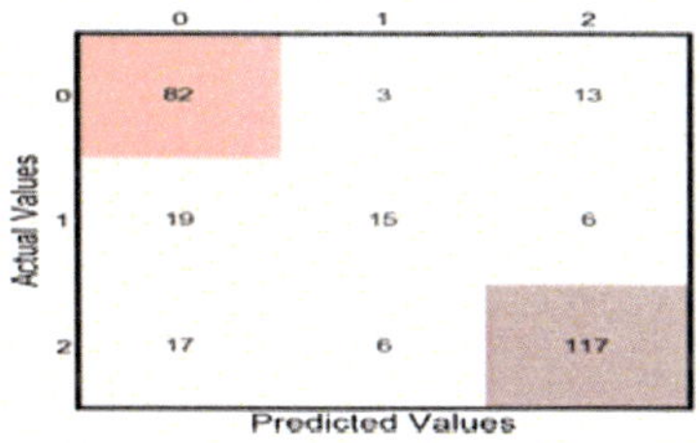

	precision	recall	f1-score	support
N	0.69	0.84	0.76	98
S	0.62	0.38	0.47	40
T	0.86	0.84	0.85	140
accuracy			0.77	278
macro avg	0.73	0.68	0.69	278
weighted avg	0.77	0.77	0.76	278

Fig. 3 Results obtained using SVM classifier

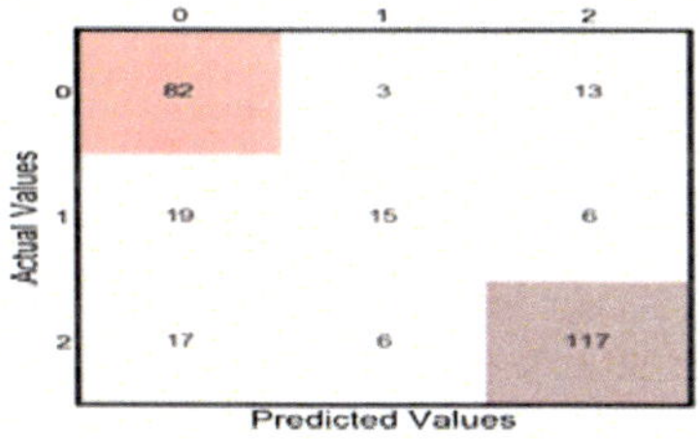

	precision	recall	f1-score	support
N	0.69	0.84	0.76	98
S	0.62	0.38	0.47	40
T	0.86	0.84	0.85	140
accuracy			0.77	278
macro avg	0.73	0.68	0.69	278
weighted avg	0.77	0.77	0.76	278

Fig. 4 Results obtained using KNN Classifier

the best possible Decision Boundary, so we can categorize data points easily. It selects extreme points that support the hyperplane imagination, in such cases called as vectors of support and algorithms used for machine learning are called Vector Support Machines (Fig. 3).

KNN Classifier: K-Nearest Neighbors (KNN) is used for regression and classification problems. Based on similarity measures it classifies new data points (Fig. 4).

Logistic Regression classifier: It is a supervised classification algorithm in which an output variable (or Y), which accepts only discrete values for a given set of inputs (features or X), is called logistic regression (Fig. 5).

Decision Tree classifier: A decision tree supports the decisions and their consequences in the form of a tree-structure model. Decision tree is used to define a structural approach that is most likely to meet an objective, especially in decision analysis (Fig. 6).

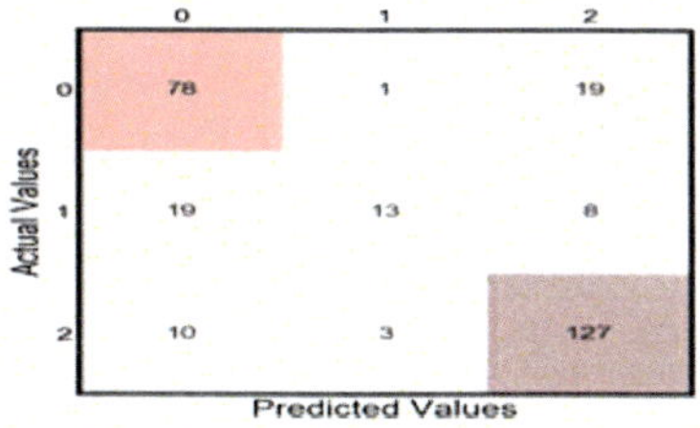

N	0.73	0.80	0.76	98
S	0.76	0.33	0.46	40
T	0.82	0.91	0.86	140
accuracy			0.78	278
macro avg	0.77	0.68	0.69	278
weighted avg	0.78	0.78	0.77	278

Fig. 5 Results obtained using logistic regression classifier

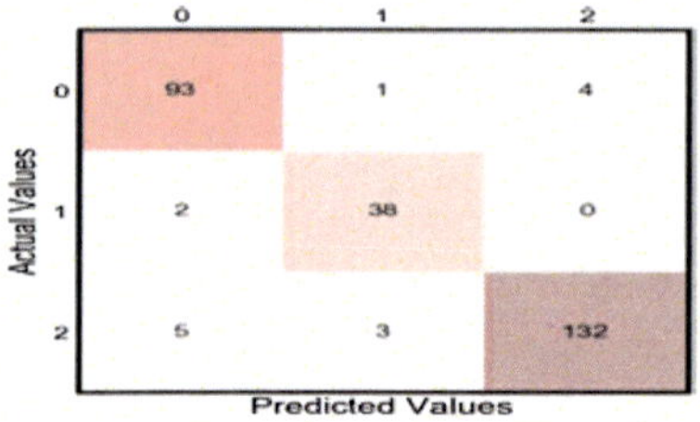

	precision	recall	f1-score	support
N	0.93	0.95	0.94	98
S	0.90	0.95	0.93	40
T	0.97	0.94	0.96	140
accuracy			0.95	278
macro avg	0.94	0.95	0.94	278
weighted avg	0.95	0.95	0.95	278

Fig. 6 Results obtained using Decision Tree

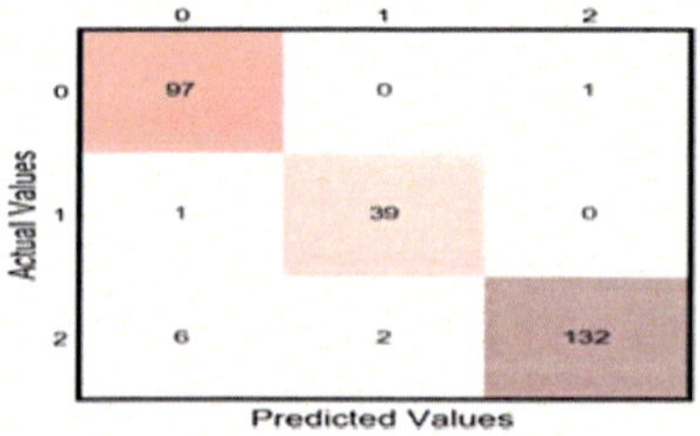

	precision	recall	f1-score	support
N	0.93	0.99	0.96	98
S	0.95	0.97	0.96	40
T	0.99	0.94	0.97	140
accuracy			0.96	278
macro avg	0.96	0.97	0.96	278
weighted avg	0.97	0.96	0.96	278

Fig. 7 Results obtained using Random Forest algorithm

Random Forest classifiers: A random forest allows a number of decision tree classifiers on various subsamples of the dataset and used for improving the predictive accuracy and avoiding overfitting (Fig. 7).

5 Results

Analysis of results:
The experimental phase has been classified into three stages:

1. Training the network using the dataset available.
2. Detection process for the normal cell, sickle cell, Thalassemia cell.
3. Calculating accuracy based on Test Predictions.

$$Accuracy = \frac{Noofcorrectpredictions}{TotalNoofpredictions} \tag{1}$$

Table 1 Performance of various classification algorithms

S. NO.	Name of the classification algorithm	Result of prediction accuracy (%)	Dataset size (No of observations X No of parameters)
1	Support Vector Machine (SVM)	76	1387 X 13
2	K-Nearest Neighbor Algorithm (KNN)	77	1387 X 13
3	Logistic Regression (L)	78	1387 X 13
4	Decision Tree Classifier (DTC)	95	1387 X 13
5	Random Forest Algorithm (RF)	96	1387 X 13

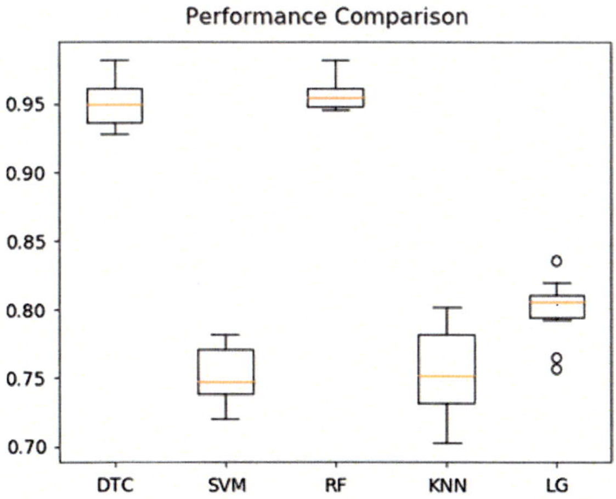

Fig. 8 Performance comparison based onaccuracy

Prediction using classifiers (Table 1):

Performance Comparison:

The above box-plot summarizes that the Decision Tree Classifier (DTC) and Random Forest (RF) classifier algorithms could predict the identification of sickle cell from the patients sample with the highest accuracy of 95% and 96% respectively in comparison to SVM, KNN, and Logistic regression (Fig. 8).

6 Conclusion

Sickle cell disease is a hematological disorder, previously it was one of the particular features of tribal people but now it is spreading to the entire world in a rapid manner and needs immediate attention. This paper has presented a description about Sickle

Cell Disease (SCD) and its history in the international scenarios and the national scenario (in the context of India). The symptoms of the disease, signs, complications, and treatment for the disease are presented. We have also characterized the blood cells in normal people with that of sickle cell and Thalassemia patients under various blood parameters. This paper also presents the identification of sickle cell disease with higher accuracy using various classifiers and to develop good prediction models that help to reduce the time and efforts in pain management systems of sickle cell disease. The results show that the classification algorithms like SVM, KNN, Logistic Regression, Decision Tree, and Random Forest give the accuracy of prediction like 76%, 77%, 78%, 95%, and 96%, respectively. Finally the results show that the Random Forest algorithm gives high accuracy of prediction in the identification of sickle cell from the blood samples.

Acknowledgements **JNTUH, TEQIP-III**: We sincerely thank Jawaharlal Nehru Technological University, Hyderabad, Technical Education Quality Improvement Programme-III (JNTUH TEQIP-III) for the award of project (proceedings No: JNTUH/TEQIP-III/CRS/2019/CSE/04 Dated 22-07-2019) with an amount of Rs 3,00,00/- for the duration of one year (Aug 2019–July 2020) as part of Collaborative Research Project Scheme. We also sincerely thank the reviewers and coordinator of JNTUH-TEQIP-III, Dr. Padmaja Rani, Professor, Department of CSE, JNTU Hyderabad for their constant support during the reviews for carrying out this project.

TSCS: We sincerely thank Thalassemia and Sickle cell society (TSCS) for accepting our request for Collaborative Research Project and their support for entering into MoU. We also thank the personal at TSCS named Mr. Chandrakant Agarwal, President-TSCS, Dr. Suman Jain, Chief Medical Research Officer, and Secretary-TSCS, Mr. Allam Ravi Kumar Reddy, Data Manager-TSCS, Dr. Saroja Kondaveeti, Medical Officer-TSCS, Dr. Padma Gunda, Research Scientist-TSCS, Mr. Mohd Abdul Tufeeq Baig, Lab InchargeTSCS, Mr. Bhargava Kalvakota, Data & Admin Officer-TSCS and Ch. Devasri, Data Entry Operator-TSCS who have helped us to understand the entire scenario of sickle cell patients, process of their work, and their services to the society in the state of Telangana, India.

References

1. https://en.wikipedia.org/wiki/Red_blood_cell
2. Sickle Cell Disease core concepts for emergency physician and nurse (https://slideplayer.com/slide/3762536/)
3. https://www.mayoclinic.org/diseases-conditions/sickle-cell-anemia/symptomscauses/syc-20355876
4. https://www.nhlbi.nih.gov/health-topics/sickle-cell-disease
5. https://www.mayoclinic.org/diseases-conditions/sickle-cell-anemia/symptomscauses/syc-20355876
6. https://www.cdc.gov/ncbddd/sicklecell/data.html
7. Akrimi, J.A., et al.:Classification red blood cells using support vector machine. In:Proceedings of the 6th International Conference on Information Technology and Multimedia. IEEE (2014)
8. Elsalamony, H.A.:Anaemia cells detection based on shape signature using neural networks.Measurement **104**, 50–59 (2017)
9. Elsalamony, H.A.: Detection of anaemia disease in human red blood cells using cell signature, neural networks and SVM. Multimed. Tools Appl. **77**(12), 15047–15074 (2018)

10. Soltanzadeh, Ramin, and Hossein Rabbani. "Classification of three types of red blood cells in peripheral blood smear based on morphology." IEEE 10th INTERNATIONAL CONFERENCE ON SIGNAL PROCESSING PROCEEDINGS. IEEE, 2010.
11. Solanki, A.V.:Data mining techniques using WEKA classification for sickle cell disease. Int. J. Comput. Sci. Inf. Technol. **5**(4), 5857–5860 (2014)
12. https://www.tscsindia.org/
13. https://towardsdatascience.com/metrics-to-evaluate-your-machine-learning-algorithm-f10ba6e38234
14. https://stackabuse.com/overview-of-classification-methods-in-python-with-scikit-learn/

Diabetic Symptoms Prediction Through Retinopathy

Ambika Shetkar, C. Kiran Mai, and C. Yamini

1 Introduction

Automatic biomedical diagnostic systems have been rapidly developed because of the advent of information technologies, boost patient care, and even support the ophthalmologists. Diabetes comes under the category of metabolic disorders where a person will have elevated sugar levels in their blood, it is because of the body's inability to generate adequate insulin and can also be due to cells fail to react to the produced insulin. Diabetes blood sugar is excessively elevated and is due to damage to smaller eye blood vessels that fuels diabetes. Diabetic retinopathy (DR) can lead to the leaking fluid or bleeding of blood vessels. This induces a loss of vision to many tiny blood vessels of the eye retina. The disease's risk increases with age, which leads to a diabetic retinopathy of the middle-aged and older diabetics.

Diabetic retinopathy has been repeatedly shown to be associated with an elevated risk of cardiovascular disease it is when hardness of the coronary arteries or atherosclerosis, which is a build-up of cholesterol in the blood vessels that provide oxygen and nutrition for the heart, is the most common cause of heart disease in a person with diabetes.

When plaques of cholesterol fall up or break up, the body tries to repair the plaque breakup by delivering platelets to patch it up. If the artery is narrow, the platelets can block blood flow, not allow oxygen to be distributed and a heart attack occurs. The

A. Shetkar (✉) · C. K. Mai
VNR VignanaJyothi Institute of Engineering and Technology, Hyderabad, India
e-mail: ambikashetkar.as@gmail.com

C. K. Mai
e-mail: ckiranmai@vnrvjiet.in

C. Yamini
PES University, Bangalore, India
e-mail: ycherukury@gmail.com

C. K. Mai et al. (eds.), *Machine Learning Technologies and Applications*,
Algorithms for Intelligent Systems,
https://doi.org/10.1007/978-981-33-4046-6_2

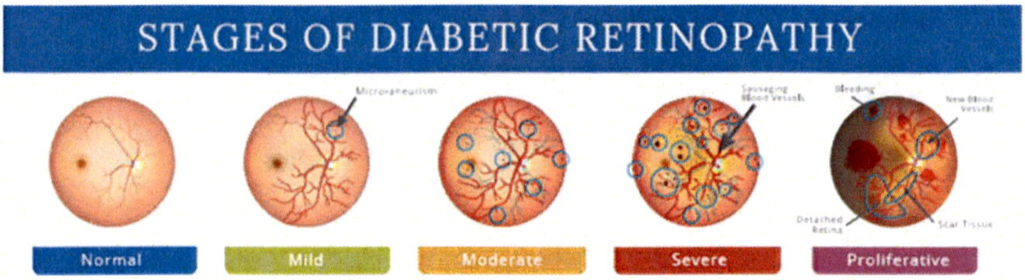

Fig. 1 Stages of diabetic retinopathy

same process can occur in all the arteries in the body, leading to a lack of blood in the brain, a stroke or a lack of blood in the feet, hands or arms that causes peripheral vascular disease.

Not only are people with diabetes at greater risk of heart disease, they are also at higher risk of heart failure, a serious medical condition in which the heart cannot effectively pump blood. This can lead to fluid build-up in the lungs that causes difficulty in breathing, or fluid retention that causes swelling in many other parts of the body (especially legs).

The purpose of this paper, is based on color fundus photography and to have an automated detection system. The project is to identify the condition of the eye's retina, provided the picture of the patient's left and right eyes. No DR, Mild, Moderate, Severe, Proliferative DR visualization is given below in Fig. 1.

The key goal of the paper presented is to use the convolutional neural network (CNN), a deep learning methodology that predicts a person's cardiovascular disorders mainly diabetic retinopathy (DR). The biggest benefit of CNN over the previous ones is that without human intervention it instantly identifies the relevant features to identify the predicted yes or no class labels. Profound machine modelling has been used to enhance the accuracy.

2 Literature

In paper [1] the authors provide an automatic process for retinal photographs to enhance and segregate blood vessels. They present a system which uses 2-D Gabor vessel wavelet because they are able to improve directional structures and have a new multilayered vessel segmentation thresholding technology. The value of the current segmentation strategy is that it works best to identify the thinnest vessels with large differences in lighting. The machine is checked with labeled pictures (i.e., DRIVE and STARE) that is available to the public in retinal pictures.

In journal paper [2] the authors surveyed on Heart Disease Prediction Using Classification with Different Decision Tree Techniques. The Decision tree, Artificial Neural Network, and Bayesian Classifier, which were some of the main methods used for constructing a classification model. The ID3, C4.5, C5.0, J48 are various decision tree techniques found in the survey. While different techniques are used extensively to predict disease, the decision tree classifier is chosen for its simplicity and accuracy.

In paper [3, 4] the author's research compares different decision tree classification algorithms providing improved diagnostic efficiency of cardiac disease with WEKA tool. The algorithms that are evaluated are the J48 algorithm, the Random Forest, and the Logistic model tree algorithms. In order to check and validate the efficiency of decision tree algorithms, the current databases of Cleveland heart disease patients from the UCI server is used. This dataset contains 303 cases and 76 attributes. The classification algorithm with optimum potential for use in large data is subsequently suggested by the authors.

In paper [5], Deep learning in retinal fundus photographs has been applied to construct an algorithm to automate diabetic retinopathy (DR) detection. The network used in this study is a neural network that uses a function which combines close pixels first with local characteristics and then aggregates them into global features. The neural network used in this research is the architecture of Inception-v3 proposed by Szegedy et al.

In paper [6], this paper gives us a deep learning (DLS) system using artificial intelligence contrasts it with professionals who use retinal images of multiethnic populations with diabetes to classify diabetes retinopathy and associated eye diseases. In DLS 90.5% sensitivity and 91.6% specific to visible diabetes retinopathy (71,896 images; 14,880 patients); 100% sensitivity and 91.1% specificity for vision-threatening diabetes retinopathy; 96.4% sensitivity and 87.2% specificity of potential glaucoma; 93.2% sensitive specificity for possible glaucoma.

In paper [7], this paper recognizes the pathological lesions of the images by convolutional neural networks (a deep learning method). This paper analyzes to compare current literature on various deep learning models for diabetic retinopathy (DR) diagnosis.

In paper [8], Diabetic retinopathy (DR) identification and sight-threatening DR (STDR) by fundus photographs taken using the mobile phone-based app Remidio "Fundus on Phone" (FOP). The images of the retinal have been tested with validated AIDR (EyeArtTM) screening software. To evaluate and validate the role of the automated fundus photography of artificial intelligence (AI)-based software for the sensing of diabetic retinopathy (DR) and sight-threatening DR (STDR) with a smartphone app and validate it against ophthalmologist's grading.

In paper [9], the authors used convolutional neural network (CNN) to distinguish between smokers and non-smoker with the use of retinal images and it also uses focus maps to enhance the perception of physiologic shifts in the retina of smokers. 165,104 retinal photographs, labeled with self-reported "smoking" or "non-smoking," were taken out of a diabetes screening system. The photos were "contrast-enhanced" or "skeletonized" in one of the two ways. For training and test sets, the dataset was split 80/20. The overall results of validation were 88.88, 93.87% for the contrast-enhanced version. In comparison, 63.63%, specificity 65.60%, were the results of the skeletonized model.

3 Existing System

Prediction of cardiovascular diseases using data mining and machine learning techniques has been used by various papers, but the accuracy to predict the diabetic retinopathy has always been the key parameter.

Most of the papers have implemented several data mining and machine learning techniques for identification and early diagnosis of cardiovascular disease with diabetic retinopathy symptoms, techniques such as 2-D Gabor wavelet method, J48 algorithm, decision tree, and convolutional neural networks showing different levels of accuracies.

4 Working Flow Model

The above block diagram in Fig. 2 is the overall process to predict the diabetic retinopathy (DR) in the patient through his/her retina image. The images are taken from the patient and are stored in the database, from the database the image is taken and some preprocessing is done to enhance the image for sending that image as an input to convolutional neural network, the network then processes the image and extracts the features, learns the features while training the dataset then it delivers the predicted classes of desired output of two classes, that is, diabetic retinopathy (DR) is present or not.

5 Proposed Work

The proposed work model with respect to the existing models and techniques, uses a deep learning technique with convolutional neural networks architecture, a deep learning model is used in order to get more accuracy from the existing systems to predict the identification of the diabetic retinopathy (DR) from the retinal color image fundus taken from various patients.

Input data is the dataset which has high-resolution photographs of color fundus retinal from five groups corresponding to five stages of the DR disease. These photographs were collected from the public database from the department of computer science of Technische Fakultat, it is a free platform for having retinopathy scanned photographs. From the dataset, 75 healthy & unhealthy retina fundus pictures are taken for the implementation.

Data preprocessing is done with the instruction images should not be used specifically for preparation, due to non-standard camera resolutions. The images taken are 512 × 512 pixels, to create a structured dataset. Training pictures with 512 × 512 pixel resolution on all three color platforms required high memory needs. The videos were transferred to a single screen, owing to this constraint. After multiple

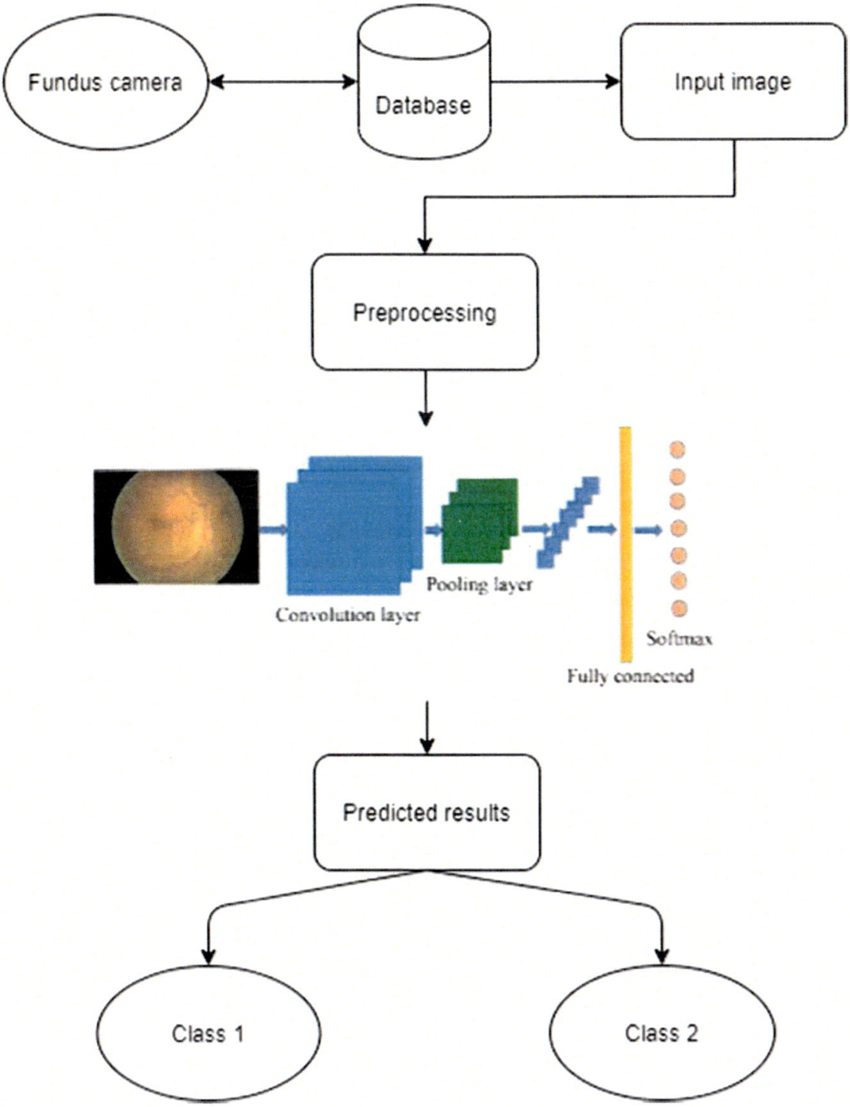

Fig. 2 Block Diagram

tests, green channel images were found to hold better information. The equalization histogram technique is used to improve the color contrast equally across pixels. To avoid the convolutional neural network from recognizing the image's intrinsic background noise, each of the images used is standardized with the Minimum–Maximum normalization. Basically preprocessing is to avoid any empty data cells and any noise before giving image data to the CNN.

CNN Architecture A Convolutional Neural Network (CNN) consists of one or more convolutional layers followed by one or more fully connected layers as in a regular multilayer neural network. A CNN's architecture is designed to take advantage of an input image's 2D hierarchical structure It is done with local connections and bound weights accompanied by some sort of pooling resulting in invariant features of the translation. Another advantage of CNNs is that they are easier to prepare with the same number of hidden units and with far fewer constraints than fully connected networks. CNNs also consider the hierarchical representation of images when practicing by piling on each other several trainable stages. Here the CNN nodes store different features that are extracted from the various retinal fundus images that are collected after data preprocessing. These features are used for prediction of diabetic retinopathy in the patients.

The final network comprises an information layer that takes the images as information with a resolution of having 512×512 pixels in them. The architecture of CNN consists of five sets of the convolution, pooling, and the dropout layers together, each one is layered above on each other. It's accompanied by the 2 sets of hidden completely linked and the pooling layers of features. That is followed by the final layer of output, where the prediction is classified into two classes that is yes class or no class, yes class means the diabetic retinopathy is present and no class means the diabetic retinopathy is not present.

6 Results

The results of this paper yield an accuracy of 89 percent with the pictures being diversified into five classes, after predicting into five classes, the resultant knowledge is procured and the patient is kept into yes class or no class based on the knowledge acquired from those five classes.

Figure 3 demonstrates the relation for each class between the number of actual test image labels and the labels expected from the ensemble model. It also indicates the number of pictures for each of the five classes which were misclassified by 1 class.

The accuracy is based on the 150 retinal fundus images taken from the database and accuracy is 89 percent. The resultant output of the implementation is given in Fig. 4.

The above Fig. 4 is the picture of the resultant output that shows the accuracy of 89% at the end of the picture. This accuracy is produced when an input of 75 healthy & 75 unhealthy retina fundus pictures are given for the implementation which used CNN architecture to predict whether diabetic retinopathy (DR) if present or not from the given retinal fundus image of the patient. The implementation of CNN is done in python environment in jupyter anaconda platform.

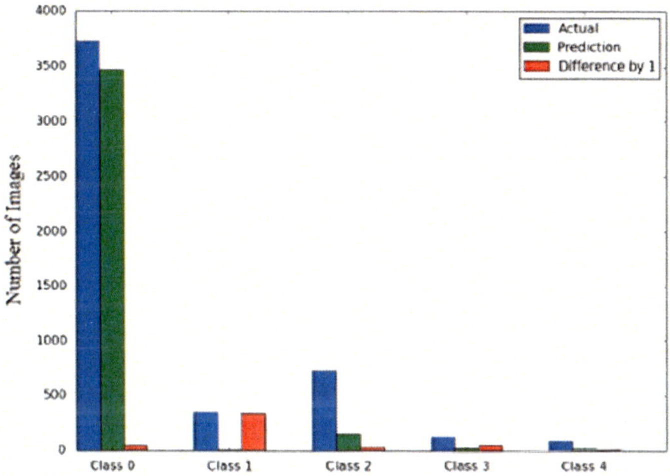

Fig. 3 Visualization of predictions

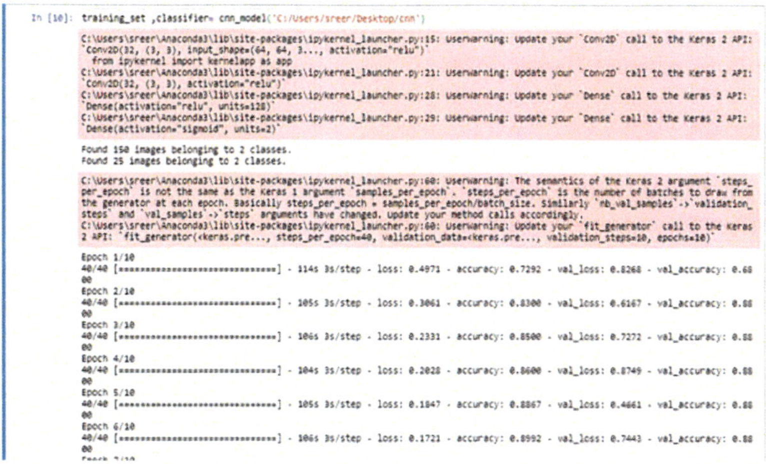

Fig. 4 Resultant output

7 Conclusion

Diabetic retinopathy has been the leading cause of blindness and visual impairment in the working-age population. Diabetic retinopathy is an end-organ reaction to a systemic disease, reflecting just one of many symptoms in microvascular and macrovascular diabetic complications. Hence the early identification and taking medication becomes more and more important than ever.

This paper proposes a deep learning model which uses the convolutional neural network architecture to identify the person has a diabetic retinopathy (DR) or not from their retinal fundus photograph. The system identifies the retinal photographs then extracts features, those features helps to get the status of the disease.

The identification of the diabetic retinopathy (DR) from the retinal color image fundus then classifying them to final two classes, represented as NO class otherwise the YES class label. The proposed system predicted with an accuracy of 89 percent, by using the deep convolutional neural networks is presented in this paper.

References

1. Akram, M.U, Khan, S.A.: Multi-layered thresholding-based blood vessel segmentation for screening of diabetic retinopathy, Springer-Verlag London Limited. Last accessed 5 Jan 2012
2. Thenmozhi, K., Deepika, P.: Heart disease prediction using classification with different decision tree techniques. Int. J. Eng. Res. Gen. Sci. (2014)
3. Patel, J., TejaUpadhyay, D., Patel, S.: Heart disease prediction using machine learning and data mining technique. Int. J. Comput. Sci. Eng. (IJCSE) (2015)
4. Ramakrishna, M., Murthy, J.V.R., Prasad Reddy, P.V.G.D., et al.: Dimensionality reduction text data clustering with prediction of optimal number of clusters. Int. J. Appl. Res. Inf. Technol. Comput. (IJARITAC), 41–49 (2011)
5. Gulshan, V., Peng, L., Coram, M., Stumpe, M.C., Wu, D., Narayanaswamy, A., Venugopalan, S., Widner, K.: Development and validation of a deep learning algorithmfor detection of diabetic retinopathy in retinal fundus photographs. Am. Med. Assoc. Last accessed 12 Jan 2016
6. Ting, D.S.W., Cheung, C.Y.L., Tan, G.S.W., Sivaprasad, S.: Development and validation of a deep learning systemfor diabetic retinopathy and related eye diseases using retinal images from multiethnic populations with diabetes. Am. Med. Assoc. Last accesssed 12 Dec 2017
7. Lakshmi, R.R., Subashini, R., Anjana, R.M., Mohan, V.: Automated diabetic retinopathy detection in smartphone-basedfundus photography using artificial intelligence. The Royal College of Ophthalmologists. Last accessed 9 Mar 2018
8. Raman, R., Srinivasan, S., Virmani, S., Siva Prasad, S., Rao, C., Raja Lakshmi, R.: Fundus photograph-based deep learning algorithms in detecting diabetic retinopathy. The Royal College of Ophthalmologists. Last accessed 7 Oct 2018
9. Bhuravane, S., Vaghef, E., Yang, S., Hill, S., Humphrey, G., Walker, N., Squirrell, D.: Detection of smoking status from retinal images. Convolutional Neural Netw. Study, Sci. Rep. Last accessed 29 April 2019

Prediction of Cervical Cancer

Oruganti Shashi Priya and Sagar Yeruva

1 Introduction

A Cancer is a disease that arises from the transformation of normal cells into tumour cells. It also spreads to other parts of the body. There are different kinds of cancers like Breast cancer, Lung cancer and cervical cancer. A cancer that occurs in the cervix cells is cervical cancer, which is the lower part of the uterus connected to the vagina. Most of the cases are due to a virus called Human Papillomavirus (HPV). Smoking, use of Oral contraceptive pills for long term and multiple pregnancies can cause cancer.Early diagnosis of cancer is very difficult. At the initial stage cervical cancer may be completely free of symptoms later we find symptoms to appear. For the presence of cervical cancer to be determined, screening techniques such as Cytology, Schiller, Hinselmann and Biopsy procedures are carried out. Similar models were evaluated on data sets, and different models performance were compared. This paper analyses the performance of Machine Learning techniques such as Artificial Neural Network (ANN) with Multi-Layer Perceptron (MLP) and Random Forest Classifier for cervical cancer prediction.

O. S. Priya (✉) · S. Yeruva
Department of CSE, VNR Vignana Jyothi Institute of Engineering and Technology, Bachupally, Hyderabad, India
e-mail: shashipriya9999@gmail.com

S. Yeruva
e-mail: sagar_y@vnrvjiet.in

C. K. Mai et al. (eds.), *Machine Learning Technologies and Applications*,
Algorithms for Intelligent Systems,
https://doi.org/10.1007/978-981-33-4046-6_3

2 Related Work

In source [1], DS Latha et al. have experimented using the data set of text format. They collected the data set using several tests like Pap smear, colposcopy, biopsy and are used for diagnosing the disease from various hospitals. Demographic data of the prospective candidates of cervical cancer using feature selection method and C4.5 classification for generation of decision trees. The data set comprises 30 instances with 7 dependent attributes and 1 decision attribute. One observation from this study is improvements in accuracy with the number of folds increasing [10]. The results obtained by the C4.5 algorithm is 100%.

In [2], provide an overall view on cancer prognosis and prediction. Therefore, for the purpose of classification there exists various kinds of methods like Artificial Neural Network (ANN), Support Vector Machine and Decision Trees were used.

In [3], R Vidya et al. have considered the input as text data and diagnosis have been done using biopsy screening tests. In this particular paper, the performance has been analysed with different Machine Learning techniques such as ID3, C4.5 and Naïve Bayesian algorithm for prediction of cervical cancer. The results for each model were obtained with stratified tenfold cross-validation. The results obtained by Naive Bayesian algorithm (81%), C4.5 (72%) and ID3 (69%).

In [4] Bargana Benazir1et al have presented various methods for controlling and preventing cervical cancer such as Pap smear, liquid-based cytology, Human Papillomavirus (HPV) and HPV vaccine. They have used different data mining techniques such as feature selection and classification. For the purpose of classification Neural Network classifier is used and patients are classified into two normal and abnormal classes.

In [5, 9], the authors have used the Feature Selection method for cervical cancer prediction. Different algorithms like Gaussian Naïve Bayes, Decision Tree, Logistic Regression, Support Vector Machine and K-Nearest Neighbour classifiers were used and comparison has been done between them for calculating the performance. The authors have preferred the Decision Tree classifier over Gaussian Naïve Bayes. A method called selection features of the Least Absolute Shrinkage and Selection Operator (LASSO) was not tested on classifiers over Gaussian Naïve Bayes.

In [6], Blessing et al. provide the Pap smear screening test, which is a popular test for detection of cervical cancer, in which they have experimented using Machine Learning Algorithms like Genetic Algorithm for data pre-processing and Support Vector Machine (SVM) algorithm for prediction. SVM doesn't perform well as they have experimented with large data sets. The overall accuracy of the model is identified as 96% for Hinselmann and 95% for Biopsy.

In [7], Parikh, Dhwaani et al. consider text as input data and diagnosis have been done using Cytology, Schiller, Hinselmann and Standard Biopsy test, Machine Learning techniques like K-Nearest Neighbour technique, Decision Trees technique and Random Forest Classifier technique are used for calculating the performance like accuracy, precision and recall. K-Nearest Neighbour is the best one with high accuracy. Higher Area under the curve (AUC) that is 0.822 compared with 0.52 for the Decision Tree, 0.5320 for Random Forest (which has very low value).

3 Proposed Framework

All the above-mentioned sources have mentioned the need for early detection of cervical cancer that may help the people to improve their life span. The following framework in Fig. 1 is used in this process.

3.1 Data Collection

It is the process of gathering data. The main objective of data collection is to obtain high-quality data which is useful for analysis. This data set is collected from UCI Machine Learning Repository.

Fig. 1 Proposed system for cervical cancer prediction

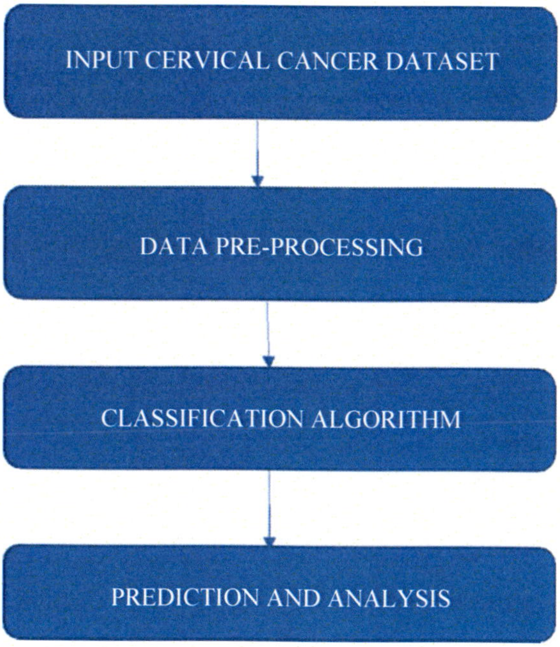

3.2 Data Pre-Processing

It is the technique of converting raw data into a more clear and understandable format. It aims at eliminating inconsistencies, missing values and errors from the data set.

Therefore Data cleaning method is used for pre-processing the data which means check for missing values, cleaning or filling missing data.

3.3 Classification Algorithms

3.3.1 Artificial Neural Network (ANN)

ANN is a method of performing tasks, instead of a computational method of programming to perform certain tasks. It is a most pragmatic model that is used to find the given data that replicates the knowledge precisely and quickly. ANN mainly contains three layers they are input layer—the raw data given as network's input, hidden layer—activities of input unit along with weight are given in between input and hidden layer, and output layer—output depends on the weights between the hidden units.

3.3.2 Random Forest

It creates multiple level Decision Trees with the usage of training dataset. The training data and testing data are created from the data for the Decision Tree. In which there is a technique called "ensemble" is used for combining multiple models to get the output.

4 Methodology

4.1 Data Description

The input data set is collected from UCI Machine Learning Repository, which consists of 836 records with 32 attributes and 4 target attributes (Hinselmann, Schiller, Cytology and Biopsy). From Fatlawi, Hayder [8], the information of each attribute is shown in Table 1

Table 1 Description of cervical cancer data

Attribute name	Attribute type	Attribute name	Attribute type
Age	Integer	STDs: Pelvic inflammatory disease	Boolean
Number of sexual partners	Integer	STDs: Genitai herpes	Boolean
Age of first sexual intercourse	Integer	STDs: Molluscum contagiosum	Boolean
No. of pregnancies	Integer	STDs: AIDS	Boolean
Smokes	Boolean	STDs: HIV	Boolean
No. of smoking years	Real	STDs: Hepatitis B	Boolean
Smokes (packs/year)	Real	STDs: HPV	Boolean
Hormonal contraceplives	Boolean	STDs: Number of diagnosis	Integer
Years of hormonal contraceplives	Real	STDs: Time since first diagnos is	Integer
1UD	Boolean	STDs: Time since last diagnosis	Integer
Years of 1UD	Real	Dx: Cancer	Boolean
STDs	Boolean	Dx: ClN	Boolean
No. of STDs	Integer	Dx: HPV	Boolean
STDs: Condylomatosis	Boolean	Dx	Boolean
STDs: Cervical condylomatosis	Boolean	Hinselmann (target)	Boolean
STDs: Vaginal condylomatosis	Boolean	Schiller (target)	Boolean
STDs: Vulvoperineal condylomatosis	Boolean	Cytology (target)	Boolean
STDs: Syphilis	Boolean	Biopsy (target)	Boolean

4.2 *Implementation*

The data set has been divided into two categories comprising of training data and test data. So there are 501 records in training data and 334 records in test data.

We test a data set using Machine Learning techniques like classification methods.

Classifiers: It is a supervised learning technique which allows computers to learn from the data. Input given to it and then uses this learning for classification of new observations. We have tested the data set with the following classifiers which include:

Multi-Layer Perceptron 2. Random Forest

4.2.1 Multi-Layer Perceptron (MLP)

It is an Artificial Neural Network where the network has many perceptrons. MLP consists of one input layer and hidden layers depending on the model we predict and how the output we want and output layers that makes the predictions about the input. An algorithm called Back Propagation is used in neural network training. Number of hidden layers used for Neural Network Architecture is two. In hidden layers, the number of neurons used is 16 or 17 that is (16, 16) or (17, 17). It is a supervised learning technique.

Supervised learning technique is used to train on the given set of inputs and output pairs. It learns the model for correlations between the given inputs and the final outputs. The concept of having multiple layers and a non-linear activation makes MLP different from a normal linear perceptron.

4.2.2 Random Forest classifier

It is an estimator (Meta) that enables with number of Decision Tree classifiers on various set of sub-samples of the dataset and uses it averages to improve the most important parameter of predictive accuracy and avoids the overfitting problem.

5 Results

5.1 Accuracies Obtained for Target Attributes Using Multi-Layer Perceptron and Random Forest Classifier

See (Table 2).

Table 2 Accuracies of the classifiers

Target attributes	Accuracy (in percentage) for multi-layer perceptron	Accuracy (in percentage) for random forest classifier
1. Hinselmann	92.3	99.0
2. Schiller	93.2	96.1
3. Cytology	91.8	93.3
4. Biopsy	91.38	96.1

Fig. 2 Precision, recall, f1-score, support of multi-layer perceptron

	precision	recall	f1-score	support
0	0.96	0.99	0.98	196
1	0.83	0.38	0.53	13
accuracy			0.96	209
macro avg	0.90	0.69	0.75	209
weighted avg	0.95	0.96	0.95	209

Fig. 3 Precision, recall, f1-score, support of multi-layer perceptron

	precision	recall	f1-score	support
0	0.97	0.97	0.97	190
1	0.70	0.74	0.72	19
accuracy			0.95	209
macro avg	0.84	0.85	0.84	209
weighted avg	0.95	0.95	0.95	209

Fig. 4 Precision, recall, f1-score, support of multi-layer perceptron

	precision	recall	f1-score	support
0	0.95	0.96	0.95	198
1	0.11	0.09	0.10	11
accuracy			0.91	209
macro avg	0.53	0.53	0.53	209
weighted avg	0.91	0.91	0.91	209

5.2 *Obtained Classification Report for Target Attributes Using Multi-layer Perceptron and Random Forest Classifiers*

5.2.1 Hinselmann

See (Fig. 2).

5.2.2 Schiller

See (Fig. 3).

5.2.3 Cytology

See (Fig. 4).

5.2.4 Biopsy

See (Fig. 5).

Fig. 5 Precision, recall, f1-score, support of multi-layer perceptron

	precision	recall	f1-score	support
0	0.97	0.96	0.97	193
1	0.59	0.62	0.61	16
accuracy			0.94	209
macro avg	0.78	0.79	0.79	209
weighted avg	0.94	0.94	0.94	209

Fig. 6 Precision, recall, f1-score, support of random forest

	precision	recall	f1-score	support
0	0.96	0.99	0.98	196
1	0.83	0.38	0.53	13
accuracy			0.96	209
macro avg	0.90	0.69	0.75	209
weighted avg	0.95	0.96	0.95	209

Fig. 7 Precision, recall, f1-score, support of random forest

	precision	recall	f1-score	support
0	0.97	0.97	0.97	190
1	0.70	0.74	0.72	19
accuracy			0.95	209
macro avg	0.84	0.85	0.84	209
weighted avg	0.95	0.95	0.95	209

Fig. 8 Precision, recall, f1-score, support of random forest

	precision	recall	f1-score	support
0	0.95	0.96	0.95	198
1	0.11	0.09	0.10	11
accuracy			0.91	209
macro avg	0.53	0.53	0.53	209
weighted avg	0.91	0.91	0.91	209

5.2.5 Hinselmann

See (Fig. 6).

5.2.6 Schiller

See (Fig. 7).

5.2.7 Cytology

See (Fig. 8).

5.2.8 Biopsy

See (Fig. 9).

Fig. 9 Precision, recall, f1-score, support of random forest

	precision	recall	f1-score	support
0	0.97	0.96	0.97	193
1	0.59	0.62	0.61	16
accuracy			0.94	209
macro avg	0.78	0.79	0.79	209
weighted avg	0.94	0.94	0.94	209

6 Conclusion

Cervical Cancer is one of the most popular effected disease especially reported in women and causes many deaths due to late identification and prediction. The main objective is to avoid the manual assessment that requires huge amount of time which may lead to wrong classification.

In this paper, we used a data set, which is obtained from UCI Machine Learning Repository for classification of cervical cancer data. We have experimented with Random Forest (RF) classifier and Multi-Layer Perceptron (MLP) algorithms to determine the existence of cervical cancer from the dataset available and also presented the performance of accuracy in prediction of cervical cancer. On the basis of experimental findings, we can infer and prove that Machine learning algorithm like Random Forest (RF) can be successfully used to predict data on cervical cancer. The average accuracy obtained for Multi-Layer Perceptron classifier for different target attributes such as Hinselmann, Schiller, Cytology and Biopsy is 92.3, 93.2, 91.8 and 91.38% is for testing dataset and the obtained accuracy using Random Forest Algorithm for different target attributes such as Hinselmann, Schiller, Cytology and Biopsy is 99.0, 96.1, 93.3 and 96.1% for testing data. Random Forest algorithm shows one of the highest results in prediction of cervical cancer data.

References

Latha, D.S., Lakshmi, P.V., Fathima, S.: Machine aided identification of risk factors of cervical cancer. IJCST **4**(3) (2013)

Kourou, K., Exarchos, T.P., Exarchos, K.P., Karamouzis, M.V.: Machine learning applications in cancer prognosis andprediction. Comput. Struct. Biotechnol. J. (2015)

Vidya, R., Nasira, G.M.: Predicting cervical cancer using machine learning techniques—an analysis. Glob. J. Pure Appl. Math. **12**(3) (2016) ISSN 0973-1768

Benazir, B., Nagarajan, A.: An expert system for predicting the cancer using data mining techniques. Int. J. Pure Appl. Math. **118**(20) (2018)

Al-Wesabi, Y.M.S., Choudhury, A.: Classification of cervical cancer dataset. In: Barker, K., Berry, D., Rainwater, C. (eds.) Proceedings of the 2018 IISE Annual Conference (2018)

Ayobami, A., Lawal, E., Abisoye, K., Opeyemi.: Prediction of cervical cancer occurence using genetic algorithm and support vector machine (2019)

Dhwaani, P., Menon, V.: Machine learning applied to cervical cancer data. Int. J. Math. Sci. Comput. **5**, 53–64 (2019). https://doi.org/10.5815/ijmsc.2019.01.05

Hayder, F.: Enhanced classification model for cervical cancer dataset based on cost sensitive classifier. Int. J. Comput. Tech. **4**(4) (2017)

Himabindu, G., Ramakrishna Murty, M., et al.: Classification of kidney lesions using bee swarm optimization. Int. J. Eng. Technol. **7**(2.33), 1046–1052 (2018)
Himabindu, G., Ramakrishna Murty, M., et al.: Extraction of texture features and classification of renal masses from kidney images. Int. J. Eng. Technol. **7**(2.33), 1057–1063 (2018)

A New 5-layer Model Approach for Pneumonia Prediction

Bandi Krishna Kanth, Gudipati Manasa, Shubham Kumar Jena, Aishwarya Kankurte, Oduri Durga Prasad, Maithre Salmon, G. Pradeep Reddy, and Swetha Namburu

1 Introduction

Infection caused in the lungs may result in pneumonia. It is serious and life threatening too. It leads to death. It is a second major cause of death after heart stroke [2]. According to a survey conducted by World Health Organization (WHO), out of all deaths occurred from different diseases, the percentage of deaths occurred due to

B. K. Kanth · G. Manasa · S. K. Jena · A. Kankurte · O. D. Prasad · M. Salmon ·
G. Pradeep Reddy (✉) · S. Namburu
Department of ECE, Gokaraju Rangaraju Institute of Engineering and Technology, 500090 Hyderabad, Telangana, India
e-mail: pradeep.19phd7025@vitap.ac.in

B. K. Kanth
e-mail: krishnakanthsrinivas@gmail.com

G. Manasa
e-mail: manasa3099@gmail.com

S. K. Jena
e-mail: jenathecool@gmail.com

A. Kankurte
e-mail: aishwaryakankurte@gmail.com

O. D. Prasad
e-mail: durgaprasadoduri888@gmail.com

M. Salmon
e-mail: salmon.vicky634@gmail.com

S. Namburu
e-mail: swethakarima@gmail.com

G. Pradeep Reddy
School of Electronics Engineering, VIT-AP University, Amaravati 522237, Andhra Pradesh, India

C. K. Mai et al. (eds.), *Machine Learning Technologies and Applications*,
Algorithms for Intelligent Systems,
https://doi.org/10.1007/978-981-33-4046-6_4

pneumonia is 14%. Every year more than one million people are affected, and more than 50,000 people die due to pneumonia in United States alone [4]. Usually, it starts with a viral or fungal infection. Later on, it increases gradually. The symptoms are continuous cough, cold, fever, short or long breath, vomiting etc., Basic test done to identify infection is blood test. To identify whether it is pneumonia or not, Chest X-rays are usually preferred [9]. Both deep learning and big data are two popular fields in the fast-growing digital world [6]. CNN using deep learning algorithms has been the popular choice for science [12]. Many methods are proposed to diagnose it. Some of them are based on laboratory tests such as Complete Blood Count (CBC), blood gases and serum electrolytes. Other methods include non-laboratory tests such as chest X-ray, Computed Tomography (CT), Bronchoscope to know the presence of pneumonia [13]. The proposed method in this paper takes chest X-ray and label it as binary (0 &1) i.e., normal and pneumonia. With the advancement in deep learning, many authors have proposed different approaches to predict Pneumonia which includes transfer learning [1, 5] on existing CNN architectures like VGG16, inceptionV3 and also by fine tuning [3]. Other methods like CheXnet which is a 121-layer convolutional architecture [10] takes large computational time. Masking R-CNN an ensemble model is also used [7] whose accuracy is less compared to the model proposed in this paper. Few authors have performed with Machine Learning techniques like SVM, Decision Trees [8], etc. The proposed model gives better accuracy with less computational time.

In this paper, Section II discusses about pre-processing the data, model summary and testing the data. Section III deals with results of the model. Section IV describes the final conclusion.

2 Proposed Model

2.1 Data Pre-Processing

The data set considered is "chest- xray- pneumonia" which was provided by Paul Mooney in Kaggle. It consists of 3 folders namely training, testing and validation. There are 5,863 x-ray images which are categorized as 2 groups: pneumonia affected chest X-rays and normal (not affected by pneumonia) chest X-rays. All radiographs were initially screened for quality control when analyzing chest X-ray images by excluding all unclear images [11]. All the images in the 3 folders are merged and divided into categories namely Pneumonia and Normal. For converting unlabeled data into labeled data, "0" is assigned to normal images and "1" is assigned to pneumonia images. Data is shuffled in order to learn all features. Before feeding to the model the images are converted to grayscale and resized to 128 X 128 and then normalized. Pneumonia chest X-ray and Normal Chest X-ray are shown in Figs. 1 and 2, respectively.

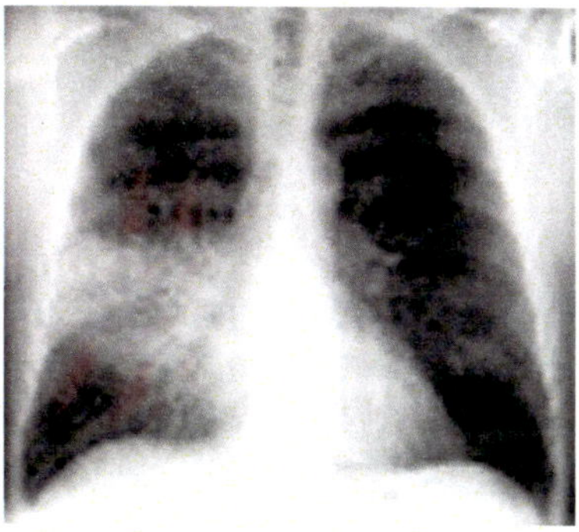

Fig. 1 Pneumonia affected X-ray

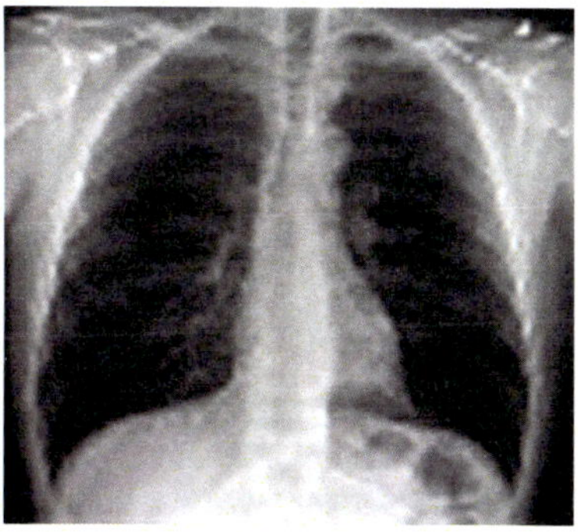

Fig. 2 Normal chest X-ray

2.2 Model Building

The proposed model is a 5-layer convolutional architecture designed to detect pneumonia using X-rays shown in Fig. 3 (generated from Spyder). This model comprises of Convolution 2D layers, flatten layers, Dense layers and dropout is also used to remove unwanted weights which reduces overfitting. At few layers, Max pooling is performed to reduce computational time and amount of parameters. Each image passes through the sequence of layers as shown in Fig. 4.

2.3 Training and Testing Model

The model is trained on 80% of the images with the parameters shown in Table 1. After training, the model is tested on remaining 20% of the images.

Table. 2 shows that different combinations of optimizers used with single activation function ReLu. Combinations are tried with transfer learning on VGG16 architecture and different metrics are listed. From the table it is clear that the maximum accuracy attained is only 94%.

Layer (type)	Output Shape	Parameters
conv2d_1 (Conv2D)	(None, 128, 128, 32)	320
activation_1 (Activation)	(None, 128, 128, 32)	0
conv2d_2 (Conv2D)	(None, 126, 126, 32)	9248
activation_2 (Activation)	(None, 126, 126, 32)	0
max_pooling2d_1 (MaxPooling2)	(None, 63, 63, 32)	0
dropout_1 (Dropout)	(None, 63, 63, 32)	0
conv2d_3 (Conv2D)	(None, 61, 61, 64)	18496
activation_3 (Activation)	(None, 61, 61, 64)	0
max_pooling2d_2(MaxPooling2)	(None,30,30,64)	0
dropout_2 (Dropout)	(None, 30, 30, 64)	0
flatten_1 (Flatten)	(None, 57600)	0
dense_1 (Dense)	(None, 64)	3686464
activation_4 (Activation)	(None, 64)	0
dropout_3 (Dropout)	(None, 64)	0
dense_2 (Dense)	(None, 2)	130
activation_5 (Activation)	(None, 2)	0

Fig. 3 5-layer model summary

The flow of the execution is described in Fig. 5. An X-ray image is given as input and preprocessed. The processed image is fed to the model. Finally, the model classifies the image as either normal or pneumonia affected, and then the output is predicted.

3 Results

From Fig. 6, It is observed that accuracy increases as the number of epochs increases. In Fig. 7, As the number of epochs increases it is clear, the loss reduces to some extent, and then it ceases.

The confusion matrix describes the performance of the model on testing data. Considering normal image classification as positive case, from Fig. 8 it is clear that

Fig. 4 Proposed network architecture

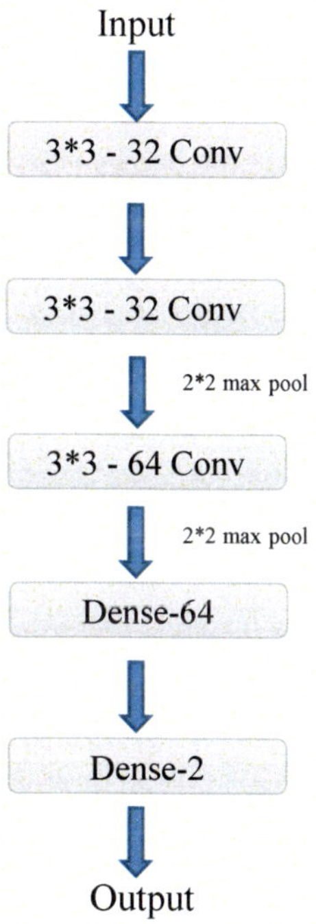

Table 1 Parameters used in proposed model

Description	Value/Type
Batch size	16
Loss function	Categorical cross entropy
Optimizer	Adam
Metric	Accuracy
Activation function	ReLu

Table 2 Different combinations of activation functions, optimizers and metrics

Activation function	Optimizer	Epoch	Batch_size	Loss	Accuracy	Val_loss	Val_acc
ReLu	Adagrad	20	16	4.1439	0.7429	6.0443	0.6250
ReLu	Sgd	20	16	4.1315	0.7381	6.0185	0.6520
ReLu	Adam	20	16	4.1336	0.9410	6.0443	0.6678
ReLu	Rmsprop	20	16	1.9728	0.6094	6.0263	0.6598

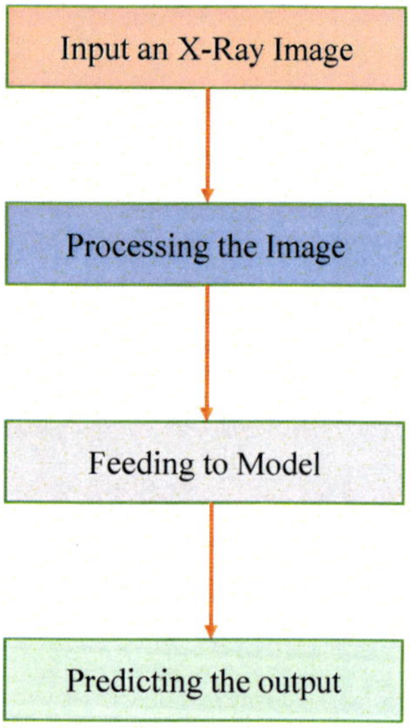

Fig. 5 Flow chart

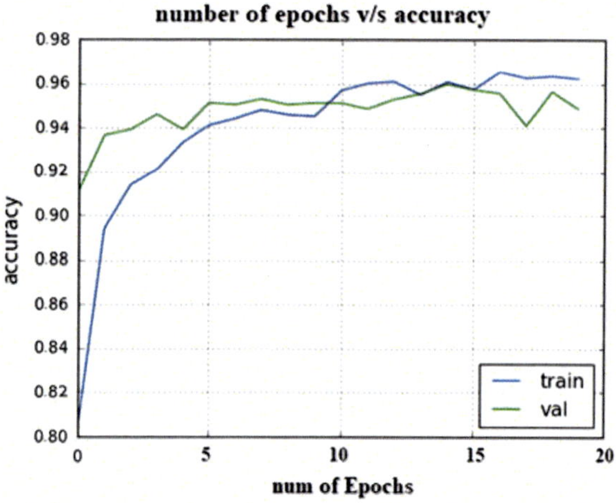

Fig. 6 Relation between training and validation accuracy

295 images are classified as normal (True Positive), whereas 15 images are classified as normal (False Negative) which are really pneumonia affected. Likewise, 45 images are classified as pneumonia (False Positive) which are actually normal images and 812 images are classified as Pneumonia (True Negative).

For the proposed model, precision, recall and f1-score values are shown in Table 3. Precision value of 0.87 for Class 0 indicates when an image is classified as normal image it is correct 87% of the time and when the value is 0.98 for Class 1 it indicates that an image is classified as pneumonia 98% of the time, respectively. Recall 0.95 of Class 0 and 0.95 of Class 1 indicates that the model is 95% sensitive towards predicting normal and pneumonia, respectively.

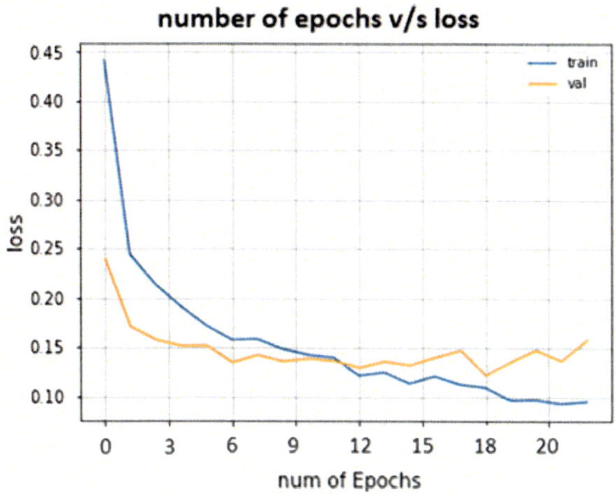

Fig. 7 Relation between training and validation loss

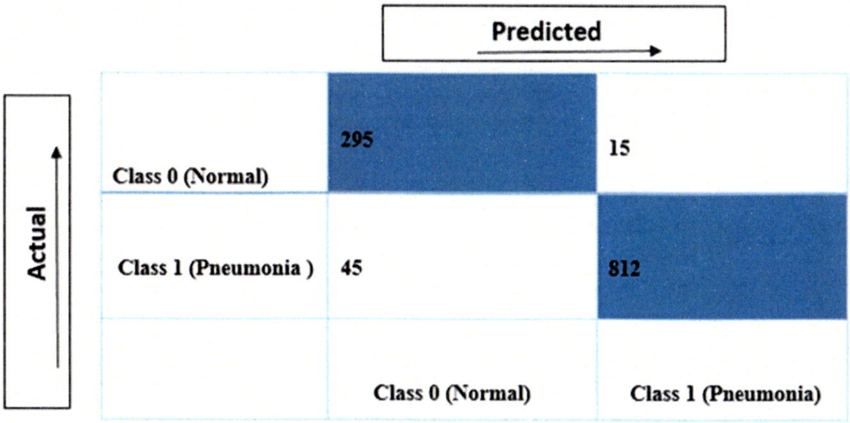

Fig. 8 Confusion matrix

Table 3 Performance of various metrics

	Precision	Recall	f1-score
Class 0 (NORMAL)	0.87	0.95	0.91
Class1 (PNEUMONIA)	0.98	0.95	0.96

4 Conclusion

In this paper, chest X-rays are taken to train the deep learning neural network. The proposed model is 5-layer convolution architecture. The process of evaluation is done using precision, recall and f1-score metrics. This method is compared with various CNN architectures of machine learning. The accuracy obtained is better than the other existing models which was worked upon Kaggle data set. Using transfer learning on VGG16 with different optimizers, the maximum accuracy obtained is 94% and the proposed model obtained 97.5%.

References

1. Allaouzi, I., Ben Ahmed, M.: A novel approach for multi-label chest X-Ray classification of common thorax diseases. IEEE Access **7**, 64279–64288 (2019). https://doi.org/10.1109/ACCESS.2019.2916849
2. Ge, Y., Wang, Q., Wang, L., Wu, H., Peng, C., Wang, J., Xu, Y., Xiong, G., Zhang, Y., Yi, Y.: Predicting post-stroke pneumonia using deep neural network approaches. Int. J. Med. Inform. **132**(May), 103986 (2019). https://doi.org/10.1016/j.ijmedinf.2019.103986
3. Guo, Y., Shi, H., Kumar, A., Grauman, K., Rosing, T., Feris, R.: Spottune: Transfer learning through adaptive fine-tuning. In: Proceedings of the IEEE Computer Society Conference on Computer Vision and Pattern Recognition, pp. 4800–4809 (2019). https://doi.org/10.1109/CVPR.2019.00494
4. Huang, S.M., Monam, M., Cortes, E.: Weakly Supervised Pneumonia Localization (n.d.)
5. Huang, Z., Pan, Z., Lei, B.: Transfer learning with deep convolutional neural network for SAR target classification with limited labeled data. Remote. Sens. **9**(9), 1–21 (2017). https://doi.org/10.3390/rs9090907
6. Jakhar, K., Hooda, N.: Big data deep learning framework using keras: a case study of pneumonia prediction. In: 2018 4th International Conference on Computing Communication and Automation, ICCCA 2018, 1–5. https://doi.org/https://doi.org/10.1109/CCAA.2018.8777571
7. Ko, H., Ha, H., Cho, H., Seo, K., & Lee, J. (2019). Pneumonia Detection with Weighted Voting Ensemble of CNN Models. 2nd International Conferences on Artificial Intelligence and Big Data, pp. 306–310 (2018). https://doi.org/10.1109/ICAIBD.2019.8837042
8. Navada, A., Ansari, A.N., Patil, S., Sonkamble, B.A.: Overview of use of decision tree algorithms in machine learning. IEEE Control Syst. Grad. Res. Colloq. **2011**, 37–42 (2011). https://doi.org/10.1109/ICSGRC.2011.5991826
9. Rajaraman, S., Candemir, S., Kim, I., Thoma, G., Antani, S.: Visualization and interpretation of convolutional neural network predictions in detecting pneumonia in pediatric chest radiographs. Appl. Sci. (Switzerland), **8**(10) (2018). https://doi.org/10.3390/app8101715
10. Rajpurkar, P., Irvin, J., Zhu, K., Yang, B., Mehta, H., Duan, T., Ding, D., Bagul, A., Langlotz, C., Shpanskaya, K., Lungren, M.P., Ng, A.Y.: CheXNet: radiologist-level pneumonia detection on chest X-Rays with deep learning, 3–9 (2017). https://arxiv.org/abs/1711.05225
11. Saul, C.J., Urey, D.Y., Taktakoglu, C.D.: Early diagnosis of pneumonia with deep learning (2019). https://arxiv.org/abs/1904.00937
12. Stephen, O., Sain, M., Maduh, U.J., Jeong, D.U.: An efficient deep learning approach to pneumonia classification in healthcare. J. Healthc. Eng. **2019**, (2019). https://doi.org/10.1155/2019/4180949
13. Toğaçar, M., Ergen, B., Cömert, Z.: A deep feature learning model for pneumonia detection applying a combination of mRMR feature selection and machine learning models. Irbm **1**, 1–11 (2019). https://doi.org/10.1016/j.irbm.2019.10.006

Prediction of Liver Malady Using Advanced Classification Algorithms

K. Sravani, G. Anushna, I. Maithraye, P. Chetan, and Sagar Yeruva

1 Introduction

The liver is the human body's second largest internal organ, playing a vital role in metabolism and performing many essential functions such as digestion of food, release of the body's toxic factor, decomposition of red blood cells, storage of nutrients, etc. These functions make the liver the 4th most important organ, where in its absence; the tissues of the body will instantly die due to lack of energy and nutrients. Liver malady is considered to be the biggest health issue in the world. Problems of liver disorders are not found until it is always too late because even partially damaged liver continues to function. Early diagnosis could potentially save lives. Thus, the findings of this study are significant from both the computer science and medical professional point of view. In this paper, comparison of 2 computer aided medical diagnostic approaches is done.

The two approaches are Artificial Neural Network and Support Vector Machine which involves training a system such that it responds to various patient features such

K. Sravani (✉) · G. Anushna · I. Maithraye · P. Chetan · S. Yeruva
Department of CSE, VNR Vignana Jyothi Institute of Engineering and Technology, Hyderabad, Telangana, India
e-mail: Sravs8820@gmail.com

G. Anushna
e-mail: ganushna99@gmail.com

I. Maithraye
e-mail: iruvurumaitreyi01@gmail.com

P. Chetan
e-mail: cherrychetan78@gmail.com

S. Yeruva
e-mail: Sagar_y@vnrvjiet.in

C. K. Mai et al. (eds.), *Machine Learning Technologies and Applications*,
Algorithms for Intelligent Systems,
https://doi.org/10.1007/978-981-33-4046-6_5

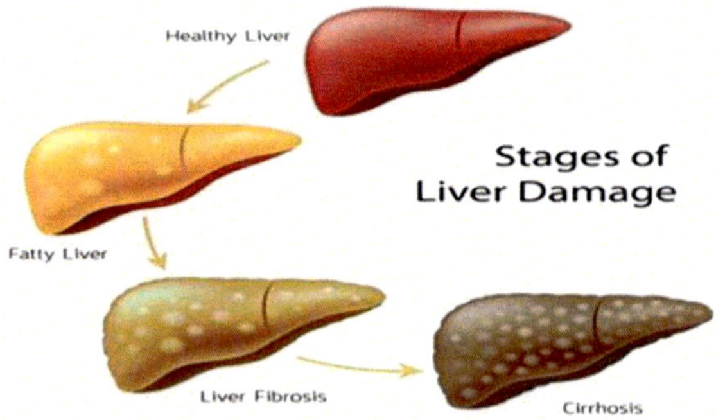

Fig. 1 Normal stages of liver damage

as age of the patient, bilirubin range, alkaline phosphatase, and aspartate aminotransferase. Machine learning has greatly impacted the prediction and diagnosis of liver disease in the biomedical field. Machine learning provides a guarantee to boost the identification and prediction of disease of concern in the biomedical field and also to increase the objectivity of decision taking. Medical problems can be quickly solved using machine learning techniques and the diagnostic costs would be reduced. For this reason two classification techniques to identify patients with or without liver disease are used (Fig. 1).

1.1 Symptoms and Signs

Liver malady shows some kind of symptoms and signs which include:

- Yellow skin and eyes
- Loss of appetite
- Pain or tenderness in the abdominal area
- Bruises easily visible than before
- Dark urine color.

1.2 Causes

Liver disease causes may include:

- Infections like Hepatitis A, B, C
- Genetics
- Cancer and other growths
- Immune system abnormality.

1.3 *Statistical Information*

Liver malady is very common nowadays and occurs worldwide irrespective of age, sex, and region. Almost 10 lakh patients of liver cirrhosis are newly diagnosed every year in India and according to WHO, Liver malady is the 10th most common cause of death in India. Out of every 5 Indians 1 is affected with liver disease. Liver cancer is the 16th leading cause of death and Cirrhosis is the 11th most common cause of death worldwide; together, they contribute to almost 3.5% of all deaths globally. According to the Global Burden of Disease (GBD) study, published in BMC Medicine, one million people died of cirrhosis in 2010, and one million suffer from hepatic cancer.

2 Literature Survey

Estimating the Surveillance of Liver Disorder Using Classification Algorithms [1]

The experiment was performed by Aneeshkumar and Jothi Venkateswaran [1] using a total of 2453 actual data consisting of 15 parameters that was collected from a Public Charitable Hospital located in Chennai. They used two classification approaches namely Naive Bayesian and C4.5 decision tree. For the total dataset, 2230 cases with liver disease and 223 cases are not. The total datasets were divided into the ratio for 50–50, 75–25, 90–10, and the accuracy was evaluated. Consequently, the highest precision (99.20%) lies in the decision tree of C4.5 with a splitting ratio of 90–10.

Comparative Prediction Performance with Support Vector Machine and Random Forest Classification Techniques [2]

Aljahdali and Hussain [2] advised that classification approaches can be employed to increase accuracy. SVM and RF are used with variations in kernel and kernel parameters to learn, identify, and compare data on cancer, heart, and liver disease. The findings are evaluated using a predictive performance measure. Results with Radial Base Function with SVM are more accurate and they can be compared with Random Forest technique.

Liver Patient Classification Using Intelligent Techniques [3]

Gulia et al. [3] proposed artificial intelligence approaches for liver patient classification. This paper deals with the chosen algorithms for liver patient's classification (J-48 algorithm, RF algorithm, Multi-Layer Perceptron algorithm, SVM algorithm, and Bayesian Network algorithm). SVM algorithm is said to give highest accuracy (71.355) prior to the application of feature selection. But, after the application of feature selection using Greedy Stepwise Technique on 11 attributes, the Random Forest algorithm is seen as the best output algorithm. Therefore, the results of the classification algorithms are compared and the Random Forest algorithm has outscored.

Liver Disease Prediction Using SVM and Naive Bayes Algorithm [4]

Vijayarani and Dhayanand [4] used the methods of data mining such as grouping, clustering, rules of association, etc. Naive Bayes and SVM are the algorithms used in this project. The former has optimum precision as compared with FT Tree algorithm and Naive Bayes algorithm. Using SVM and Wrapper approach, an improved precision rate was observed in the medical lab test cost with minimal processing time from experimental tests. Changed Particle Swarm Optimization and SVM Least Squares resulted in the highest accuracy in classification. Naive Bayes algorithm accuracy is 61.28, and SVM is 79.66.

Threshold Doses and Prediction of Visually Apparent Liver Dysfunction After Stereotactic Body Radiation Therapy in Cirrhotic and Normal Livers Using Magnetic Resonance Imaging [5]

Doi et al. [5] examined the MRI of cirrhotic and normal liver patients after stereotactic body radiation therapy (SBRT) to find out the threshold dose for focal liver damage. For these 54 cirrhotic patients with hepatocellular carcinoma (HCC) and 10 non-cirrhotic patients with liver metastases a total of 64 patients underwent SBRT for liver tumors. The threshold dose for hepatic dysfunction was calculated by determining individually.

Diagnosis of Liver Diseases Using Machine Learning [6]

Sontakke et al. [6] in an aim to show the benefits of Machine Learning, explained the techniques of predicting liver diseases from genetic microarrays to the computer science community. For these two approaches were used one is a symptomatic approach to diagnosis and the other genetic approach. The first method involves ANN model to classify the patients as having liver malady and not having liver malady. The second one is applying Micro-Array Analysis to ANN and MLP.

Analysis of Classification Algorithms for Liver Disease Diagnosis [7]

Ghosh and Waheed [7] suggested that the main aim of this research work should be focused on predicting liver disease using classification algorithms namely Naive Bayes algorithm, Bagging algorithm, Logistic Regression algorithm, and REP Tree algorithm based on data collected from the liver patient. The dataset consisted 583 records with 11 features collected from Andhra Pradesh. Compared to other algorithms Bagging algorithm was found to perform better as it had highest performance with lowest error rate.

Liver Disease Classification Using Deep Learning Algorithm [3]

Anand and Neelanarayan (2019) proposed to separate the helpful data of the patient using the ANN algorithm for the liver issue report. A Hybridized Artificial Neural Network is used to predict liver disease Function using C4.5 estimation, SVM, KNN, and Neural Network to find out patient has liver disease using ILPD dataset. This paper uses 11 different features from the dataset for liver disease prediction. From all the algorithms, ANN obtained better results.

A Comparative Study on Liver Disease Prediction Using Supervised Machine Learning Algorithms [8]

Rahman et al. [8] developed an successful diagnostic method for patients with chronic liver infection using six distinctive supervised machine learning classifications. Logistic Regression LR classifier offers the highest order accuracy 75% depending on the F1 test to predict liver disease and Naive Bayes NB gives 53% the least accuracy.

Diagnosis of Liver Disease Using Machine Learning Techniques [9]

Jacob et al. [9] conducted the project with machine learning approaches by comparing different classification algorithms based on different performance metrics [10]. Four classification algorithms Logistic Regression, SVM, KNN, and ANN were considered in this study. Of all the algorithms used ANN has outscored with an accuracy of 92% and is considered as the best model for the prediction.

3 Dataset Description

3.1 ILPD (Indian Liver Patient Dataset)

The ILPD dataset is collected from the UCI Machine Learning Repository. This data set is categorical and includes two classes that is patient has liver disease or patient does not have liver disease. It has a total of 583 records which includes 416 records of liver patients and 167 records of non-liver patients. The dataset was obtained from northeast Andhra Pradesh, India. In this dataset 75.64% that is 441 are male and 24.36% that is 142 are female patients. Table 1 contains dataset of 11 different features while further study 10 parameters and 1 parameter as target class is chosen.

3.2 Correlation of Dataset

Correlation value indicates how strongly two different characteristics move in conjunction. Its value lies between −1 and 1 and it can be positive or negative (Fig. 2).

In positive correlation, the features are directly proportional, i.e. if one feature increases the other increases. In negative correlation, the features are indirectly proportional, i.e. if one feature increases the other decreases. Correlations nearer to −1 or 1 indicate a strong relationship whereas closer to 0 indicate a weak relationship. Correlation 0 indicates no relationship.

Table 1 Dataset parameters and its description

No	Attributes	Attribute type	Description	Range
1	Age	Numeric	Patient's age	51 ± 10
2	Sex	Nominal	Patient's gender	M/F
3	Total bilirubin	Numeric	Bilirubin breaks down heme in vertebrates	0.1–1.2 (mg/dL)
4	Direct bilirubin	Numeric	Bilirubin is graded direct after gluconic acid conjugation in the liver	0.1–1.2 (mg/dL)
5	Alkaline phosphatase	Numeric	It helps down proteins in the body	40–129 U/L
6	Alanine aminotransferase	Numeric	It is blood test to find damage to liver	7–55 (U/L)
7	Aminotransferase aspartate	Numeric	Higher level suggests liver damage	8–48 U/L
8	Total proteins	Numeric	Albumin and globulin	6–8 (g/dl)
9	Albumin	Numeric	It nourishes tissues	3.5–5.0 (g/dL)
10	Albumin and globulin ratio	Numeric	It measures concentration of proteins	8–61 U/L
11	Result	Numeric	0 indicates no liver disease and 1 indicates no liver disease	0/1

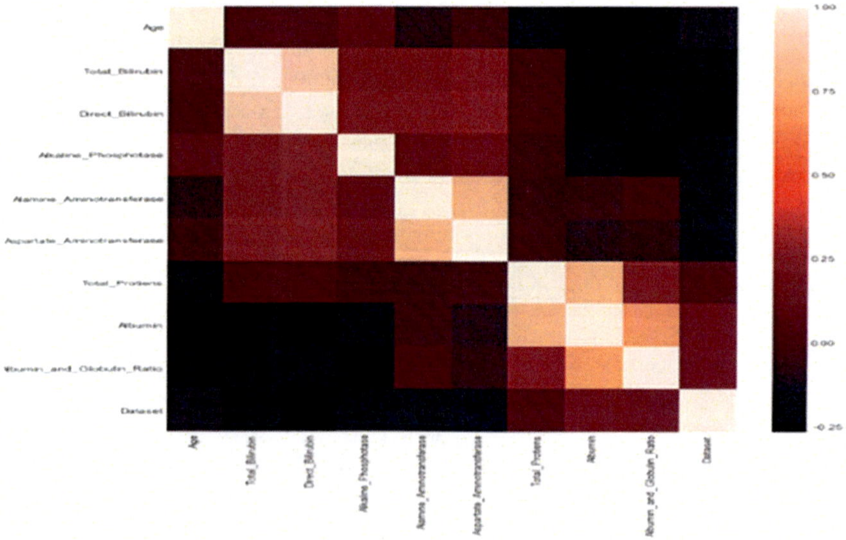

Fig. 2 Correlation heat map of dataset values

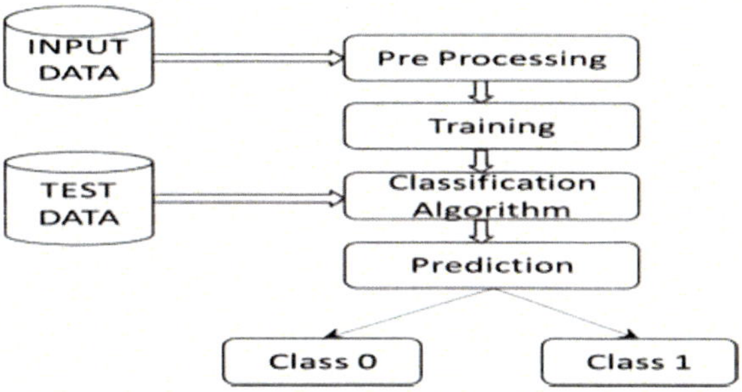

Fig. 3 System architecture of prediction algorithm

4 System Architecture

Data is the foundation for any real time project. Data Gathering is very important process as the data that is going to be collected will directly influence the model that is going to build (Fig. 3).

After the data collection process is complete, need to pre-process the data to eliminate any potential data imbalances. Now the collected data is split into training data and validation data. Majority of data is used for Training and is generally called training data and the leftover data is used for testing and is called Testing data. Training the model well results in good results. Final step involves evaluating the performance of our model using the testing data this performance measure represents how model actually performs in the real world. The next step involves selecting the model among many models scientists have created that best suits our data and requirements. After choosing the model, training the model where data is used in incremental manner to improve the model's ability to predict. Once the training process is completed, testing process is done.

5 Algorithms

5.1 Support Vector Machine

A support vector machine (SVM) is a supervised type of machine learning model that uses classification algorithms for classification problems in two classes. Integration of Support vector machine with collection of connected, supervised learning methods are used for regression and classification. The SVM is the advanced technology used in mathematical learning theory, with full classification algorithms (Fig. 4).

These model methods can be applied in both linear data classification and nonlinear data classification. The classification function is performed by SVM by maximizing the margin by classifying both classes while reducing classification errors. SVM algorithm is often used as a regression and classification prediction tool

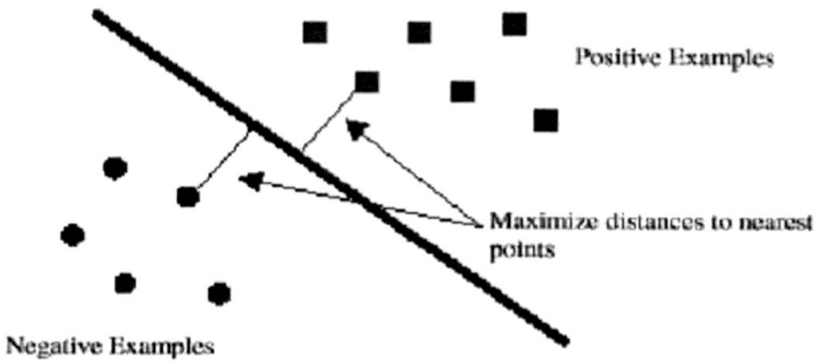

Fig. 4 Working method of SVM

which applies machine learning principles for optimization of accuracy predicted while avoiding over fitting. A better methodology would also avoid over fitting of data.

5.2 Artificial Neural Network

Artificial Neural Network (ANN) is called as the connectionist system. This is a method of information processing which is influenced by the way information is processed by the biological neural network which is the human brain. The main goal is to build a system that performs specific computational tasks faster than the conventional systems. These tasks include the identification and classification of patterns, approximation, optimization, and clustering of data. This includes a huge number of highly interconnected processing units that integrate to solve a particular problem. Messages passing through the network can influence the configuration of ANN when a neural network changes or learns depending on the I/O.

Artificial neural networks are known to be nonlinear mathematical data pre-processing methods where the hierarchical interactions between I/O are recreated or correlations are discovered. ANN has the ability to self-learn which enables them to produce better results as more data becomes available. The neurons are organized into several layers in the Neural Network. Each layer is connected to the other layers on both sides where the layers on one side receive input signals that the network requires to know or process and the layers on the other side deal with feedback and information responses. There are hidden units in between those two layers. The inter-connection in between the hidden units and the output unit with all other units in the layers are known as weights. In this paper, Jupyter notebook and Kaggle notebook as a tool and python 3.7 as a programming language is used (Fig. 5).

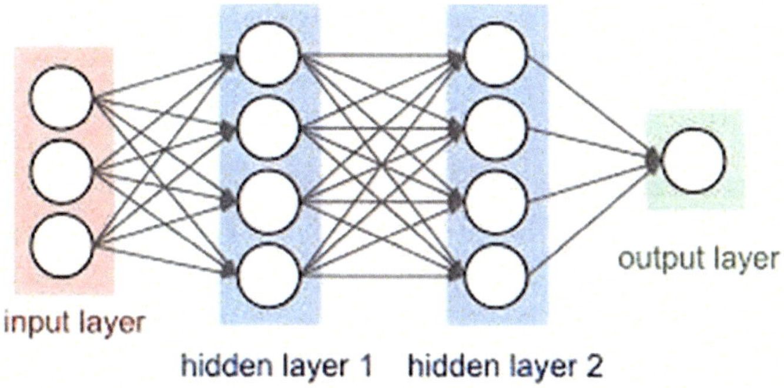

Fig. 5 Working method of ANN

Table 2 Results obtained using SVM algorithm

Accuracy	Precision	F1 score
0.78%	0.78%	0.86%

6 Results and Analysis

In our data collection all the attributes reflect the most favorable signs of patients leading to liver disease. Choosing the best attribute among all the attributes becomes a difficult task, as only effectively produce our results based on this attribute. The class label in present dataset which is categorical is also in the test dataset as label data have collected. Then converted nominal values in the class label to binary values 0 or 1 by importing label encoder and applied cross validation in order to divide dataset into testing and training dataset where 80% is for training and 20% is for testing. Then applied the Artificial Neural Network and the support vector machine to compute the result.

6.1 Metrics

Accuracy = (True Positives + True Negatives)/Total

Precision = True positives/(True positives + False Positives)

F1 Score = 2 * ((Precision * Recall)/(Precision + Recall)).

Table 3 Results obtained using ANN algorithm

Accuracy	Precision	F1 score
0.94%	1.0%	0.96%

6.2 Results Obtained Using SVM Algorithm

Table 2 and Table 3 present the results obtained from SVM and ANN algorithms respectively.

7 Conclusion

The use of predicting systems is necessary for prescient examination of the disease as it enables us to foresee the disease in advance and also save individuals' lives through the expectation of fix. So, here in this paper presented two methods for predicting liver disease in patients using classification algorithms. Support Vector Machine (SVM) and Artificial Neural Network (ANN) were implemented to predict if the patient is suffering from liver disease or not suffering from liver disease. The data from the report after the liver test is fed into the model to predict the disease. The dataset used was the Indian Liver Patient Dataset taken from the UCI repository. Thus, based on the characteristics of the dataset that is utilized, here make use of selected algorithms in order to better predict the desired outcomes. The ANN model is generally a good choice when the data is categorical and when dealing with huge datasets. While, SVM model tends to give better results when the data is categorical. The performance of SVM and ANN was evaluated and ANN performed better with an accuracy of 95% whereas SVM obtained 78% accuracy. The metrics like accuracy, precision can be enhanced by using various mixtures of algorithms and by tuning the hyper parameters of the model. Therefore, Artificial Neural Network is considered to be the best model with better accuracy and least execution time

References

1. Aneeshkumar, A.S., Jothi Venkateswaran, C.: Estimating the surveillance of liver disorder using classification algorithms. Int. J. Comput. Appl. 0975-8887 (2012)
2. Aljahdali, S., Hussain, S.N.: Comparative prediction performance with support vector machine and random forest classification techniques. Int. J. Comput. Appl. **69**(11) (2013)
3. Gulia, A., Vohra, R., Rani, P.: Liver patient classification using intelligent techniques. Int. J. Comput. Sci. Inf. Technol. **5**(4), 5110–5115 (2014)
4. Vijayarani, S., Dhayanand, S.: Liver disease prediction using SVM and Naïve Bayes algorithms. Int. J. Sci. Eng. Technol. Res. (IJSETR) **4**(4), 816–820 (2015)
5. Doi, H., et al.: Threshold doses and prediction of visually apparent liver dysfunction after stereotactic body radiation therapy in cirrhotic and normal livers using magnetic resonance imaging. J. Radiat. Res. **57**(3), 294–300 (2016)

6. Sontakke, S., Lohokare, J., Dani, R.: Diagnosis of liver diseases using machine learning. In: 2017 International Conference on Emerging Trends & Innovation in ICT (ICEI). IEEE (2017)
7. Ghosh, S.R., Waheed, S.: Analysis of classification algorithms for liver disease diagnosis. J. Sci. Technol. Environ. Inform. **05**(01)
8. Rahman, A.K.M.S., et al.: A comparative study on liver disease prediction using supervised machine learning algorithms. Int. J. Sci. Technol. Res. **8**(11), 419–422 (2019)
9. Jacob, J., et al.: Diagnosis of liver disease using machine learning techniques. Int. Res. J. Eng. Technol. **5**(04) (2019)
10. Himabindu, G., Ramakrishna Murty, M., et al.: Extraction of texture features and classification of renal masses from kidney images. Int. J. Eng. Technol. **7**(2.33), 1057–1063 (2018)

A Fuzzy Based Approach for Indian Standard Classification of Soils

A. Sujatha, L. Govindaraju, N. Shivakumar, and V. Devaraj

1 Introduction

Expert Systems are systematic computer programs which provide the solution of problems depending on task specific knowledge and inference techniques at the level of a human expert [1]. These systems are initiated from the study of artificial intelligence, which is a user-friendly interactive computer program that integrates the prowess of an expert or a group of experts in a well-defined domain. The data used in various judgment models is affected by unpredictability leading to ambiguity. A first source of unpredictability comes from the uncertainty of the data, due to the subtle nature of social and natural attributes. Another type of unpredictability is indefinite that may be due to the values obtained from a measuring instrument or the observer who undertakes the task. Fuzzy expert system has proved to deal with such unpredictabilitys to solve the real-world complex problems [2].

Fuzzy expert system is one such system introduced by Kandel and Schneider [3]. This is an intelligent tool capable of making decisions and also deals with ambiguous data. FES has improved the excellence, effectiveness and quality in recent times.

A. Sujatha (✉) · N. Shivakumar
Department of Mathematics, RV College of Engineering, Bengaluru, Karnataka, India
e-mail: sujathaa@rvce.edu.in

N. Shivakumar
e-mail: shivakumarn@rvce.edu.in

L. Govindaraju
Department of Civil Engineering, UVCE, Bangalore University, Bengaluru, Karnataka, India
e-mail: lgr_civil@yahoo.com

V. Devaraj
Department of Civil Engineering, Ambedkar Institute of Technology, Bengaluru, Karnataka, India
e-mail: professor.devaraj@gmail.com

C. K. Mai et al. (eds.), *Machine Learning Technologies and Applications*,
Algorithms for Intelligent Systems,
https://doi.org/10.1007/978-981-33-4046-6_6

FES has been applied in numerous real world problems such as numerical classification of soil and mapping, land evaluation, slope stability, rock engineering, tunneling, project management, waste water treatment, online scheduling, performance indexing, computer security, gesture recognition, medical diagnosis, agricultural problem to deal the vagueness by mimicking the human way of thinking [4]. Expert System was applied in classification of soil for agricultural purposes where in computer-aided soil classification was developed involving substantial number of rules, inter-parametric associations and subjective considerations [5–7]. Research in this area then developed on two frontages, a computer algorithm for the rule-based inference process was developed [8], and a representation of subjective evaluations was obtained by the application of fuzzy logic [9–11]. Fetz et al. have developed a research work for application of fuzzy models in geotechnical engineering based on interval analysis on α-cut set [12]. In 1998, Hayo M. G. et al., suggested a fuzzy expert system to compute an index of "Ipest" which contemplates an expert insight of the potential environmental impact by applying pesticides in the field [13]. Many other applications on FES include: tunnel boring machine performance modeling [14] prediction of liquefaction [15], interpretation of a model footing response [16], compacted soils swelling potential, modeling of soil shearing resistance angle using soft computing systems [17]. In 2011 Adoko reviewed the applications based on fuzzy inference technique in Geotechnical Engineering [18]. In 2014, Muhammad Akram et al., presented a detailed review on the application of expert system and its potent advantages in the Civil Engineering field [19]. In 2014, T. S. Umesha., et.al, developed a fuzzy model for contaminated parameters of soil [20].

The goal of this study is to develop an interactive, user-friendly fuzzy rule-based system using fuzzy IF–THEN rules to quantify Indian Standard classification of soils for engineering purpose in qualitative terms considering the index properties of soils. Fuzzy rules are generated based on the membership functions defined for index properties for classification of soils.

1.1 Review of Fuzzy Set Theory

Fuzzy set theory is a flexible tool that can be tolerated for imprecise data. The most important advantage is that unlike other modeling techniques, the opinion of experts can be utilized in fuzzy set theory.

1.1.1 Definition

Let D be a nonempty set. A fuzzy set Q in D is characterized by its membership function $A : D \to [0, 1]$ and A is interpreted as the degree of membership of element l in fuzzy set Q for each $l \in D$. In fuzzy set theory, fuzzy sets are denoted by membership functions. Membership functions are selected randomly in practice. The most widely used membership functions are usually represented in triangular, trapezoidal, Gaussian forms. The triangular function defined by parameters a, m and b, where$a \leq \mathrm{m} \leq \mathrm{b}$

$$A(g) = \begin{cases} 0 & g \leq e \\ \frac{g-e}{n-e} & e \leq g \leq n \\ 1 & g = n \\ \frac{f-g}{f-n} & n \leq g \leq f \\ 0 & g \geq f \end{cases} \quad (1)$$

1.1.2 Operations on Fuzzy Sets

Let J and K be two fuzzy sets of the universe of discourse Y with membership function $J(g)$ and $K(g)$, respectively.

- The union of the fuzzy sets $J\&K$ is defined by $J \cup K(g) = \max[J(g), K(g)]$, $\forall g \in Y$.
- The intersection of J and$K$, is defined by $J \cap K(x) = \min[J(g), K(g)], \forall g \in Y$.
- The complement of J, denoted by $\bar{J}$, is defined by $\bar{J}\ (g) = 1 - J(g), \forall g \in Y$.

1.1.3 Fuzzy Classification System

Rule-based fuzzy classification systems require the input factors to be defined through fuzzy sets and partitions. In the preceding part of the rules, the fuzzy variables represent the input factors, and the consequent part is a class to which the system belongs. A standard fuzzy classification principle can be expressed by

R_l: *If* P_l*is* Y_l*and* P_2*is* Y_2*and .... and* P_m*is* Y_m*then Class* $= D_l$

where R_j is the rule identifier, $P_1, \ldots, P_m$ are the input variables of the example considered in the problem (represented by fuzzy sets), $P_{1,\ldots,} P_m$ are the linguistic values used to describe the input values, then $D_i \in D$ is the class to which the system belongs to. Figure 1 represents a flow diagram of a fuzzy rule-based system. An

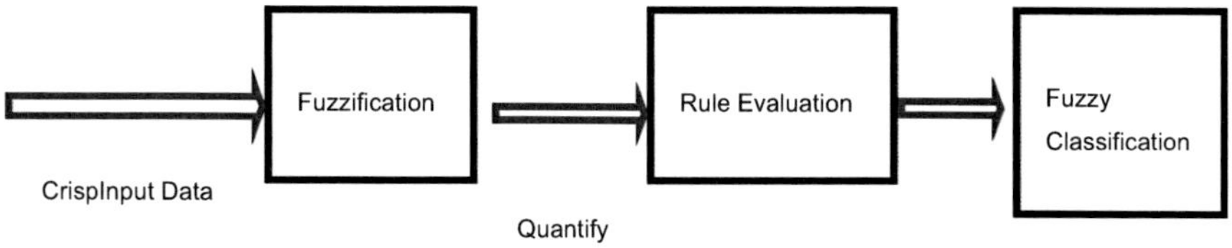

Fig. 1 Flow diagram of fuzzy expert system

example is categorized by comparing each rule in the Fuzzy rule base to a particular class to which it belongs [21]. The sum of compatibility degrees for a set of fuzzy rules for each class is calculated by using the fuzzy reasoning method and the class with highest sum is used to classify the example considered.

2 Indian Standard Classification System

Vagueness exists in the input parameters as their relation to classification is not crisp. In Fuzzy approach system, fuzzy sets are defined for the input parameters. A fuzzy set represents the set of values that a parameter can take. All the data contained in a fuzzy set is described by its membership function. The parameters are represented in the form of membership functions covering the probable range of values that a parameter can take in most of the situations. The inputs are taken to discover the degree to which they belong to each of the suitable fuzzy sets through membership functions. In this study, a triangular function is used for fuzzification of the inputs.

This study aims to construct fuzzy rule-based system for Indian Standard (IS) classification system in qualitative terms considering the following six index properties of soils namely; Particle size smaller than 4.75 mm, Particle size smaller than 0.075 mm, uniformity coefficient, coefficient of curvature, liquid limit, plastic limit and the output is Indian Standard classification of soil consisting of twenty five subgroups as shown in Table 1 [22, 23].

3 Procedure for Fuzzy Classification

The fuzzy decision tree induction method [24–27] comprises of the following steps.

1. Fuzzification of the input parameters
2. Induction of a fuzzy decision tree
3. Conversion of the decision tree into a set of rules
4. Application of the fuzzy rules for classification.

Fuzzification is the process of making a crisp quantity fuzzy. In this study, the input parameters to the system are particle size smaller than 4.75 mm, particle size smaller than 0.075 mm, uniformity coefficient, coefficient of curvature, liquid limit, and plastic limit. Ambiguity or vagueness exists in all these parameters as their relation to classification is not crisp. In Fuzzy rule-based system the parameters representing inputs are represented in the form of fuzzy sets. A fuzzy set represents the set of values that a parameter can take. All the data contained in a fuzzy set is described by its membership function. The parameters are represented in the form of membership functions covering the probable range of values that a parameter can take in most of the situations. The inputs are taken to ascertain the degree to which they belong

Table 1 Outputclasses

Type	Soil subgroup
1	Well graded gravel (GW)
2	Poorly graded gravel (GP)
3	Well graded sand (SW)
4	Poorly graded sand (SP)
5	Silty gravel (GM)
6	Silty sand (SM)
7	Clayey gravel (GC)
8	Clayey sand (SC)
9	Well graded gravel with clay (GW-GC)
10	Well graded gravel with silt (GW-GM)
11	Poorly graded gravel with clay (GP-GC)
12	Poorly graded gravel with silt (GP-GM)
13	Well graded sand with clay (SW-SC)
14	Well graded sand with silt (SW-SM)
15	Poorly graded sand with clay (SP-SC)
16	Poorly graded sand with silt (SP-SM)
17	Silty gravel–clayey gravel (GM-GC)
18	Silty sand–clayey sand (SM-SC)
19	Inorganic/organic silt with low plasticity (ML/OL)
20	Clay with low plasticity (CL)
21	Silty clay with low plasticity (CL-ML)
22	Inorganic/organic silt with intermediate plasticity (MI/OI)
23	Clay with intermediate plasticity (CI)
24	Inorganic/organic silt with high plasticity (MH/OH)
25	Clay with high plasticity (CH)

to each of the appropriate fuzzy sets through membership functions. In this study, a triangular function is used for fuzzification of the inputs.

In this study, Triangular membership functions are defined for each of these fuzzy sets and the membership values for any given input parameters are obtained. Fuzzy decision tree algorithm is used to construct fuzzy rules for IS classification of soils.

The membership function for particle size smaller than 4.75 mm size is divided into five ranges denoted by fuzzy descriptors namely Very Low (VL), Low (L), Medium (M), High (H) and Very High (VH).

The membership function developed for the above fuzzy descriptors are

$$A_{VL}(p) = \begin{cases} \frac{25-p}{25} & 0 \le p \le 25 \\ 0 & p \ge 25 \end{cases} \tag{2}$$

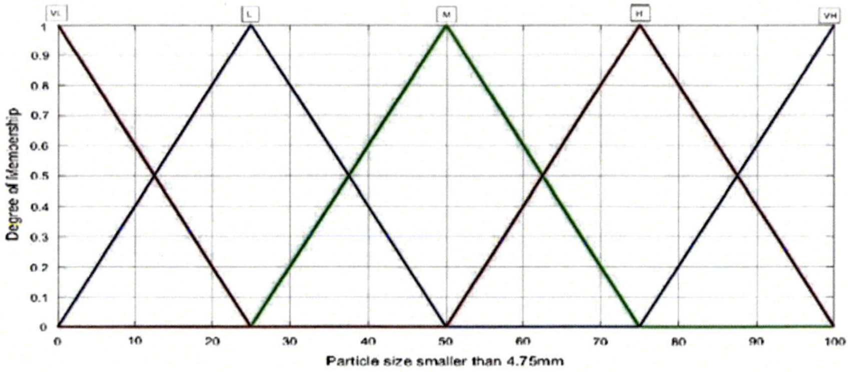

Fig. 2 Fuzzy membership function for particle size smaller than 4.755 mm

$$A_L(p) = \begin{cases} \frac{p}{25} & 0 \le p \le 25 \\ \frac{50-p}{25} & 25 \le p \le 50 \end{cases} \tag{3}$$

$$A_M(p) = \begin{cases} \frac{p-25}{25} & 25 \le p \le 50 \\ \frac{75-p}{25} & 50 \le p \le 75 \end{cases} \tag{4}$$

$$A_H(p) = \begin{cases} \frac{p-50}{25} & 50 \le p \le 75 \\ \frac{100-p}{25} & 75 \le p \le 100 \end{cases} \tag{5}$$

$$A_L(p) = \left\{ \frac{p-75}{25} \; 75 \le p \le 100 \right. \tag{6}$$

Figure 2 represents the triangular fuzzy membership function developed. The fuzzy membership value for any given coarse fraction passing 4.75 mm IS Sieve can be obtained.

The membership functions for the input variables namely particle size smaller than 0.075 mm, uniformity coefficient, coefficient of curvature, liquid limit, plasticity index are divided into nine ranges denoted by fuzzy descriptors namely very-very-low (VVL), very-low (VL), low (L), low-medium (LM), medium (M), medium–high (MH), high (H), very-high (VH), very-very-high (VVH) as shown in Table 2. The triangular membership functions developed for the input variables in terms of fuzzy descriptors are shown in Figs. 3, 4, 5, 6 and 7. The membership values for any of the input variables can be obtained [28].

4 Fuzzy Rules Constructed for IS Classification System

1. If (particle size smaller than 4.75 mm is VL, L) Λ (particle size smaller than 0.075 mm is VVL, VL) Λ (uniformity coefficient is M) Λ (coefficient of curvature is M) ⇒ (the soil is classified as GW).

Table 2 Fuzzy sets for the input variables

Linguistic variables (membership functions) and its parameters	Input variables					
	Particle size smaller than 4.75 mm	Particle size smaller than 0.075 mm	Uniformity coefficient	Coefficient of curvature	Plasticity index	Liquid limit
VVL	–	[0, 2.5]	–	–	–	–
VL	[0, 25]	[0, 5]	[0, 2]	–	[0, 2]	[0, 17.5]
L	[0, 50]	[2.5, 8.5]	[0, 4]	[0, 1]	[0, 4]	[0, 35]
LM	–	[5, 12]	[2, 5]	[0, 5.2]	[2, 5.5]	[17.5, 42.5]
M	[25, 75]	[8.5, 31]	[4, 6]	[1, 3]	[4, 7]	[35, 50]
MH	–	[12, 50]	[5, 8]	[2, 6.5]	[5.5, 33.5]	[42.5, 75]
H	[50, 100]	[31, 75]	[6, 10]	[3, 10]	[7, 60]	[50, 100]
VH	[75, 100]	[50, 100]	[8, 10]	[6.5, 10]	[33.5, 60]	[75, 100]
VVH	–	[75, 100]	–	–	–	–

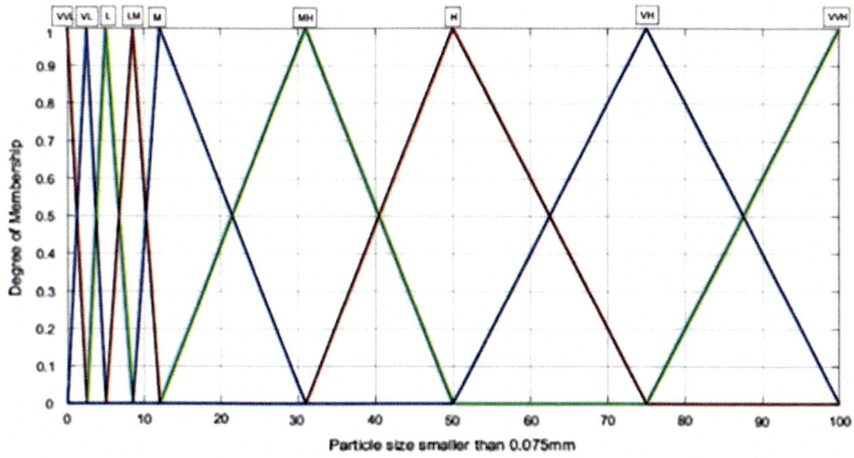

Fig. 3 Fuzzy membership function for particle size smaller than 0.075 mm

2. If (particle size smaller than 4.75 mm is VVL, L) Λ (particle size smaller than 0.075 mm is VVL, VL) Λ (uniformity coefficient is VL, L) Λ (coefficient of curvature is L, H, VH) ⇒ (the soil is classified as GP).

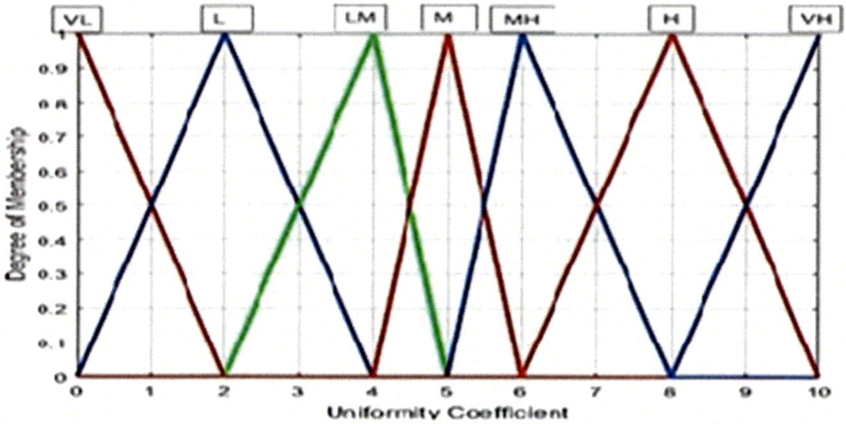

Fig. 4 Fuzzy membership function for uniformity coefficient

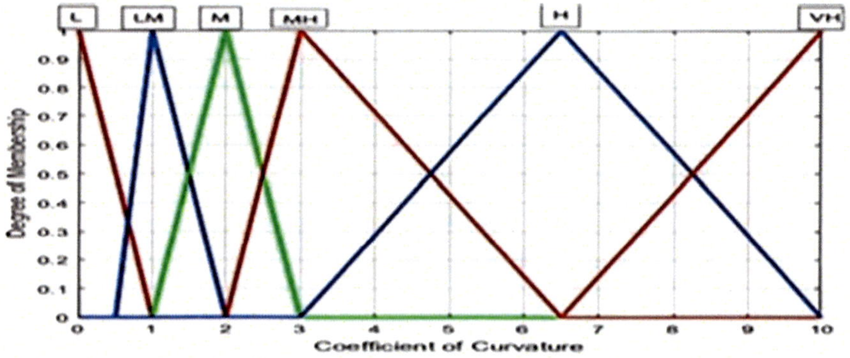

Fig. 5 Fuzzy membership function for coefficient of curvature

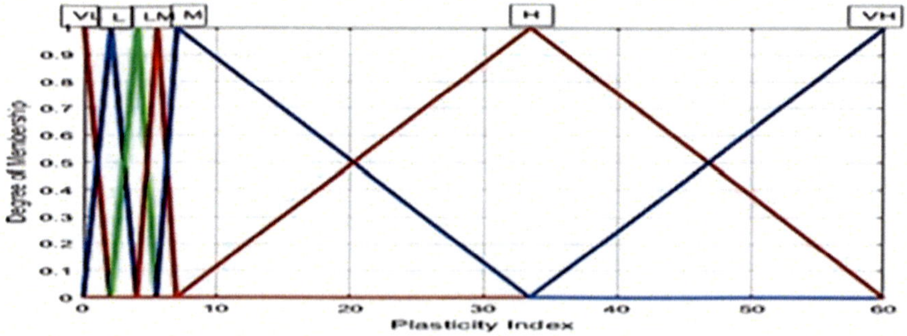

Fig. 6 Fuzzy membership function for plasticity index

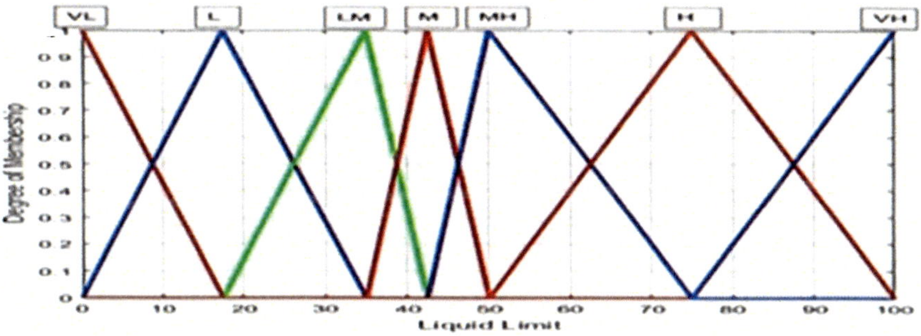

Fig. 7 Fuzzy membership function for liquid limit

3. If (particle size smaller than 4.75 mm is H, VH) Λ (particle size smaller than 0.075 mm is VVL, VL) Λ (uniformity coefficient is H, VH) Λ (coefficient of curvature is M) ⇒ (the soil is classified as SW).
4. If (particle size smaller than 4.75 mm is H, VH) Λ (particle size smaller than 0.075 mm is VVL, VL) Λ (uniformity coefficient is VL, M) Λ (coefficient of curvature is L, H, VH) ⇒ (the soil is classified as SP).
5. If (particle size smaller than 4.75 mm is VL, L) Λ (particle size smaller than 0.075 mm is MH, H, VH, VVH) Λ (plasticity index is VL, L) ⇒ (the soil is classified as GM).
6. If (particle size smaller than 4.75 mm is H, VH) Λ (particle size smaller than 0.075 mm is MH, H, VH, VVH) Λ (plasticity index is VL, L) ⇒ (the soil is classified as SM).
7. If (particle size smaller than 4.75 mm is VL, L) Λ (particle size smaller than 0.075 mm is MH, H, VH, VVH) Λ (plasticity index is H, VH) ⇒ (the soil is classified as GC).
8. If (particle size smaller than 4.75 mm is H, VH) Λ (particle size smaller than 0.075 mm is MH, H, VH, VVH) Λ (plasticity index is H, VH), ⇒ (the soil is classified as SC).
9. If (particle size smaller than 4.75 mm is VL, L) Λ (particle size smaller than 0.075 mm is LM) Λ (uniformity coefficient is M) Λ (coefficient of curvature is M) Λ (Plasticity index is H, VH) ⇒ (the soil is classified as GW-GC).
10. If (particle size smaller than 4.75 mm is VL, L) Λ (particle size smaller than 0.075 mm is LM) Λ (uniformity coefficient is M) Λ (coefficient of curvature is M) Λ (plasticity index is VL, L) ⇒ (the soil is classified as GW-GM).
11. If (particle size smaller than 4.75 mm is VL, L) Λ (particle size passing 0.075 mm is LM) Λ (uniformity coefficient is VL, L) Λ (coefficient of curvature is L, H, VH) and (plasticity index is H, VH) ⇒ (the soil is classified as GP-GC).
12. If (particle size smaller than 4.75 mm is VL, L) Λ (particle size passing 0.075 mm is LM) and (uniformity coefficient is VL, L) Λ (coefficient of curvature is L, H, VH) Λ (plasticity index is VL, L) ⇒ (the soil is classified as GP-GM).

13. If (particle size smaller than 4.75 mm is high, VH) Λ (particle size passing 0.075 mm is LM) Λ (uniformity coefficient is VH, H) Λ (coefficient of curvature is M) Λ (plasticity index is high, VH) ⇒ (the soil is classified as SW-SC).
14. If (particle size smaller than 4.75 mm is high, VH) Λ (particle size passing 0.075 mm is LM) Λ (uniformity coefficient is H, VH) Λ (coefficient of curvature is M) Λ (plasticity index is VL, L) ⇒ (the soil is classified as SW-SM).
15. If (particle size smaller than 4.75 mm is high, VH) Λ (particle size passing 0.075 mm is LM) Λ (uniformity coefficient is VL, L, M) Λ (coefficient of curvature is L,H, VH) Λ (plasticity index is H, VH) ⇒ (the soil is classified as SP-SC).
16. If (particle size smaller than 4.75 mm is H, VH) Λ (particle size passing 0.075 mm is LM) Λ (uniformity coefficient is VL, L, M) Λ (coefficient of curvature is L, H, VH) Λ (plasticity index is VL, L) ⇒ (the soil is classified as SP-SM).
17. If (particle size smaller than 4.75 mm is VL, L) Λ (particle size passing 0.075 mm is LM) Λ (plasticity index is M) ⇒ (the soil is classified as GM-GC).
18. If (particle size smaller than 4.75 mm is high, VH) Λ (particle size passing 0.075 mm is LM) Λ (plasticity index is M) ⇒ (the soil is classified as SM-SC).
19. If (particle size smaller than 0.075 mm is VH, VVH) Λ (liquid limit is VL, L) Λ (plasticity index is VL, L) ⇒ (the soil is classified as ML/OL).
20. If (particle size smaller than 0.075 mm is VH, VVH) Λ (liquid limit is VL, L) Λ (plasticity index is H, VH) ⇒ (the soil is classified as CL).
21. If (particle size smaller than 0.075 mm is VH, VVH) Λ (liquid limit is VL, L) Λ (plasticity index is M) ⇒ (the soil is classified as CL-ML).
22. If (particle size smaller than 0.075 mm is VH, VVH) Λ (liquid limit is M) Λ (plasticity index is VL, L) ⇒ (the soil is classified as MI/OI).
23. If (particle size smaller than 0.075 mm is VH, VVH) Λ (liquid limit is M) Λ (plasticity index is H, VH) ⇒ (the soil is classified as CI).
24. If (particle size smaller than 0.075 mm is VH, VVH) Λ (liquid limit is high, VH) Λ (plasticity index is VL, L) ⇒ (the soil is classified as MH/OH).
25. If (particle size smaller than 0.075 mm is VH, VVH) Λ (liquid limit is H, VH) Λ (plasticity index is H, VH) ⇒ (the soil is classified as CH).

5 Validation of Fuzzy Expert System

Twenty soil samples are considered for the validation of the developed fuzzy model. For validation of the developed fuzzy model, first the validation set (20 soil samples) was classified by fuzzy model intended for this aim and then the same set was classified using laboratory test results.

The proposed fuzzy rule-based system is considered for the IS classification of soil which is implemented in MATLAB. In the proposed work, twenty soil samples are considered for the illustration of the developed fuzzy rule-based system. The index values of the twenty soil samples considered are as shown in Table 3. The

Table 3 Outputs of the fuzzy rule-based model for soil samples

	Soil samples (test dataset)						Highest degree of possibility value	Active rule number	Soil group
	Particle size smaller than 4.75 mm	Particle size smaller than 0.075 mm	Uniformity coefficient	Coefficient of curvature	Plasticity index	Liquid limit			
Soil 1	25	10	–	–	20	15	0.8333	7	GC
Soil 2	10	3	4.5	3.5	–	–	0.64	1	GW
Soil 3	10	35	–	–	25	22	0.5989	5	GM
Soil 4	70	8	7	3	32	29	0.7529	14	SW-SM
Soil 5	100	100	–	–	40	32	0.8962	22	OI
Soil 6	20	8	4.8	2.5	40	36	0.8329	10	GW-GM
Soil 7	10	70	–	–	20	14	0.8038	21	CL-ML
Soil 8	35	2.6	4.8	2.3	0	0	0.8040	1	GW
Soil 9	85	80	–	–	28	25	0.71	19	ML
Soil 10	75	3	4.3	6	–	–	0.8314	4	SP
Soil 11	90	78	–	–	26	16	0.7710	20	CL
Soil 12	80	25	–	–	40	37	0.7284	6	SM
Soil 13	70	40	–	–	45	22	0.7334	8	SC
Soil 14	30	8	5.5	1.5	40	22	0.7198	9	GW-GC
Soil 15	100	75	–	–	42	36	0.9467	22	OI
Soil 16	78	10	4.5	4	45	39	0.8379	16	SP-SM
Soil 17	90	78	–	–	75	45	0.9148	24	OH
Soil 18	78	10	4.5	4	26	16	0.8819	15	SP-SC
Soil 19	70	40	–	–	45	22	0.9223	8	SC

(continued)

Table 3 (continued)

	Soil samples (test dataset)						Highest degree of possibility value	Active rule number	Soil group
	Particle size smaller than 4.75 mm	Particle size smaller than 0.075 mm	Uniformity coefficient	Coefficient of curvature	Plasticity index	Liquid limit			
Soil 20	85	80	–	–	33	30	0.7957	19	OL

index properties of soil are fuzzified and their membership values are calculated for the samples 1–20 and the values are as shown in Table 3. It uses the achieved definitive fuzzy rules to classify the samples and the degree of possibility for the output attributes that are shown in Table 3.

By using a numerical example illustration, the working method of the proposed model can be explained as follows:

Laboratory test results of a soil sample 1 are as shown in Table 3. At the first step, The Fuzzification of the input parameters yields the following fuzzy inputs (Figs. 3, 4, 5, 6 and 7) for the next step in the inference process:

Input 1: particle size smaller than 4.75 mm is L with membership degree 1.

Input 2: particle size smaller than 0.075 mm is L with membership degree 0.5714 and LM with membership degree 0.4286.

Input 3: Plasticity index is LM with membership degree 0.3333 and M with membership degree 0.6667.

Then, the fuzzified values were applied by the proposed model to initiate suitable rules, depending on the weight factor, the fuzzy output of each rule was calculated using MIN operator and grouped into one fuzzy output using MAX operator and the degree of possibility are calculated. Fuzzy rule 7 gets the highest degree of possibility (0.8333) among the fuzzy rules 1, 2, 3, 4, 5……0.25. Therefore, Sample 1 is classified as Clayey Gravel. This classification result coincides with the laboratory classification system.

The similar procedures were executed concurrently by the model for the crisp values in the dataset. The result of this procedure for the soil samples is summarized in Table 3.

The analysis done for sample 1 is repeated for sample 2. Fuzzy rule 1 gets the highest degree of possibility (0.64) among the fuzzy rules 1, 2, 3, 4, … 25. Therefore, Sample 2 is classified as *Well graded Gravel*. This classification coincided with the laboratory classification.

The analysis is repeated for samples 3, 4, 5 … 0.20. From Table 3, it is observed that soil samples 2 and 8 are classified as GW, sample 3 as GM, sample 4 as SW-SM, sample 5 and 15 as OI, sample 6 as GW-GM, sample 7 as CL-ML, sample 9 as ML, sample 10 as SP, sample 11 as CL, sample 12 as SM, sample 13 and 19 as SC, sample 14 as GW-GC, sample 16 as SP-SM, sample 17 as OH, sample 18 as SP-SC, sample 20 as OL. The soil class obtained from the developed fuzzy rule-based model coincides with the laboratory classification.

6 Conclusion

Mathematical models in deterministic form are used to solve the qualitative problems in engineering. But there are uncertainties due to complex nature of problem. In this paper, fuzzy rule-based system is developed using triangular membership function and 25 definitive fuzzy rules to quantify IS classification of soil in qualitative terms. The developed model is validated with laboratory test results indicating that the developed fuzzy rule-based system can be effectively used for IS classification of soil.

References

1. Bolloju, N., Schneider, C., Sugumaran, V.: A knowledge-based system for improving the consistency between object models and use case narratives. Expert Syst. Appl. (2012). https://doi.org/10.1016/j.eswa.2012.02.126
2. Zadeh, L.A.: Fuzzy sets. Inf. Control **8**(3), 338–353 (1965). https://doi.org/10.1016/S0019-9958(65)90241-
3. Kandel, A., Schneider, M.: Fuzzy sets and their applications to artificial intelligence. Adv. Comput. (1989). https://doi.org/10.1016/S0065-2458(08)60046-7
4. Zadeh, L.: Soft computing and fuzzy logic. IEEE Softw. **11**(6), 48–56 (1994)
5. McCracken, R.J., Cate, R.B.: Artificial intelligence, cognitive science, and measurement theory applied in soil classification. Soil Sci. Soc. Am. J.**50**(3), 557–561 (1986). https://doi.org/10.2136/sssaj1986.03615995005000030003x.
6. Soil Survey Division Staff: Soil survey manual. Soil Conservation Service. U.S. Department of Agriculture Handbook 18 (1993)
7. Soil Survey Staff: Soil survey manual agriculture. Handbook 18 (2017)
8. Galbraith, J.M., Bryant, R.B., Ahrens, R.J.: An expert system for soil taxonomy. Soil Sci. **163**(9), 748–758 (1998). https://doi.org/10.1097/00010694-199809000-00008
9. McBratney, A.B., Odeh, I.O.A.: Application of fuzzy sets in soil science: fuzzy logic, fuzzy measurements and fuzzy decisions. Geoderma **77**(2–4), 85–113 (1997). https://doi.org/10.1016/S0016-7061(97)00017-7
10. Hu, Z., Chan, C.W., Huang, G.H.: A fuzzy expert system for site characterization. Expert Syst. Appl. (2003). https://doi.org/10.1016/S0957-4174(02)00090-8
11. Burrough, P.A., Macmillan, R.A., van Deursen, W.: Fuzzy classification methods for determining land suitability from soil profile observations and topography. J. Soil Sci.**43**(2), 193–210 (1992). https://doi.org/10.1111/j.1365-2389.1992.tb00129.x
12. Fetz, T., Jäger, J., Köll, D., Krenn, G., Lessmann, H., Oberguggenberger, M., Stark, R.F.: Fuzzy models in geotechnical engineering and construction management. In: Analyzing Uncertainty in Civil Engineering, pp. 211–239 (2005)
13. Van Der Werf, H.M.G., Zimmer, C.: An indicator of pesticide environmental impact based on a fuzzy expert system. Chemosphere **36**(10), 2225–2249 (1998). https://doi.org/10.1016/S0045-6535(97)10194-1
14. Alvarez Grima, M., Bruines, P.A., Verhoef, P.N.W.: Modeling tunnel boring machine performance by neuro-fuzzy methods. Tunn. Undergr. Sp. Technol.**15**(3), 259–269 (2000). https://doi.org/10.1016/S0886-7798(00)00055-9
15. Rahman, M.S., Wang, J.: Fuzzy neural network models for liquefaction prediction. Soil Dyn. Earthq. Eng. **22**(8), 685–694 (2002). https://doi.org/10.1016/S0267-7261(02)00059-3
16. Provenzano, P., Ferlisi, S., Musso, A.: Interpretation of a model footing response through an adaptive neural fuzzy inference system. Comput. Geotech. **31**(3), 251–266 (2004). https://doi.org/10.1016/j.compgeo.2004.03.001
17. Kayadelen, C., Taşkiran, T., Günaydin, O., Fener, M.: Adaptive neuro-fuzzy modeling for the swelling potential of compacted soils. Environ. Earth Sci. **59**(1), 109–115 (2009). https://doi.org/10.1007/s12665-009-0009-5
18. Adoko, A.C., Wu, L.: Fuzzy inference systems-based approaches in geotechnical engineering—a review. Electron. J. Geotech. Eng. (2011)
19. Mayadevi, N., Vinodchandra, S.S., Ushakumari, S.: A review on expert system applications in power plants. Int. J. Electr. Comput. Eng. (2014). https://doi.org/10.11591/ijece.v4i1.5025

20. Umesha, M.A.J.T.S., Dinesh, S.V.: Fuzzy modeling for contaminated soil parameters. Int. J. Fuzzy Syst. Adv. Appl. **1**, 66–73 (2014)
21. Cordón, O., Del Jesus, M.J., Herrera, F.: A proposal on reasoning methods in fuzzy rule-based classification systems. Int. J. Approx. Reason. **20**(1), 21–45 (1999). https://doi.org/10.1016/S0888-613X(00)88942-2
22. Budhu, M.: Soil Mechanics and Foundations (2010)
23. Mishra, B.: Indian system of soil classification: a way forward. Agric. Res. Technol. Access J. **3**(1), 1–9 (2016). https://doi.org/10.19080/artoaj.2016.03.555606
24. Chen, S.M., Lee, S.H., Lee, C.H.: A new method for generating fuzzy rules from numerical data for handling classification problems. Appl. Artif. Intell. **15**(7), 645–664 (2001). https://doi.org/10.1080/088395101750363984
25. Schwartz, D.G., Klir, G.J., Lewis, H.W., Ezawa, Y.: Applications of fuzzy sets and approximate reasoning. Proc. IEEE (1994). https://doi.org/10.1109/5.282229
26. Bohlender, G., Kaufmann, A., Gupta, M.M.: Introduction to fuzzy arithmetic, theory and applications. Math. Comput.**47**(176), 762 (1986). https://doi.org/10.2307/2008199
27. Ross, T.J.: Fuzzy Logic with Engineering Applications, 3rd edn. Wiley (2010)
28. Sujatha, A., Govindaraju, L., Shivakumar, N.: Application of fuzzy rule based system for highway research board classification of soils. Int. J. Fuzzy Log. Syst.**10**(2), 1–14 (2020). https://doi.org/10.5121/ijfls.2020.10201

Blocking Mobile Based Games and Nullifying the Search String Containing Inappropriate Words

Tadivaka Sai Swetha, V. Baby, Chouda Sowsheel, Rasamsetti Himabindu, Bhukya Rohith, and Abdul Azeez Ahmed

1 Introduction

In this modernized world, everything right from education to shopping, payments, and professions are being digitalized. As a result, the usage of computers, mobile phones, and various digital devices has increased tremendously. In this era of digitalization, it is observed that every problem is being solved effectively using the technology and mobile phones, on the other side of the coin, there are many disadvantages too. Few of them are as follows.

1. Excessive usage of the devices
2. Misuse of the resources
3. Variations in behavioral patterns
4. Effects of mental health
5. Lack of physical activity.

T. Sai Swetha (✉) · V. Baby · C. Sowsheel · R. Himabindu · B. Rohith · A. A. Ahmed
Department of CSE, VNRVJIET, Secunderabad, Hyderabad, India
e-mail: tsaiswetha03@gmail.com

V. Baby
e-mail: baby_v@vnrvjiet.in

C. Sowsheel
e-mail: choudasowsheel98@gmail.com

R. Himabindu
e-mail: himabindu.rasamsetti@gmail.com

B. Rohith
e-mail: bhukyarohith491@gmail.com

A. A. Ahmed
e-mail: azeez0784@gmail.com

C. K. Mai et al. (eds.), *Machine Learning Technologies and Applications*,
Algorithms for Intelligent Systems,
https://doi.org/10.1007/978-981-33-4046-6_7

Due to these disadvantages, many people can get affected. Majority of the people who are prone to more usage of mobile phones are children and teenagers.

1.1 Children and Mobile Based Games

As education is being digitalized, instead of notes and textbooks every resource is being provided to the student in digital form in devices like tablets (tabs). A child is spending more time with the digital devices. In earlier days an individual used to spend leisure time or relax by reading books, playing outdoor games, developing arts such as music, painting, dance, etc. In the current scenario, it is being observed that not only for education even for the purpose of relaxation or spending the leisure time mobile phones or other forms of digital devices are being used. Most of the children are playing mobile-based games. Most popularly played games include PUBG, Clash of Clans, Mini Militia, Blue Whale Game, Pokémon Go, etc. Although playing these games is fun and entertaining, there are many drawbacks to it. Many horrifying incidents occurred due to addiction to those games. Not only addiction, there are many other side effects of playing mobile-based games. They are as follows:

1. The blue light that emits from the mobile phone damages the eyes of the individual.
2. Valuable time of the individual is wasted.
3. There will be lack of hobbies in the individuals.
4. Addiction to games may slow down the brain growth.
5. It can also cause sleeping disorders like Insomnia.
6. There can be damage to mental health as well as the physical health.
7. It can also lead to evolution of suicidal thoughts in an individual.

1.2 Searches Containing Inappropriate Terms

Due to the evolution of internet, accessing various kinds of resources have become easier. There is more chance of distraction if the resources are used in a wrong manner. Nowadays teenagers and kids are no less in using mobile phones when compared to youngsters and adults. If the individuals of such age groups search about the content that is not suitable to their age it might result in change of behavior and mindset. The age of teenagers is likely to know and explore about everything even though it is not relevant or good for them. It is observed that there are more chances for them to include bad or inappropriate words in their searches. Teenagers are much sensitive people to deal with, thus they have more chances in ending their lives for small things. Suicide, sleeping pills, porn, sex, death, depression are some of the irrelevant and inappropriate searches for their age. It is the responsibility of the parents, schools, colleges, and society to see that a kid or a teenager is behaving in a right way and also have a right mindset and attitude.

2 Literature Survey

Luo et.al. [1] proposed a framework to detect whether the android applications in the children category contain inappropriate content in the form of videos, audio, and pictures. This framework is based on the policies of maturity rating. It works with the help of three modules namely Simulation Operation Module, Capture Module, and Content Inspection Module. Thus, this framework helps the parents to differentiate Android applications that should not be applied for the kids under the age of 12 years.

Moreno and Salazar [2] proposed a system that reduces the risks and prevention of threats against children. This system is based on WhatsApp conversations. These conversations are sent to a central server and then they are analyzed and classified. If any threats were detected, an alert will be sent to the parent.

Elmogy and Elkhowiter [3] proposed a system which could monitor and control the usage of mobile phone by the child. This system allows the parent to view the overview of statistics of the usage of the mobile device. The results are displayed in graphical manner so that it will become easy for a parent to observe the child.

Haz et al. [4] proposed a system which is a web browser software that has features like controlling from single computer access, restriction of adult content pages, browsing schedules, blocking of pop-ups, etc.

Kharad and Kulkarni [5] proposed that usage of URL analysis and Keyword analysis together would filter the content effectively because even if the address of the website is changed by the owner, with the help of keywords inappropriate contents can be blocked.

Lochtefeld et al. [6] proposed a system which helps in decreasing the addiction to the mobile applications and games by allowing the parents to create three types of rules. First rule helps to prevent the usage at specific times of the day. Second rule helps to restrict the application forever. Third rule helps to restrict the usage of the application only for certain amount of time.

Ding et al. [7] proposed a system that uses URL analysis on the top of content analysis, text-based content analysis, and single pass content analysis to filter and block the websites.

Du et al. [8] proposed a system which performs web filtering by using text classification. The samples of banned web pages are used to block the web pages. The class of the web pages that must be blocked is first characterized and then websites are blocked on the basis of similarity.

Aristofany et al. [9] proposed a system which uses data mining algorithms on the internet browsing history of an individual to reduce the errors in identification of negative content websites.

Baishya and Kakoty [10] proposed a model to reduce the limitations such as decreased productivity, incorrect usage of legal issues, and network resources by integrating Expert filtering technology in content base domain area. The authors also discussed about various techniques for web filtering such as Keyword analysis, URL analysis, packet analysis, Rating System, Black listing and White listing, novel filters technique, etc., and concluded that most of these approaches have limitations and few of them have not fully filtered as per the requirement.

Dinh et al. [11] proposed a model that works on the basis of Naïve Bayes algorithm to filter and block the webpages that contain pornographic content. The content of those webpages is in Vietnamese. In this system, when the users enter the URL, its presence in the black list database is checked. If the URL is not present in that database, it's checked using the classifier and then added to the white list or black list and then the entire database is updated and again the URL is checked, if it's present in the black list database, it's blocked.

3 Existing Systems and Its Limitations

From the limitations of all the above-mentioned systems, Table 1, it can be inferred that the below are the common limitations. They are as follows

1. Child can uninstall the blocked application.
2. There is no nullification of the search string.
3. There is no option to block multiple applications for different time periods.
4. There is no option to send alerts to the parent about child's browsing history.

4 Proposed System

In order to solve the above-mentioned limitations, an application named "LaveBract" is designed. The word Lave means "clean." The proposed system has the capability to

1. Restrict the screen time of the applications.
2. To prevent the child from uninstalling the restricted application.
3. To set different time periods for different applications for restricting.
4. Restrict the child from using the websites based on keywords given by parent.
5. Parents can setup the keywords that should not be in search string.
6. Alert the parent about the search activity of the child.

From the limitations of existing systems, it can be understood that URL analysis and firewalls are most commonly used methods to block the websites. There is no option to nullify the search string containing inappropriate words before sending it to the bowser and also there is no system to alert the parent about child's browsing activity. This the most unique feature of the proposed application "Lave Bract."

There are three modules used in this application. They are

1. **Parent Module**: This module facilitates registration of the user, login of the user, display of installed applications in the device, and setting the time limit.

Table 1 Existing systems and its limitations

Name	Features	Limitations
Monitoring the child manually	Parent monitors the child manually and checks on the browser history	Delay in detecting the mental health of the child, child may feel the loss of freedom, manual effort is involved
Setting up parental controls on Google play	The type of content that can be downloaded is restricted on basis of maturity level, blocks few features of games, back up the data, allows the parent to manage kid's device, helps in monitoring the screen time	Unavailability in some of the countries, visibility of applications and movies that are downloaded before parental control is assed even if they are not in the range of the rating set, it does not filter 100% of sexually explicit books
Fami safe app	Requires of installation of app on both child and parent devices, tracks location, monitors the activity of the child, availability of screen time tracking and control, blocks websites	No frequent update in location, existing geo-fence settings cannot be edited, cannot set different periods of screen time to limit the phone usage, websites cannot be blocked on basis of words. The restricted application can be uninstalled
Parental control secure kids app	It has facility to block the webpages and applications, has a feature to set alarms on child device, locates and blocks the child devices, displays the statistics of usage of apps on the device	Child can uninstall the restricted application, there is no word-based website blocking, the application crashes on the kid's device
Google family link app	Requires two separate applications, i.e. one for parent and other for child device respectively, allows the parent to add another parent to supervise the child, set password rules, monitor the number of screens unlock attempts and incorrect passwords typed, lock the screen control, disabling camera, etc.	Allows the child to uninstall the blocked application, no keyword-based website blocking, does not allow the parent to remotely see the child usage of a particular app, can't show the past searches or history and also don't notify the parent about irrelevant searches

2. **Child Module**: This module facilitates restricting/blocking/hiding of the selected application and prevents the uninstall of the restricted application.
3. **Application Module**: This module facilitates nullification of search string, sending alerts to the parent, identify and find the synonyms of inappropriate words.

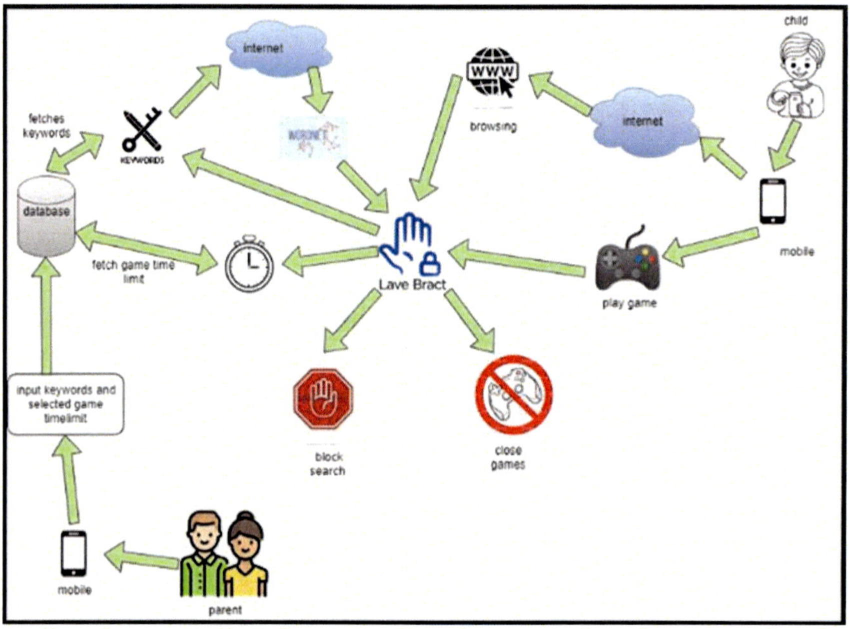

Fig. 1 System architecture of Lave Bract

Figure 1 depicts the system architecture of the application.

5 Implementation

Lave Bract is implemented with the help of XML, Java [12], Natural Language Processing [13], Android [14] Studio, IBM Rational Rose, and the following concepts of python.

- File concepts are used to store the search history of the individual.
- Google search module is used to search the given query if it is appropriate.
- Web browser module is used to open the URL in the browser of the device.
- Nltk package helps to find synonyms of the inappropriate words.
- Smtplibmodule is used to alert the parent through email.

While implementing this application, standards of Clean Code [15] are followed and this application is tested using various kinds of testing [16, 17] techniques.

5.1 Working

1. As shown in Fig. 1, first the parent uses Lave Bract application to give input (keywords/timer).
2. All the given details are stored in the database.
3. Now, whenever child starts playing games, this application fetches the timer details. If the time exceeds the limit set by the parent, the application that is being used by the child is closed and hidden, otherwise go to step (4).
4. Allow the child to play the game until the time limit is over.
5. When the child browses about anything, this application searches for inappropriate words or their respective meanings.
6. If such words are found then this application nullifies the search string and sends alert to the parent, otherwise goto step (7).
7. Display the results for the query searched by the child.
8. Stop.

5.2 Experimental Results

After successful registration and login into Lave Bract, parent views the list of applications installed in the device. From that list parent selected an application which is a game named "2048." The selection of the application and setting time limit is depicted in Fig. 2. The time set by the parent is 2 min. When child plays the game, after completion of 2 min, the game "2048" is hidden. As the application is hidden, it helps to prevent the uninstallation by the child. This time setting remains intact and can be modified by the parent whenever needed. An option is given to the parent for unhiding the restricted application also.

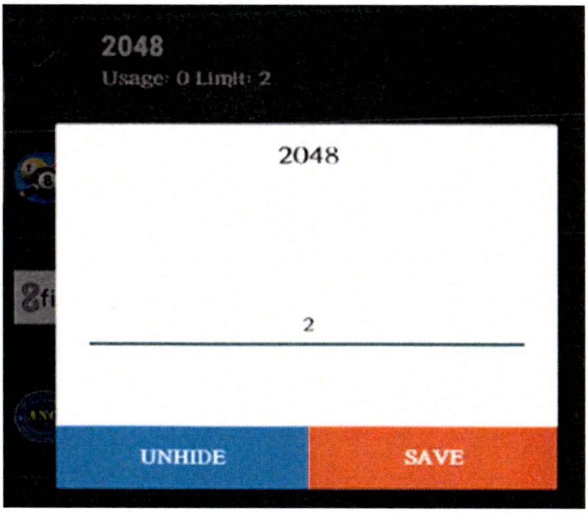

Fig. 2 Parent setting time limit in minutes

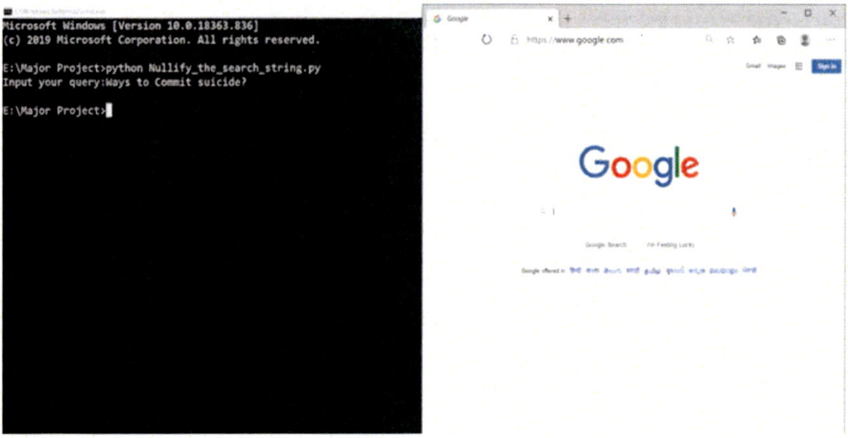

Fig. 3 Result of query containing inappropriate words in CLI application

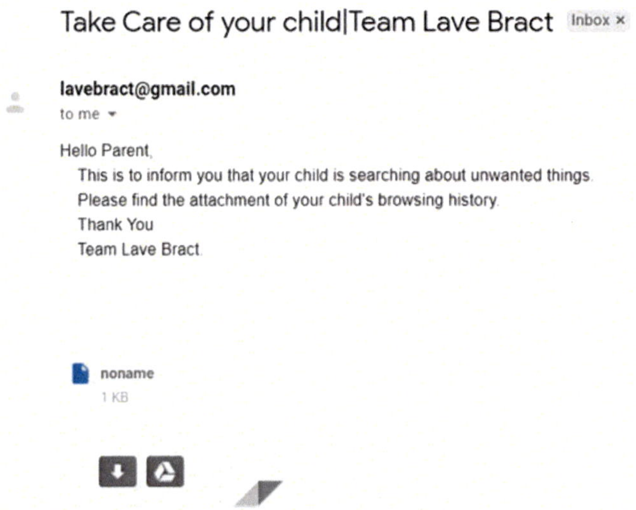

Fig. 4 Email alert that is sent to the parent

If the child opens the browser and searches for any query. Searching of query using CLI application is depicted in Fig. 3 and Searching of query using GUI application is depicted in Fig. 6. Here the search query is "Ways to commit suicide?" The following activities occur in the background before showing the results.

1. Firstly, all the punctuations are removed from the search query.
2. Then the search string is converted into lowercase and leading and trailing spaces are removed.

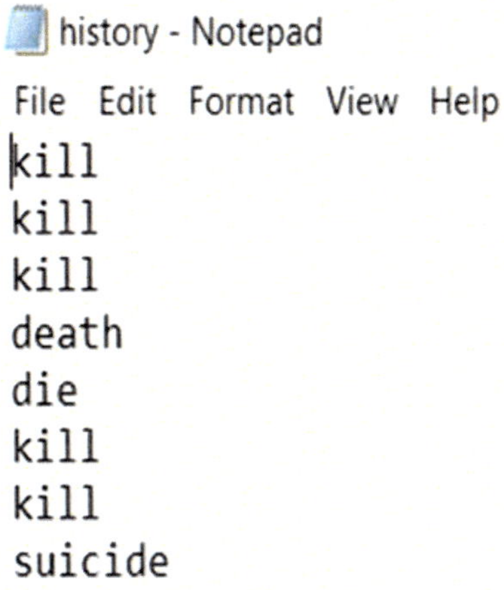

Fig. 5 Browsing history of child

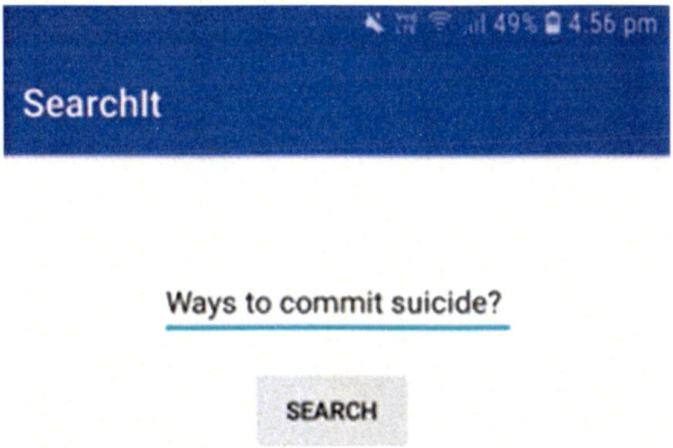

Fig. 6 Child's search containing inappropriate words

3. Now, the input query is tokenized. The tokens for the given input are as follows. "ways," "to," "commit,", "suicide."
4. Now each token is parsed and checked for the presence of inappropriate words or meanings of an inappropriate words. So, in the query that is taken
   - "ways," "to," "commit" are not inappropriate words
   - suicide is an inappropriate word.

1. As the inappropriate word is present in the query, it is added to a file as shown in Fig. 5 and then the search string is nullified and a blank page is displayed to the child and an email alert is sent to the parent. Result of searching this query in CLI application is depicted in Fig. 3 and the result of searching this query in GUI application is depicted in Fig. 7. The email alert that is sent to the parent via

Fig. 7 Result of child's query

CLI application is depicted in Fig. 4 and the email alert that is sent to the parent via GUI application is depicted in Fig. 8.

As of now the following words and its synonyms are considered as inappropriate. They are kill, killed, killing, sex, porn, death, die, suicide, molest, molestation, molested, suicidal, coupling, intimacy, liquor, terrorism, and names of harmful drugs.

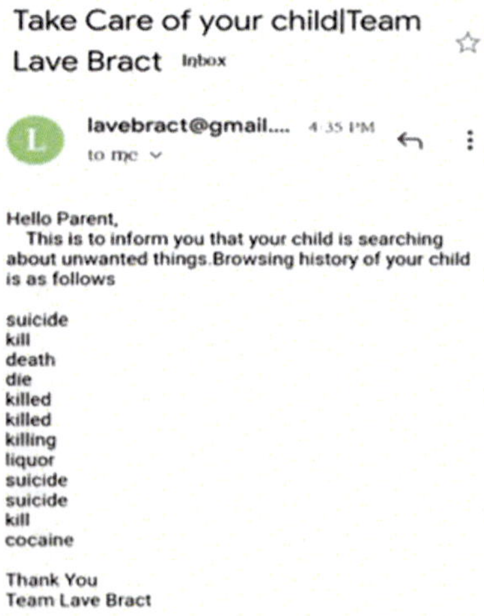

Fig. 8 Email alert that is sent to the parent

6 Conclusion and Future Scope

Hence, by using the concepts of Android "Lave Bract" is able to restrict the app usage after the time limit that is set by the parent and also prevent the uninstall of the restricted application by the child and by using the concepts of Natural Language Processing this application is able to nullify the search string, if it contains any inappropriate words before sending it to google and alert the parent regarding the activity of the child successfully. With this application a parent can identify the mental status of the child and steer him or her toward right path. So, If Lave Bract is used with the existing approaches, the parental control can be done in most effective way without making child feel that his or her freedom is lost. This application is further enhanced by using Machine Learning [18] Algorithms to identify the inappropriate words, adding the support of wordnet to the GUI application and deploying onto different platforms by converting it into a cross platform application using Ionic [19] framework.

Reference

1. Luo, Q., Liu, J., Wang, J., Tan, Y.: Automatic content inspection and forensics for children android apps. IEEE Internet Things J. 1–12. IEEE(2020)
2. Moreno, E.F.C., Salazar, G.K.C.: Development of an application for parental control of WhatsApp on android mobile devices. In: 2019 International Conference on Information Systems and Software Technologies (ICI2ST), pp. 16–23. IEEE(2019)
3. Elmogy, A.M., Elkhowiter, K.: Parental control system for mobile devices. Int. J. Comput. Appl. 16–23 (2017)
4. Haz, L., Guarda, T., Zambrano, I., Sánchez, C.: Internet Based parenting control application on teenagers. In: 2017 12th Iberian Conference on Information Systems and Technologies (CISTI). IEEE (2017)
5. Kharad, V.S., Kulkarni, S.S.: Design model on website filtering and blocking. Int. J. Adv. Res. Comput. Sci. Softw. Eng. 247–249 (2015)
6. Lochtefeld, M., Bohmer, M., Ganev, L.: AppDetox: helping users with mobile app addiction. In: MUM '13: 12th International Conference on Mobile and Ubiquitous Multimedia, pp. 1–2 (2013)
7. Ding, C., Chi, C.-H., Deng, J., Dong, C.-L.: Centralized content-based web filtering and blocking: how far can it go? In: IEEE SMC'99 Conference Proceedings. 1999 IEEE International Conference on Systems, Man, and Cybernetics (Cat. No. 99CH37028), pp. 115–119. IEEE(1999)
8. Du, R., Safavi-Naini, R., Susilo, W.: Web filtering using text classification. In: The 11th IEEE International Conference on Networks, 2003, ICON 2003, pp. 325–330. IEEE(2003)
9. Aristofany, A., Saptawati, G.A.P., Asnar, Y.: Internet browsing history data analysis for automatic negative content website identification (Case Study: TRUST+™ Positif). In: 2018 5th International Conference on Data and Software Engineering (ICoDSE). IEEE (2018)
10. Baishya, A., Kakoty, S.: A review on web content filtering, its technique and prospects. Int. J. Comput. Sci. Trends Technol. (IJCST) 37–40 (2019)
11. Dinh, T.-A., Ngo, T.-B., Vu, D.-L.: A model for automatically detecting and blocking pornographic websites. In: 2015 Seventh International Conference on Knowledge and Systems Engineering (KSE), pp. 244–249. IEEE
12. Schildt, H.: Java, The Complete Reference, 7th edn. McGraw Hill Professional (2006)

13. Klein, E., Loper, E.: Natural Language Processing with Python, 1st edn. O'Reilly Media (2009)
14. Jerome, J.F., Marzio, D.: Android, A Programmer's Guide, 1st edn. McGraw Hill (2010)
15. Martin, R.C.: Clean Code: A Handbook of Agile Software Craftsmanship, 1st edn. Prentice Hall PTR (2008)
16. Pressman, R.S.: Software Engineering—A Practitioner's Approach, 5th edn. McGraw Hill Education (2001)
17. Naik, K., Tripathy, P.: Software Testing and Quality Assurance Theory and Practice, 2nd Revised edn. Wiley-Blackwell (2016)
18. Bishop, C.M.: Pattern Recognition and Machine Learning, 1st edn. Springer (2006)
19. Griffith, C.: Mobile App Development with Ionic, 1st edn. O'Reilly Media (2017)

A Methodology to Retrieve Information from Ontologies with the Application of D2R Mapping and SPARQL

Mahavadi Meghana, R. Kranthi Kumar, Vadaguri Srikala, N. Sahithi, and M. Jyothsna

1 Introduction

Earlier the information retrieval was performed from database that takes complex queries for retrieval and may not bring result with respect to user interests. To overcome this drawback, people started using "Resource Description Framework (RDF)" ontologies that describes the data as relationship between the concepts in data. In simple words Ontology is a information tank which are interlinked and describes it in the form of concepts, properties, and attributes of various other concepts. An ontology containing distinct concepts with its instances of classes forms base for knowledge in [1] RDF is a form of representing ontologies in the form of graph. RDF is a graphical representation of Ontologies whereas TURTLE is its textual syntax. The ambiguous nature of the retrieval of information incorporates the necessity to the user to enable meaningful information retrieval. With information being the foremost important part of the many applications, projects the utilization of the technologies that use semantic web, ontologies to try and do this is often becoming more and more attractive. A set of methods and tools are developed that are used for information retrieval where semantics can be observed automatically from the results. In our work, we adopt a methodology to retrieve information of user interest from ontologies using SPARQL for querying and D2R for mapping data to RDF format.

M. Meghana (✉) · R. K. Kumar · V. Srikala · N. Sahithi · M. Jyothsna
V.N.R Vignana Jyothi Institute of Engineering and Technology, Hyderabad, India
e-mail: meghanasony1998@gmail.com

R. K. Kumar
e-mail: kranthikumr.rudrarapu@gmail.com

V. Srikala
e-mail: srikala98@gmail.com

N. Sahithi
e-mail: nunnasahithi777@gmail.com

M. Jyothsna
e-mail: mandepudijyothsna@gmail.com

C. K. Mai et al. (eds.), *Machine Learning Technologies and Applications*,
Algorithms for Intelligent Systems,
https://doi.org/10.1007/978-981-33-4046-6_8

2 Literature Survey

Information Retrieval is a capability of retrieving meaningful data from large amounts of data. The meaningful data refers to information. Ontology-driven information retrieval can be carried out by retrieving knowledge from Ontologies which may be in the form of RDF or TURTLE. For retrieval, we perform querying where semantically processed queries are sent to the Information Retrieval engines for relevant information retrieval. The most recent work in the domain is the work that explores the structure, execution, and use of a semantic methodology driven for retrieving the real-word data concerning to water supply management network [2]. This work aims to the process of design, implementation, and usage of ontology to record, store, integrate data was done by Escobar et al. [2]. The semi-automated retrieval is performed to retrieve information from reference collected data from which ontologies are constructed. Semantic search is similar searching by understanding the semantics of data based on keywords and other inputs [3]. When it comes to accuracy of retrieval, it is well performed in [4] but the execution is done using complex queries. Similarly, in [5], the searching application is built but query complexities are a major drawback. The work in [6] include a semantically designed search engine for XML. Here, advanced retrieval techniques are used to retrieve documents semantically relevant to user's request in ranking order. Here, we can observe retrieval performed bymany techniques and ranking them which is time taking [7]. The semi-automated ontologies-based IR systems were implemented based on keywords of documents where a graphical tool was developed for the process [8]. In [9] ontologies are built and are visualized for geological data. Exploratory analysis is also performed on these ontologies for querying and comparing information from multiple sources. Similarly, Kara et al. [10] performed semantic querying on ontologies for information retrieval by applying "semantic indexing". Guan et al. [11] proposed a mechanism where user was given privilege of processing queries which was later modified by semantic query processing module. Similar works include [12–14].

3 Limitations in Existing Systems

In the existing systems, we see how the technology helps in current world and how it has wide applications.

- Not fully automated IR system.
- Aids information retrieval which is not fully relevant but is based on reference documents which are previously collected the semi-automate IR system is built. Built simple ontology model where one can analyze the data structures of insurance data stored in ontologies but not the advanced features like retrieval and prediction.
- Using ontologies, the data exchange took place where the ranking for various techniques haven't took place, but retrieval from ontologies is performed by taking

many techniques and ranking of techniques into the account, which is time taking and process becomes with no ease.
- The exploration of various techniques and ranking of those techniques took place which performs retrieval of information but does not account for fastness and quick results are not expected.

4 Proposed System

Information retrieval from ontologies is a process where relevance of retrieved Information plays an important role. By considering the above drawbacks, we are proposing a system that will adopt process of information retrieval from ontologies where information retrieval is performed after databases are mapped to ontologies and then later the retrieval page is plugged to HTML web page. D2R Mapping maps data from the relational databases to RDF format. SPARQL is a query language that plays an important role in querying information from ontologies, i.e., RDF or TURTLE format [15]. We first take databases and map them to ontologies using D2R mapping where we get mapped data in RDF format then SPARQL query is passed on the RDF data in order to retrieve desired output. Finally, SPARQL end point is connected to HTML web page where it becomes user friendly for querying and viewing information. SPARQL Endpoint is a access point for an HTTP protocol extension where HTTP network will be able to receive and process SPARQL protocol requests. Here the process is fully automated, no relevant data is previously collected. The process here undergoes is a single process that brings quick and accurate results. Here, our main aim lies in ease of the process and bringing user-friendly platform.

5 Methodology

The process of information retrieval from ontology is performed by following a few sub-processes. The sub-processes include the following.

- Mapping databases to ontologies, i.e., RDF format.
- Retrieving information from ontologies.

5.1 Mapping Databases to Ontologies

The data present in the form of tables, i.e., rows and columns format can be transformed to RDF graphical format as well as TURTLE, i.e., textual syntax of RDF. "RDF is resource description framework" represents ontologies in graphical format. RDF follows tree-structured graphical data representation. "Terse Resource Description Framework" (TURTLE) is textual representation , and "Resource Description Framework" (RDF) data model is graphical. Mapping of data is done by D2R mapping for which first we should install D2R server in our system and also download Java MySQL connector.jar file. After performing the above operations, we need to go to command prompt from the path where D2R server is installed and its respective files are stored. Its advised to store the D2R server downloaded folder in D drive of system, i.e., data drive and make sure Java MySQL connector jar file exists inside D2R server folder, i.e., "D:/d2rq0.8.1/mysql-connector-java-5.1.39". Before all this Java must be installed in the system. Path setup should be done where we add D2R server directory to windows path, and it must be verified too. This is the preliminary setup required to start the actual process. Next step is to map database, i.e., to create a RDF mapping for MySQL database using D2R server. With reference to [16], we execute command "generate-mapping -o outputfilename.ttl -u youruserid -p yourpassword jdbc:mysql://localhost/yourdatabasename" The output of mapping is stored in .ttl format in "D:/d2rq0.8.1/outputfilename.ttl". After executing the above steps now we have to start D2R server web interface by following the below steps: Run command in command prompt "d2r-server outputfilename.ttl". We can view D2R web interface at "http://localhost:2020". Finally, we dump databases by creating a RDF mapping of a MySQL database using D2R server. To perform this in Windows command prompt, we run command "dumprdf -f RDF/XML -b http://localhost:2020/outputfilename.ttl ¿ outputfilename.rdf". After executing we can view our RDF dump file in "D:/d2rq-0.8.1/outputfilename.rdf".

5.2 *Retrieving Information from Ontologies*

To retrieve information from ontologies, we execute command in SPARQL format. SPARQL is query language which is used to retrieve and manipulate data in RDF format [17].

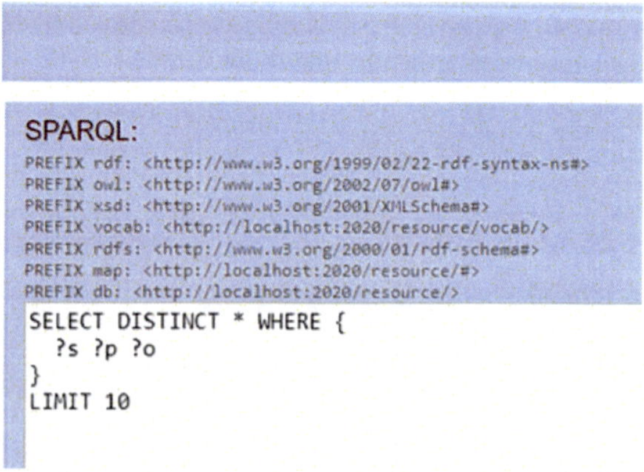

Finally, we connect SPARQL end point to HTML web page where it becomes user friendly to retrieve information based on entered ID of each data. HTML form connects the entered values to a Python script that plugs the values to SPARQL query and formats the result as HTML. The steps for the the above process include Firstly, Write a SPARQL query that requests for specific information Next, Create a web page in HTML form where the end user can enter the value with respect to the query. Finally, Add the SPARQL query to a Python common gateway interface script that takes the values passed from the web form, connects them into the respective places in the query, sends the query to endpoint, and later displays the result as HTML. With this one can view results of our information retrieval in html web page.

6 Results

After data of relational database undergoes into D2R mapping the data is transformed to RDF format and TURTLE format. The data in the form of rows and columns is transformed first to TURTLE(.ttl) format by D2R mapping where we create RDF mapping to for MySQL database using D2R server. The start page of D2R server is viewed as follows:

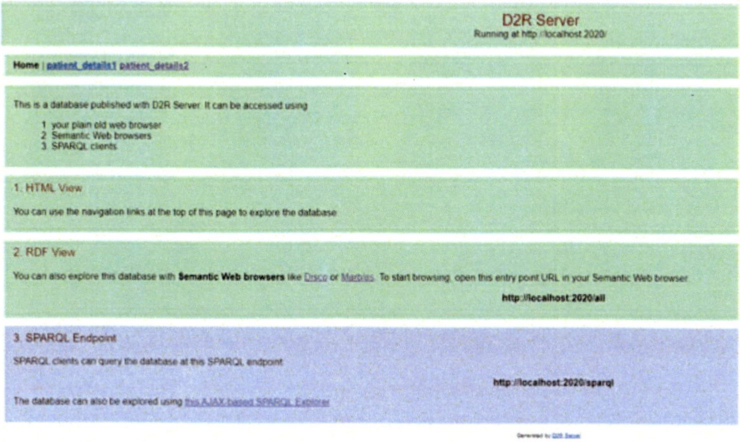

The TURTLE format of data can be represented as following patient details database which was previously of relational database format.

```
@prefix map: <#> .
@prefix db: <> .
@prefix vocab: <vocab/> .
@prefix rdf: <http://www.w3.org/1999/02/22-rdf-syntax-ns#> .
@prefix rdfs: <http://www.w3.org/2000/01/rdf-schema#> .
@prefix xsd: <http://www.w3.org/2001/XMLSchema#> .
@prefix d2rq: <http://www.wiwiss.fu-berlin.de/suhl/bizer/D2RQ/0.1#> .
@prefix jdbc: <http://d2rq.org/terms/jdbc/> .

map:database a d2rq:Database;
        d2rq:jdbcDriver "com.mysql.jdbc.Driver";
        d2rq:jdbcDSN "jdbc:mysql://localhost/patient_database?useSSL=false";
        d2rq:username "root";
        jdbc:autoReconnect "true";
        jdbc:zeroDateTimeBehavior "convertToNull";
        .
    # Table patient_details
map:patient_details a d2rq:ClassMap;
        d2rq:dataStorage map:database;
        d2rq:uriPattern "patient_details/@@patient_details.id@@";
        d2rq:class vocab:patient_details;
        d2rq:classDefinitionLabel "patient_details";
        .
map:patient_details__label a d2rq:PropertyBridge;
        d2rq:belongsToClassMap map:patient_details;
        d2rq:property rdfs:label;
        d2rq:pattern "patient_details #@@patient_details.id@@";
```

For the above TURTLE format data in RDF graph is

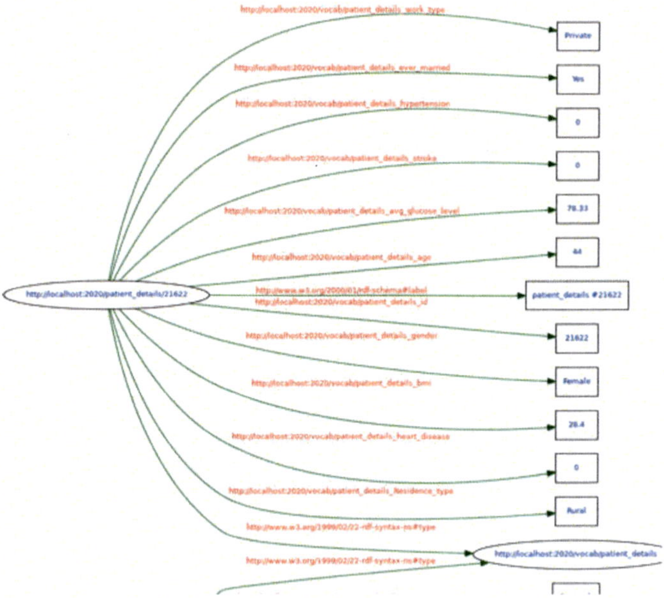

To retrieve information from ontologies we execute command We execute command "SELECT DISTINCT ?hasValue ?isValueOf WHERE ¡ http://localhost:2020/resource/patientdetails/36306 ¿property?hasV alueUNION http://localhost:2020/resource/patientdetails/36306 ¿property ?hasValue ORDER BY (!BOUND(?hasValue)) ¿property?hasValue?isValueOf"ž

SPARQL results:

s	p	o
db:patient_details2/36306	vocab:patient_details2_Residence_type	"Urban"
db:patient_details2/36306	vocab:patient_details2_hypertension	"0"
db:patient_details2/36306	vocab:patient_details2_ever_married	"Yes"
db:patient_details2/36306	vocab:patient_details2_age	80
db:patient_details2/36306	vocab:patient_details2_gender	"Male"
db:patient_details2/36306	vocab:patient_details2_smoking_status	"formerly smoked"
db:patient_details2/36306	vocab:patient_details2_work_type	"Private"
db:patient_details2/36306	rdfs:label	"patient_details2 #36306"
db:patient_details2/36306	vocab:patient_details2_id	36306
db:patient_details2/36306	vocab:patient_details2_avg_glucose_level	83.84

After performing this we connected this information retrieval process to HTML page and create a user friendly platform to retrieve information from ontologies.

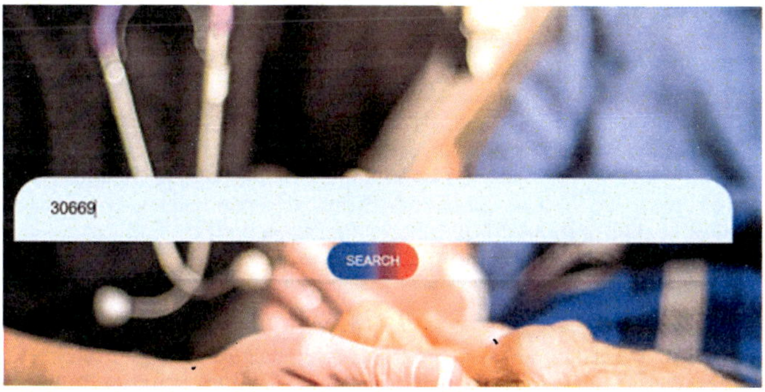

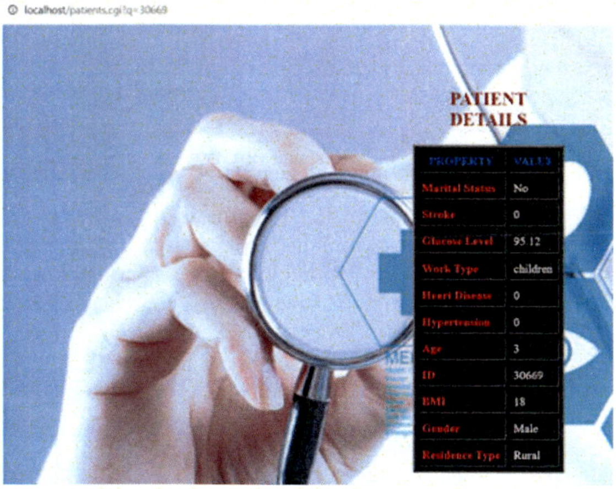

7 Conclusion

A simple retrieval system is proposed where a web page in front end takes input from user and provides required information in the output. The back-end process include the patient's databases that are converted to ontologies using D2R mapping and after getting RDF data, and these data are queried using SPARQL query language. Then finally SPARQL endpoint is connected to HTML web page using python script. This process is performed on hospital patient database. The proposed system is a general retrieval system, and it can be more advanced in future when technologies like machine learning and deep learning are integrated. This paper gives us a glimpse of how semantic web works, can how queries are processed using SPARQLE. Finally, the complete process is done in back-end where results are viewed in front-end web page which serves as retrieval system.

References

1. Noy, N.F., McGuinness, D.L.: Ontology development 101: a guide to creating your first ontology. Stanford Knowledge Systems Laboratory Technical Report KSL-01-05 and Stanford Medical Informatics Technical Report SMI-2001-0880, March 2001
2. Escobar, P., Roldán-García, M. del M., Peral, J., et al.: An ontology-based framework for publishing and exploiting linked open data: a use case on water resources management. J. Appl. Sci. **10**, 779 (2020). https://doi.org/10.3390/app10030779
3. Chi, N.-W., Jin, Y.-H., Hsieh, S.-H.: Developing base domain ontology from a reference collection to aid information retrieval. J. Automat. Construct. **100**, 180–189 (2019). https://doi.org/10.1016/j.autcon.2019.01.001
4. Alkahtani, N., et al.: A semantic multi-agent system to exchange information between hospitals. J. Procedia Comput. Sci. **109**, 704–709 (2017). https://doi.org/10.1016/j.procs.2017.05.381
5. Mutiara, A., Putri, T., Silfianti, W., Muslim, A., Oswari, T.: Semantic-web-based searching application for doctors schedule and facilities in hospital. J. Theoret. Appl. Inf. Technol. **59**, 189–196 (2014)
6. Cohen, S., Mamou, J., Kanza, Y., Sagiv, Y.: XSEarch: a semantic search engine for XML. In: Proceedings of the 29th International Conference on Very Large Data Bases, vol. 29. Berlin, Germany (2003)
7. Jimeno-Yepes, A., et al.: Ontology refinement for improved information retrieval. Inf. Process. Manage. **46**, 426–435. Pergamon Press Inc. https://doi.org/10.1016/j.ipm.2009.05.008 (2010)
8. Vallet, D., Fernández, M., Castells, P.: An ontology-based information retrieval model. In: Gómez-Pérez, A., Euzenat, J. (eds.) The Semantic Web: Research and Applications. ESWC: Lecture Notes in Computer Science, vol. 3532. Springer, Berlin, Heidelberg (2005)
9. Wang, C., Ma, X., Chen, J.: Ontology-driven data integration and visualization for exploring regional geologic time and paleontological information. J. Comput. Geosci. **115**, 12–19 (2018). https://doi.org/10.1016/j.cageo.2018.03.004
10. Kara, S., Alan, Ö., Sabuncu, O., Akpinar, S., Çiçekli, N.K., Alpaslan F.N.: An ontology-based retrieval system using semantic indexing. Paper presented at the 2010 IEEE 26th International Conference on Data Engineering Workshops (ICDEW 2010), Long Beach, CA, 2010, 1–6 March 2010, pp. 197–202 (2010). https://doi.org/10.1109/ICDEW.2010.5452700.
11. Guan, J., Zhang, X., Deng, J., Qu, Y.: An ontology-driven information retrieval mechanism for semantic information portals. In: 2005 First International Conference on Semantics, Knowledge and Grid, Beijing, p. 63 (2005). https://doi.org/10.1109/SKG.2005.42.

12. Patil, S.M., Jadhav, D.M.: Semantic search using ontology and RDBMS for cricket. Int. J. Comput. Appl. **46**, 26–31 (2012)
13. Yadav, P., Singh, R.P.: Ontology-based intelligent information retrieval method for document retrieval. Int. J. Eng. Sci. **4** (2012). https://doi.org/10.13328/j.cnki.jos.004622
14. Konstantinou, N., Spanos, D.-E., Chalas, M., Solidakis, E., Mitrou, N.: VisAVis: an approach to an intermediate layer between ontologies and relational database contents. Paper presented in Proceedings of the CAISE'06 Third International Workshop on Web Information Systems Modeling (WISM '06), Luxemburg, 5–9 June 2006
15. Michel, F., Montagnat, J., Zucker, C.: A survey of RDB to RDF translation approaches and tools. In: Hal.archives-ouvertes.fr. https://hal.archives-ouvertes.fr/hal-00903568v2 (2014). Accessed 16 Feb 2020
16. Cyganiak, R., et al.: Getting Started D2RQ Mapping Language. http://d2rq.org/getting-started (2012). Accessed 12 Jan 2020
17. Prud'hommeaux, E., Seaborne, A.: Sparql query language for rdf. Available via DIALOG. http://www.w3.org/TR/2008/REC-rdf-sparql-query-20080115/ (2008). Accessed 11 Jan 2019

Fuzzy Logic Controller for Accurate Diagnostics in X-Ray Film

Rakesh Kumar Tripathi and Javaid Ahmad Shah

1 Introduction

The concept of Fuzzy logic helps computers in decision-making in such a way that coincides with human behavior. Productivity gets tremendously increased in industries with the help of fuzzy logic. It makes production most convenient and helps industries in economical terms [1, 2]. Professor Lotfi A. Zadeh (in 1965) foremost proposed the concept of fuzzy logic. It was presented in one of his research papers under the heading Fuzzy logic or Fuzzy sets. It was E. H. Mamdani in 1974 who prepared the first fuzzy logic-based controller experiment for a steam engine. With this experiment, E. H. Mamdani showed that it becomes easy for a computer to process linguistic statements, which was given by Zadeh [3]. Fuzzy Logic Control (FLC) presents an important part of fuzzy reasoning derived from control theory based on mathematical models [4]. Fuzzy logic seems to have a promising future in the design of the controller mechanism in various domestic appliances that are used in day-to-day life [5]. In 1987, the Sendai railway started its operations, which was the first exciting application of Fuzzy logic. Thus, Fuzzy logic reaches its successful development in industry and in the commercial market rather than in universities [6]. The characteristics of a controller system are determined by the way/method by which it transforms the input quantities into output quantities. When an intelligent control system faces problem stimuli, it emits appropriate problem-solving responses, which are usually imprecise. Moreover, an intelligent system learns and generalizes from a limited/minimal number of experiences, which are mostly imprecise, and thus give

R. K. Tripathi (✉) · J. A. Shah
Dr. APJ Abdul Kalam University, Indore, MP, India
e-mail: drrakeshkumartripathi@gmail.com

J. A. Shah
e-mail: jsjavaidjs@gmail.com

C. K. Mai et al. (eds.), *Machine Learning Technologies and Applications*,
Algorithms for Intelligent Systems,
https://doi.org/10.1007/978-981-33-4046-6_9

rise to new input–output relationships. The processing of imprecise information is the domain of Fuzzy logic.

In the present day world, human health gets special caring [7]. As for as the human body is concerned, it is the most complicated and complex structure, and thus any error in radiographic positioning or diagnosis of any ailment can easily occur unless technicians have a common set of rules which are used to describe the human body and its various types of movements [8, 9]. Scientists are trying to find new ways for supporting technicians/clinicians in the diagnosis of diseases [7]. In this paper, a Fuzzy logic controller is used by choosing various parameters, which helps the radiologist to get better-developed X-ray films for an accurate diagnosis.

2 Preliminaries

X-rays (electromagnetic energy) are formed when electrons at high speed are bombarded on an anode made of tungsten. These rays possess the properties of both waves and particles and have a short wavelength than visible light, which allows them to pass/penetrate the matter [8, 9]. Now, another concept is how X-rays can produce images of internal parts of the body?

Different body tissues have different densities, which allow us to see inside the body by creating a shadow gram. The body is composed of tissues containing many different elements, which have different atomic numbers. The density of any element depends upon its atomic number. High dense elements block the X-rays more effectively than less dense elements. Therefore shadows of internal body structures become visible because they are made of different types of elements, e.g., when an X-ray strikes a bone made of calcium, it is blocked and the bone will appear white on the X-ray film. When an X-ray strikes a less dense element, it passes through it easily. Therefore, the lungs will appear darker on the X-ray film. Thus, a fracture will appear dark while the intact bone will remain white [8]. The five basic radiographic densities are.

1. Metal (Bright White), 2 Mineral (White), 3 Fluid/Soft tissues, 4 Fat (Dark gray) and 5 Air (Black).

Thus we conclude that the densities of the body tissues depend upon the weight and height of the body; the more the weight or height of the body, the more is the density of the tissue and more X-ray exposure is needed in order to obtain a better X-ray film. Thus, the main concept here is the required amount of X-ray exposure in order to obtain a suitable radiograph. Each time, a fixed set of exposure factors has to be chosen to obtain a required radiograph (X-ray image). The choice of these factors will depend on the thickness, density and pathology of the region to be examined [10]. The exposure factors are

1. Mille ampere second (mAs), 2. Kilo voltage (Kv) and 3. Focus to Film distance (FFD).

3 Fuzzy Logic in X-Ray Film

A radiologist works hard for developing his skills so that his work will be appreciated. A radiologist can make selections of different exposure levels and different parts of the body for which the X-ray film has to be taken. It will help him/her to get better-developed films for accurate diagnostics. With regard to the exposure, the exposure levels such as low exposure, normal exposure and high exposure are selected. For the parts of the body, the following types of tissues based on density are selected, which are low dense tissue, normal tissue and high dense tissue. After making those selections, the satisfaction score for the part of the body is computed based on the fuzzy inference system. If the radiologist is unsatisfied with the satisfaction score, he/she can return to selections [10]. Therefore, this system helps the radiologist to make better decisions. The system has three inputs and one output.

These input linguistic variables are a range of exposure (E) {low exposure (LE), normal exposure (NE), high exposure (HE)}, age of the person (A) {child (C), very young (VY), young (Y), middle aged (MA), old (O) and very old (VO)}, and weight of the body (W) {very underweight(VUW), underweight (UW), normal weight (NW), overweight (OW) and very overweight (VOW)}. Also, satisfaction score (S), i.e., Output variables are {very dark (VD), medium dark (MD), dark (D), light (L), very light (VL), bright (B), medium bright (MB) and very bright (VB)}, i.e., the nature of X-ray film. The Fuzzy rule base consists of ninety rules, some of them are

R1:- If "E" is LE, "A" is C and "W" is VUW, then S is a very light film.
R2:- If "E" is LE, "A" is C and "W" is UW, then S is a light film.
R3:- If "E" is LE, "A" is C and "W" is NW, then S is a bright film.
R4:- If "E" is LE, "A" is C and "W" is OW, then S is a medium bright film.
R5:- If "E" is LE, "A" is C and "W" is VOW, then S is a very bright film.
R6:- If "E" is LE, "A" is VY and "W" is VUM, then S is a very light film.
R7:- If "E" is LE, "A" is VY and "W" is UW, then S is a light film.
R8:- If "E" is LE, "A" is VY and "W" is NW, then S is a bright film.
R9:- If "E" is LE, "A" is VY and "W" is OW, then S is a medium bright film.
Moreover, R90:- If "E" is HE, "A" is VO and "W" is VOW, then S is a very light film.
All the ninety rules (relations) are mentioned in the matrix representation table.

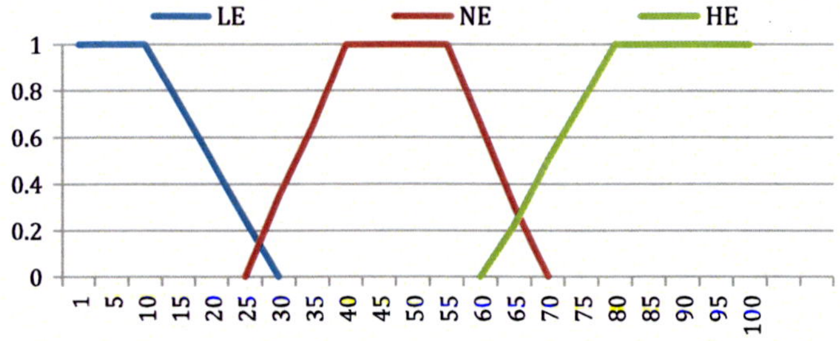

Fig. 1 Input variable of exposure level

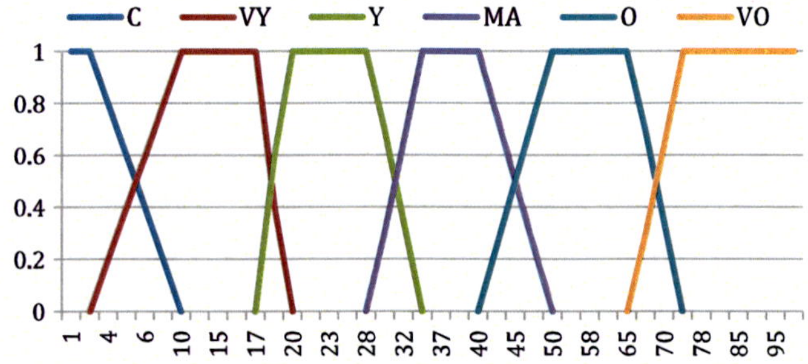

Fig. 2 Input variable of age level

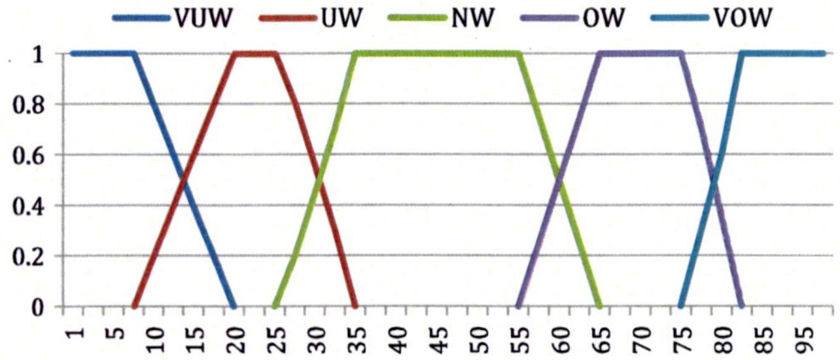

Fig. 3 Input variable of weight level

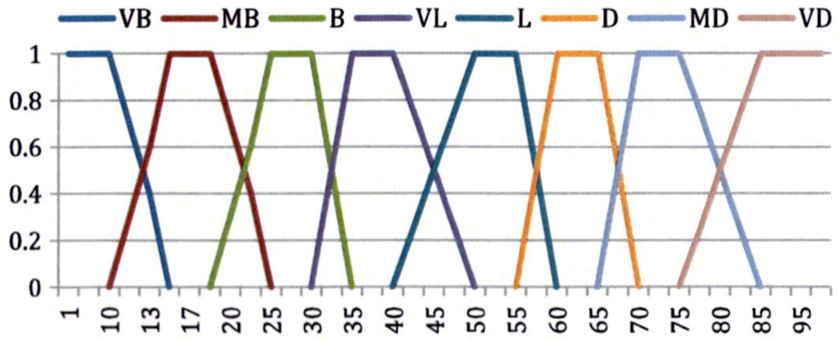

Fig. 4 Output variable of X-ray radiograph

Table 1 Tabulated form of input–output variables

Input	Input	Input	Output
Age range for given matrix is 0–6 (Exposure level)	(Age level)	(Weight level)	(X-ray radiograph level)
LE	C	VUW	VL
LE	C	UW	L
LE	C	NW	B
LE	C	OW	MB
LE	C	VOW	VB
Age range for given matrix is 5–21			
LE	VY	VUW	VL
LE	VY	UW	L
LE	VY	NW	B
LE	VY	OW	MB
LE	VY	VOW	VB
Age range for given matrix is 15–35			
LE	Y	VUW	VL
LE	Y	UW	L
LE	Y	NW	B
LE	Y	OW	MB
LE	Y	VOW	VB
Age range for given matrix is 30–50			
LE	ME	VUW	VL
LE	ME	UW	L
LE	ME	NW	B
LE	ME	OW	MB
LE	ME	VOW	VB
Age range for given matrix is 40–75			
LE	O	VUW	VL
LE	O	UW	L
LE	O	NW	B
LE	O	OW	MB
LE	O	VOW	VB
Age range for given matrix is 65–100			
LE	VO	VUW	VL
LE	VO	UW	L
LE	VO	NW	B
LE	VO	OW	MB

(continued)

Table 1 (continued)

Input	Input	Input	Output
LE	VO	VOW	VB
Age range for given matrix is 0–6			
NE	C	VUW	MD
NE	C	UW	D
NE	C	NW	L
NE	C	OW	B
NE	C	VOW	MB
Age range for given matrix is 5–21			
NE	VY	VUW	VD
NE	VY	UW	D
NE	VY	NW	L
NE	VY	OW	B
NE	VY	VOW	MB
Age range for given matrix is 15–35			
NE	Y	VUW	VD
NE	Y	UW	D
NE	Y	NW	L
NE	Y	OW	B
NE	Y	VOW	MB
Age range for given matrix is 30–50			
NE	ME	VUW	VD
NE	ME	UW	D
NE	ME	NW	L
NE	ME	OW	B
NE	ME	VOW	MB
Age range for given matrix is 40–75			
NE	O	VUW	VD
NE	O	UW	D
NE	O	NW	L
NE	O	OW	B
NE	O	VOW	MB
Age range for given matrix is 65–100			
NE	VO	VUW	VD
NE	VO	UW	D
NE	VO	NW	L
NE	VO	OW	B

(continued)

Table 1 (continued)

Input	Input	Input	Output
NE	VO	VOW	MB
Age range for given matrix is 0–6			
HE	C	VUW	VD
HE	C	UW	MD
HE	C	NW	D
HE	C	OW	L
HE	C	VOW	VL
Age range for given matrix is 5–21			
HE	VY	VUW	VD
HE	VY	UW	MD
HE	VY	NW	D
HE	VY	OW	L
HE	VY	VOW	VL
Age range for given matrix is 15–35			
HE	Y	VUW	VD
HE	Y	UW	MD
HE	Y	NW	D
HE	Y	OW	L
HE	Y	VOW	VL
Age range for given matrix is 30–50			
HE	ME	VUW	VD
HE	ME	UW	MD
HE	ME	NW	D
HE	ME	OW	L
HE	ME	VOW	VL
Age range for given matrix is 40–75			
HE	O	VUW	VD
HE	O	UW	MD
HE	O	NW	D
HE	O	OW	L
HE	O	VOW	VL
Age range for given matrix is 65–100			
HE	VO	VUW	VD
HE	VO	UW	MD
HE	VO	NW	D
HE	VO	OW	L
HE	VO	VOW	VL

3.1 *Membership Functions of Input–Output Linguistic Variables are shown in Figs. 1, 2, 3 and 4*

3.2 *The Matrix Representation of Input (Exposure, Age, Weight) and Output (Quality of X-Ray Film) Linguistic Variables is Shown in Table 1*

4 X-Rays Exposure Data Collection

Here, given below are X-ray exposure data collections of different patients with varying age groups and different weights where we obtain developed X-ray films which depend upon the range of X-ray exposure and the density of the body tissue. We collect this data from different hospitals, and it is given in Table 2.

After evaluating the above data, we come to the conclusion that by changing the exposure range, density of tissue or FFD, different types of X-ray radiograph can be developed which may be underdeveloped, developed and overdeveloped. Let us take some examples from the mentioned data and change any one of the factors to prove that the X-ray radiograph will be different each time we change the factors and shown in Table 3.

5 Conclusion

This paper shows that rules based on Fuzzy logic help in developing accurate/required X-ray radiograph. They help radiologists in obtaining well-developed X-ray radiograph by considering different input parameters such as weight, age and level of the exposure. Based on input parameters, different types of X-ray radiograph can be achieved as output parameters. Due to this adjustment of input parameters, a person can be saved from the side effects of using higher exposure in place of required exposure. This in turn will result in the well-being of the patient. It will save the patient from other disorders like vomiting, bleeding, fainting, hair loss, etc., which may arise when the exposure level is high. This method will show better results than the traditional methods which are used by the radiologists.

Table 2 Data collection from different hospitals

Exposure mAs/kvp	Body part to be examined	Age	Weight	X-ray film
180–80	Lumbar spine (lat.)	35	75	Developed
23–50	Nasal bone	7	22	Developed
25–55	Knee joint (lat.)	32	70	Developed
25–55	Knee joint (AP)	32	70	Developed
175–80	Lumbar spine (AP)	40	75	Developed
180–80	Lumbar spine (lat.)	40	75	Developed
20–55	Knee joint (lat.)	37	60	Developed
20–55	Knee joint (AP)	37	60	Developed
40–50	Ankle joint (AP)	45	67	Developed
55–65	Pelvis	6	17	Developed
63–70	Ankle joint (lat.)	45	67	Developed
35–65	Chest (AP)	70	75	Developed
35–65	Chest(lat.)	35	65	Developed
30–60	Chest (AP)	35	65	Developed
175–70	Lumbar spine (AP)	59	72	Developed
185–70	Lumbar spine (lat.)	59	72	Developed
50–75	Foot (AP)	50	65	Developed
30–70	Chest (AP)	78	75	Developed
52–76	FOOT (lat.)	50	70	Developed
30–60	Chest (PA)	32	65	Developed
178–65	Lumbar spine (AP)	48	70	Developed
180–68	Lumbar spine(lat.)	48	70	Developed
25–40	Fore arm (AP)	9	25	Developed
25–40	Fore arm (lat.)	9	25	Developed
50–65	Foot (AP)	35	55	Developed

Table 3 Representation of X-ray film with varying factors

Exposure mAs/kvp	Body part to be examined	Age	Weight	X-ray film
250–150	Lumbar spine (lat.)	35	75	Overdeveloped
150–200	Nasal bone	7	22	Overdeveloped
100–200	Knee joint (AP)	37	60	Overdeveloped
150–250	Ankle joint (lat.)	45	67	Overdeveloped
29–15	Lumbar spine (lat.)	59	72	Underdeveloped
20–15	Cervical spine (lat.)	45	80	Underdeveloped
50–20	Pelvis	65	50	Underdeveloped
40–20	Lumbar spine (AP)	45	85	Underdeveloped
150–200	Fore arm (lat.)	9	25	Overdeveloped

References

1. Zadeh, L.A.: Fuzzy Sets. Inf. Control **8**(3), 338–353 (1965)
2. Demetgul, M., Ulkir, O., Waqar, T.: Washing machine using fuzzy logic. Autom. Control Intell. Syst. **2**(3), 27–32 (2014).https://doi.org/10.11648/j.acis.20140203.11
3. Mamdani, E.H.: Application of fuzzy algorithms for control of simple dynamic plant. Proc. IEEE **121**(12), 1585–1588 (1974)
4. Zadeh, L.A.: Fuzzy sets as a basis for a theory of a possibility. Fuzzy Sets Syst. **1**(1), 3–28 (1978)
5. Lootsma, F.A.: Fuzzy Logic for Planning and Decision-Making, vol. 8 (1997). ISBN 978–1–4419–4779–6 ISBN 978–1–4757–2618–3 (eBook). https://doi.org/10.1007/978-1-4757-2618-3
6. Tuan, T.M., Duc, N.T., Hanoi, P.V., Son, L.H.: Dental diagnosis from X-ray images using fuzzy rule-based systems. Int. J. Fuzzy Syst. Appl. **6**(1) (2017)
7. Benseler, J.S.: The Radiology Handbook: A Pocket Guide to Medical Imaging. Ohio University Press, Athens 45701 (1954). ISBN-13:978-0-8214-1708-9 (pbk: alk. paper)
8. Iancu, I.:AMamdani type fuzzy logic controller. In: Fuzzy Logic—Controls, Concepts, Theories and Applications (2012). ISBN: 978-953-51-0396-7. https://doi.org/10.5772/2662
9. Whitley, A.S., Sloane, C., Hoadley, G., Moore, A.D., Alsop, C.W.: Clarks Positioning in Radiography, 12th edn (2005). First published in Great Britain in 2005 by Arnold. ISBN: 0340763906
10. Himabindu, G., Ramakrishna Murty, M., et al.: Image preprocessing of abdominal CT scan to improve visibility of any lesions in kidneys. J. Theor. Appl. Inf. Technol. **96**(8) (2018). (E-ISSN 1817-3195 / ISSN 1992-8645)

Heart Attack Classification Using SVM with LDA and PCA Linear Transformation Techniques

S. Vamshi Kumar, T. V. Rajinikanth, and S. Viswanadha Raju

1 Introduction

In the present situation, a heart attack takes place based on many factors. In day-to-day life, stress and living habits play a major role in the functionality of the heart of a human body. In the heart, there are two main blood vessels for the supply of blood through coronary arteries. If the artery gets completely blocked, then it leads to a heart attack. In the medical industry, there are many cases where there is an availability of data related to many diseases apart from problems leading to the storage of huge data. The cardiac data is taken for analysis in order to know the causes of heart-related issues and problems. Data mining is useful for the extraction of useful information from medical data with the help of Data mining techniques. It helps in diagnosing heart-related problems in an effective manner. In this paper, there are many classifier techniques involved in machine learning. Among those classifiers, mostly used techniques are Random Forest (RF), Naïve Bayes (NB), Logistic Regression (LR), Support Vector Machine (SVM), KNN, Decision Tree (DT), and many more for finding better results.

S. Vamshi Kumar (✉)
Department of CSE, SNIST, Hyderabad, Telangana, India
e-mail: vamshik670@gmail.com

S. Viswanadha Raju
Department of CSE, JNTUHCEJ, Jagityal, Telangana, India
e-mail: svraju.jntu@gmail.com

T. V. Rajinikanth
Professor and Dean R & D, Department of CSE, SNIST, Hyderabad, Telangana, India
e-mail: rajinitv@gmail.com

C. K. Mai et al. (eds.), *Machine Learning Technologies and Applications*,
Algorithms for Intelligent Systems,
https://doi.org/10.1007/978-981-33-4046-6_10

1.1 Support Vector Machine (SVM)

In an N-dimensional area to know a hyperplane, the data points are distinctly classified based on the number of features (N). The main target of SVM is to plot the data points and to know the maximum distance between the class label and also the maximum margin. This helps in future data points to be classified in a better manner [1]. There are different kernels involved in finding the margin which is maximizing. To solve non-linear problem a linear classifier method is used with the help of kernel tricks.

- **Linear kernel (linear)**: For Linearly separable data, the data is separated using a single line. It is the most commonly used kernel by everyone for a large number of features in a dataset.
- **Polynomial kernel (poly)**: It is the line drawn in a zig-zag way where the data points are located to differentiate the class labels based on the maximized margin.
- **Radical Basis Function (rbf)**: It is a real value function that depends only on the distance between input and some fixed data points. In this, the margin is created where there is a difference in class label where the margin line is created.

1.2 Decision Tree (DT)

DT is like a flow of data in a step-by-step manner in a tree structure, where each node represents different attributes in each level of the tree as the sub-tree represents the test attributes, the edges represent the test data outcome, and child node (leaf node) represents the label of the class [2].

The attributes used are Age, Serum Cholesterol mg/dl (chol), Thal, Fasting Blood Sugar (fbs), Sex, Slope of Peak (slope), Chest Pain type (cp), Resting ECG (restecg), Maximum Heart Rate achieved (thalach), Resting Blood Pressure (trestbps), Exercise-induced Angina (exang), Oldpeak, Number of major vessels (ca), and Target. Dimensionality reduction is done for the attributes to reduce the dataset from higher dimensionality to lower dimensionality using linear transformation techniques like Linear Discriminant Analysis (LDA) and Principle Component Analysis (PCA) [3].

1.3 Principle Component Analysis (PCA)

It is an unsupervised learning algorithm which ignores the class labels when the transformation of data is done that helps in increasing the dataset difference to know the directions.

1.4 Linear Discriminant Analysis (LDA)

It is a supervised algorithm which uses the class label into consideration when the transformation is done. It is a way to reduce higher dimensionality to lower dimensionality based on preserving the class discrimination information. The reduction is done as per the number of n-components (n represents the number of dimensionalities).

2 Literature Review

The data from previous papers can be used for knowing what the major components involved in heart disease prediction are. Cardiac disease was the major cause in nations such as India and the United States. Data mining and Machine learning classifiers like Linear Regression, Logistic Regression, Random Forest, Naïve Bayes, KNN, Support vector machine (SVM), and other classified results, for example, SVM encounter various types of heart-related issues. Some of the papers referred to show the selection of attributes from the dataset. Many authors have classified dataset attributes on some parameters for training and testing the accuracies.

Mythili et al. [4] The rule-based model has been used for comparing the accuracy among the SVM (linear or RBF), LR, and DT using the Cleveland dataset in the predicting the best model for heart disease prediction. There are different types of rules used for classification, regression, and association tasks that can be implemented using C-rules, A-rules, F-rules, T-rules, and P-rules. They have used the classification rule (C-rule) for this model to analyze the basis of sensitivity, specificity, and accuracy.

Hasan et al. [5] In this, the unused attributes of feature sets are removed using the information gain feature selection technique and are reduced to 10 and compared with Decision Tree (ID3), Gaussian Naïve Bayes, Random Forest, KNN, and Logistic Regression techniques for measuring accuracy, precision, recall, F1-score, sensitivity, and specificity. Among those, Logistic Regression accuracy (92.76%) is better.

Beena Bethel et al. [6] has stated in their that feature reduction techniques are used to increase the accuracy a little bit for the SVM classifier with various kernels on the Cleveland dataset that is compared with other classification techniques. Using the Thumb rule data from dataset is being separated as training and testing then SVM classification is applied.

Dinesh Kumar et al. [7] have used R software to implement ML techniques like SVM, Naïve Bayes classifier, Gradient Boosting, Logistic Regression, and Random Forest. Data visualization visualization is done to know which model gives better accuracy for the prediction.

Mohan et al. [8] have calculated the accuracy by selecting number of feature selection and model is generated. They have proposed a new model HRFLM to accomplish the best results without any restriction in feature selecting. HRFLM is the combination of Random Forest and Linear Method; this hybrid technique has given better accuracy.

Beulah Christalin Latha et al. [9] in their work they have used the ensemble techniques are bagging, boosting, stacking, and majority voting are used for the prediction of heart disease. Where their increase in accuracies of 6.92, 5.94, 7.26, 6.93% and the comparison results shows that majority voting as more improvement in accuracy. The results are measured using the Cleveland dataset (UCI repository).

Chitra et al. [10] in their work have used SVM with 'rbf' kernel for classification where the accuracy obtained is 82% and this result has been compared with CNN. A computerized algorithm is created for efficient heart attack prediction in an accurate manner with a CNN classifier. This classification is done on a dataset obtained from 270 patients, which shows CNN has better accuracy than the SVM classifier.

Beena Bethel et al. [11] in their work have used Co-clustering approach for heart disease analysis depends on weighted based technique. They have proposed the Weighted ITCC technique for Cleveland dataset for prediction of heart disease and also made performance comparisons. They observed an improvement in the accuracy performance for the weighted co-clustering techniques when compared with other techniques.

Benjamin Fredrick David et al. [12] have done a comparative analysis on 3 algorithms such as DT, Random Forest, and Naïve Bayes on the UCI repository dataset. This analysis shows that Random Forest has given the best results as compared with other results. For the results of PRC area, Precision, F-measure, Recall, and ROC area, Comparative analysis is done.

Abdar et al. [13] have used data mining algorithms such as C5.0, Logistic Regression, KNN, SVM, and Neural Network will produce results based on the UCI dataset, and comparison is done. That shows DT has got better accuracy as compared to other techniques. DT can be performed without any computation classification and generate efficient rules that clearly show which features are most significant for prediction.

Ghumbre et al. [14] has stated that SVM and RBF network structures are used for the diagnosis of heart diseases. This is done using samples collected from different patients, and the performance measures of SVM with SMO (Sequential minimize optimization) are equally the same as ANN and some other techniques in the prediction. The results of the SVM model accuracy, sensitivity, and specificity are high thus making a better option in heart disease prediction.

Beena Bethel et al. [15] has stated that CFS, MFS and their combination of two results are measured for continuous and discrete sets of features. From this, by controlling the number of discrete features the accuracy of the CFS or MFS+CFS are increased, and the efficiency of classification is improved by continuous features.

Joshi et al. [16] have used Weka tool for data mining classifiers like Decision Tree, Naïve Bayes, and KNN as per the classifier creations the performances show KNN has given better accuracy.

In this, the prediction is basically measured using the different ML algorithms to generate the results and some of the hybrid algorithms for the improvement of results. Among all these techniques, the Support Vector Machine algorithm is used with only a single kernel in many of the research papers, and there was less comparison done between the kernels. In some cases, SVM kernels are selected based on the most used kernel (linear) or by comparing the results with the remaining kernel which gives better results that have been used for prediction. But, there is no proper use of all the kernels of SVM in any of previous research predictions and also no proper comparison between the kernels.

As such, this paper represents the usage of the SVM model with all the three different kernels like Linear, Polynomial and Radical Basis Function. All these SVM kernels' results are calculated for the dimensionality reduced data using LTT methods. So, this presents the unique model, i.e. SVM with PCA and SVM with LDA. This hybrid combination model helps in generating the output in less time without giving a huge burden to the processor and gives better results as compared with the original dataset.

3 Proposed System

In the proposed method, initially, data was preprocessed to remove the noisy data, filling the values using a measure of central tendencies and classifying the number of attributes for prediction. From the literature survey, it is known that SVM and DT classifiers have given better results in terms of accuracy when compared to other classification techniques. In the dataset, there are 14 attributes among them; 13 attributes were related to input data values and the 14th attribute is a target value. Those data values cannot be included in the SVM at a time so dimensionality reduction techniques PCA and LDA were used to reduce the dimensions. SVM with PCA and SVM with LDA techniques were used. Before that, the preprocessed data was split into train and test data in the ratio 80:20 and after that, the classifier techniques were used initially on Train and followed by Test.

A comparison was made between these two classifier techniques SVM and DT for knowing which among these two algorithms is giving the best results and accuracy. In SVM, three different kernels have been used, i.e. Linear, Polynomial (poly), and Radical Basis Function (RBF). The results of SVM with three kernels were compared with that of DT.

These findings helps in knowing the performance of the algorithms and how the dataset is classified with the techniques, and how the SVM kernel will produce the three different results and their comparisons with the DT. SVM with LDA showed better results when compared with other classifiers. Along with these, the prediction of a heart attack is also known when a new data record is given to the model which will show the outcome whether a heart attack occurred or not. This prediction is done based on SVM as shown in Fig. 1.

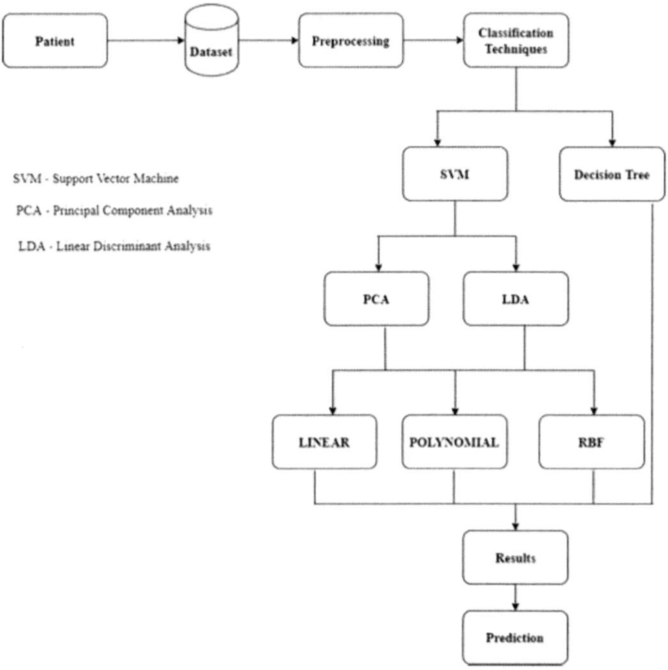

Fig. 1 Data flow of proposed system

3.1 Steps for the Proposed Method

- Initially, Patient data is collected as per the need for the prediction of the cardiac attack and that will be stored in a dataset.
- Preprocessing is done on the collected dataset, i.e.the Cleveland dataset. The unwanted data is removed and the missing values are filled with the measures of the central tendencies method such as mean, median, and mode, and also the redundant data is also removed.
- The acquired dataset is processed using classification techniques such as SVM and DT in our proposed system.
- In SVM, all attributes are used for finding the results. But to plot a graph a 2 dimensional attribute structure is needed. The dimensionality reduction for the attributes is done using the two techniques PCA and LDA which help in reducing the data to n-component dimensions that transformed data is used for the classifier.
- The transformed data is now carried to the different kernels available, the SVM Linear, Polynomial, and RBF; here, the results are generated for these three kernels individually.
- The whole data is split into training and testing parts; then the results are calculated for the kernels of SVM and the comparison is done between those kernels.

- Next, the preprocessed data is again split into trained and test parts for the Decision tree classifier for finding the results of trained and tested data.
- Overall, the results are clubbed together as training data results, testing data results, and whole dataset as training data results. Then a comparison is made between the PCA results of SVM kernels, LDA results of SVM kernels, and Decision Tree results.
- All results are stored in separate files for the future comparisons.
- Then the prediction is carried out for the newly added record to the dataset which will predict the occurrence of a heart attack. This prediction is done using the SVM classifier as it given better results.

3.2 Linear Transformation Techniques (LTT)

Linear Transformation Techniques are basically used for dimensionality reduction in the form of data matrix factorization. PCA and LDA are the two types of the LTT method to reduce the dimensionality of the space of variables.

PCA: Principle Component Analysis

PCA [3] is simply based on the eigenvector for multivariate analysis, and it is mostly used as a method to know the internal structure of the data that helps in getting the maximum variance.

PCA is used for Noise Filtering, Visualization, Feature Extraction, Stock Market Predictions, and Gene data analysis.

The aim of PCA is to identify samples in a dataset and find the correlation between independent variables.

The dimensions of independent variables, i.e. n-dimensionality dataset, are reduced by projecting it into K-subspaces where $k < n$.

Steps of PCA:

1. Scale the data to standardize.
2. Obtains the Eigenvectors and Eigenvalues from the correlation or covariance matrix, or using the Singular Vector Decomposition (SVD).
3. Now the eigenvalues are sorted in decreasing order and then choose the ‘k’ eigenvectors that correspond to the ‘k’ largest eigenvalues for the new feature subspace ($k \leq n$);‘k’ is the number of dimensions.
4. For the selected ‘k’ eigenvectors, construct the projection matrix ‘W’.
5. Transform the original independent data X via W to obtain a k-dimensionality features subspace Y as shown in Fig. 2.

LDA: Linear Discriminant Analysis

LDA [17] is one of the dimensionality reduction techniques used in the preprocessing steps of the related pattern classification. It is like a supervised learning which reduces the dimensions to n-component dimensions. LDA has the aim of creating lower-dimensionality space for the original dataset.

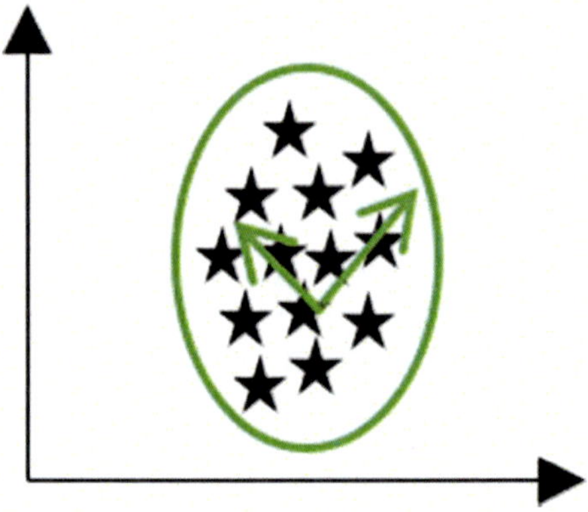

Fig. 2 PCA (the variance is maximized using component axes)

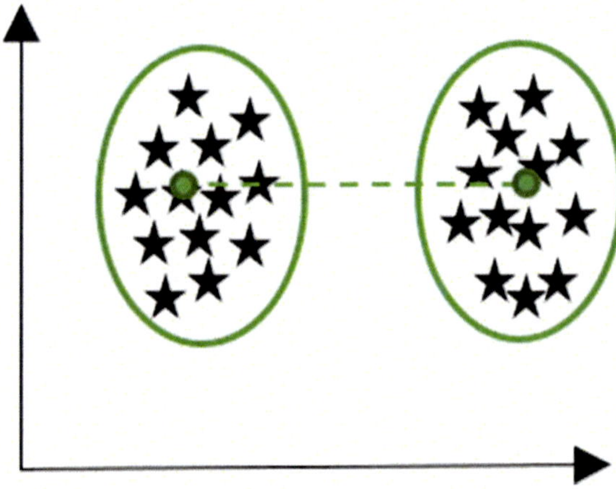

Fig. 3 LDA (for class-separation maximizing the component axes)

LDA differs from PCA as it maximize the component axes in order to separate in multiple classes with multiple separations.

The aim of LDA is to project the independent variables, i.e. feature elements (an n-dimensionality sample in the dataset) into subspace or lesser-dimensionality space: subspace K (where $k \leq n - 1$) by maintaining the label discriminatory information as shown in Fig. 3.

Steps of LDA:

1. Finding the n-dimensionality mean vectors.
2. Finding the Scatter Matrices for the mean vectors.
3. For the obtained matrix, the generalized eigenvalue problem is solved.
4. For the new feature subspace, linear discriminants are selected.
5. Using the highest eigenvalues, the 'k' eigenvectors are selected.
6. Then these samples are transformed into the new subspace.

This Linear Transformation Technique is used for the preprocessing of the dataset by removing the noisy data. The independent data is reduced to the n-component

Table 1 Trained results of SVM with PCA and SVM with LDA

SVM	PCA			LDA		
	Linear	Poly	RBF	Linear	Poly	RBF
Accuracy	71.95	66.01	70.96	84.16	82.18	85.15
Precision	71.79	73.36	69.3	91.78	91.78	89.13
Recall	64.03	35.25	64.03	79.14	64.75	79.14
F1-score	67.68	48.76	66.92	82.09	76.92	83.02

variables. The reduced data is trained on the SVM algorithms, and the results are calculated for SVM with PCA and SVM with LDA.

This hybrid combination model helps in generating the output in less time that means the time complexity will be reduced and gives less burden to the processor. As compared to the results of the original dataset with the reduced dataset, the results are much better using SVM with LDA.

4 Results and Analysis

In this section, the results like accuracies, precision, recall, and F1-score are generated and displayed. In this, we have shown the results for different classification techniques, i.e. SVM using PCA and LDA with different kernels and DT.

4.1 *Trained Results*

Table 1 includes the results obtained for SVM using PCA and SVM using LDA for trained dataset as follows.

Figure 4 represents that Trained results of SVM with LDA are better then SVM with PCA as the percentage of the RBF kernel of SVM with LDA is 84.71%.

4.2 *Tested Results*

Table 2 includes the results obtained for SVM with PCA and SVM with LDA for tested dataset as follows.

Figure 5 represents that the results of the Test Data show that SVM with LDA with Linear and RBF kernels has produced the highest accuracy of 91.8%.

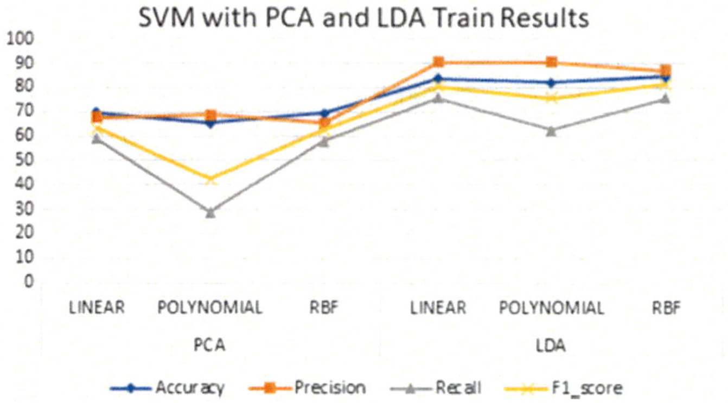

Fig. 4 Train results of SVM with PCA and LDA

Table 2 Tested results of SVM with PCA and SVM with LDA

SVM	PCA			LDA		
	Linear	Poly	RBF	Linear	Poly	RBF
Accuracy	68.85	68.85	70.49	91.8	80.33	91.8
Precision	84.57	85.41	84.21	95.03	95.03	97.49
Recall	56.25	43.75	59.38	90.62	65.62	87.5
F1-score	65.45	59.57	67.86	92.06	77.78	91.8

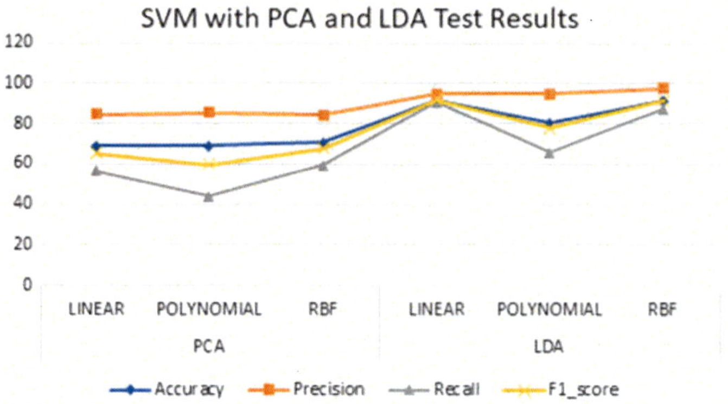

Fig. 5 Test results of SVM with PCA and LDA

Table 3 Whole dataset as trained results of SVM with PCA and SVM with LDA

SVM	PCA			LDA		
	Linear	Poly	RBF	Linear	Poly	RBF
Accuracy	71.95	66.01	70.96	84.16	82.18	85.15
Precision	71.79	73.36	69.3	91.78	91.78	89.13
Recall	64.03	35.25	64.03	79.14	64.75	79.14
F1-score	67.68	48.76	66.92	82.09	76.92	83.02

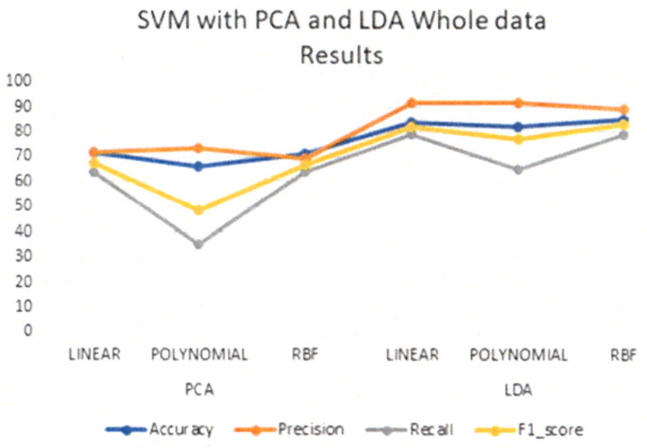

Fig. 6 Whole data results of SVM with PCA and LDA

4.3 *Whole Dataset as Trained*

Table 3 includes the results obtained for SVM with PCA and SVM with LDA for the whole dataset as trained as follows.

Figure 6 represents that results of SVM with LDA are better than the PCA when the whole dataset is taken as trained data shows that RBF kernel of SVM with LDA has the highest accuracy of 85.15%.

4.4 *Decision Tree Results*

Table 4 shows the results generated for the train, test, and whole trained results as follows.

Figure 7 represents that DT has given 100% results in Trained data and 75% accuracy for Test data, which shows that decision tree gives less error rate in trained data so the results of trained data are 100%.

Table 4 Whole dataset as trained results of SVM with PCA and SVM with LDA

J48	Train	Test	Whole trained
Accuracy	100	75	100
Precision	100	71.76	100
Recall	100	71.88	100
F1-score	100	75.41	100

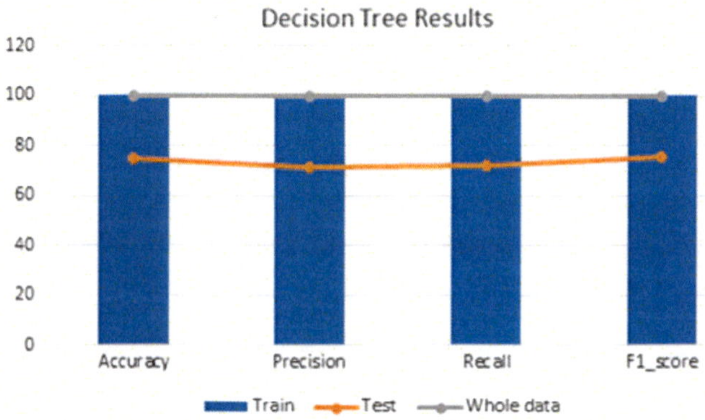

Fig. 7 Train, test, and whole data results of decision tree

The results are generated from the attributes collected from the dataset. These tables show the variation in results of trained, tested, and whole dataset of SVM with PCA, SVM with LDA and DT. All these acquired results are compared and show that SVM with LDA with linear and RBF kernels has given 91.8% highest accuracy among other techniques. As per the results obtained from the techniques, a better model which gives the best results is used for predicting heart attacks.

5 Conclusions

In this, classification techniques of machine learning namely SVM and DT were used for better outcomes of the results and compared between those two techniques. In SVM, various kernels such as Linear, Poly, RBF were tried and the results were shown. These kernels have given different outcomes for transformed data that is obtained using dimensionality reduction techniques PCA and LDA. SVM with LDA showed better results between these two. It is observed from the results that SVM with LDA with linear and RBF kernels gave the highest accuracy of 91.8% when compared to other classifiers. This hybrid model helps in achieving the results with less time complexity and gives better results when compared with the original dataset.

Our findings also help in heart attack prediction using the best classifier among the SVM kernels and DT classifiers to know whether there is a chance of heart attack or not on providing new details to our proposed system. When the whole datasets were considered under training, both techniques SVM and Decision Tree have given better accuracy in results.

Acknowledgements Thanks to providers of UCI Machine Learning Repository [18] for providing the Dataset.

References

1. https://towardsdatascience.com/support-vector-machine-introduction-to-machine-learning-algorithms-934a444fca47
2. https://en.wikipedia.org/wiki/Decision_tree
3. https://sebastianraschka.com/faq/docs/lda-vs-pca.html
4. Mythili, T., Mukherji, D., Padalia, N., Naidu, A.: A heart disease prediction model using SVM-decision trees-logistic regression (SDL). Int. J. Comput. Appl. (0975-8887) **68**(16) (2013)
5. Hasan, S.M.M., Mamun, M.A., Uddin, M.P., Hossain, M.A.: Comparative analysis of classification approaches for heart disease. In: International Conference on Computer, Communication, Chemical, Material and Electronic Engineering (IC4ME2), 20 September 2018
6. Beena Bethel, G.N., Rajinikanth, T.V., Viswanadha Raju, S.: An efficient feature reduction technique for an improved heart disease diagnosis. Int. J. Appl. Eng. Res. **10**(1), 2081–2090 (2015)
7. Dinesh Kumar, G., Santhosh Kumar, D., Arumugaraj, K., Mareeswari, V.: Prediction of cardiovascular disease using machine learning algorithms. In: IEEE International Conference on Current Trends toward Converging Technologies, Coimbatore, India (2018)
8. Mohan, S., Thirumalai, C., Srivastava, G.: Effective Heart Disease Prediction Using Hybrid Machine Learning Techniques. IEEE Access (2019)
9. Beulah Christalin Latha, C., Carolin Jeeva, S.: Improving the accuracy of prediction of heart disease risk based on ensemble classification techniques. Elsev. Inform. Med. Unlocked **16** (2019)
10. Chitra, R., Seenivasagam, V.: Heart disease prediction system using supervised learning classifier. Bonfring Int. J. Softw. Eng. Soft Comput. **3**(1) (2013)
11. Beena Bethel, G.N., Rajinikanth, T.V., Viswanadha Raju, S.: A knowledge driven approach for efficient analysis of heart disease dataset. Int. J. Comput. Appl. (0975-8887) **147**(9) (2016)
12. Benjamin Fredrick David, H., Antony Belcy, S.: Heart disease prediction using data mining techniques. ICTACT J. Soft Comput. **09**(01) (2018)
13. Abdar, M., Niakan Kalhori, S.R., Sutikno, T., Subroto, I.M.I., Arji, G.: Comparing performance of data mining algorithms in prediction heart diseases. Int. J. Electr. Comput. Eng. (IJECE) **5**(6) (2015)
14. Ghumbre, S.U., Ghatol, A.A.: Heart disease diagnosis using machine learning algorithm. In: Proceedings of the InConINDIA 2012, AISC, vol. 132, pp. 217–225. Springer, Berlin, Heidelberg (2012)
15. Beena Bethel, G.N., Rajinikanth, T.V., Viswanadha Raju, S.: Weighted co-clustering approach for heart disease analysis. In: Proceedings of the First International Conference on Computational Intelligence and Informatics, Advances in Intelligent Systems and Computing, vol. 507 (2017)

16. Joshi, S., Nair, M.K.: Prediction of heart disease using classification based data mining techniques. In: Jain L.C., et al. (eds.) Computational Intelligence in Data Mining—Volume 2, Smart Innovation, Systems and Technologies, vol. 32. Springer, India (2015)
17. https://sebastianraschka.com/Articles/2014_python_lda.html
18. Dua, D., Graff, C.: UCI Machine Learning Repositor. University of California, School of Information and Computer Science, Irvine, CA (2019). http://archive.ics.uci.edu/ml

Distributed Training of Deep Neural Network for Segmentation-Free Telugu Word Recognition

Koteswara Rao Devarapalli and Atul Negi

1 Introduction

In traditional Telugu optical character recognition (TOCR) systems [1–4], the basic unit of recognition is the connected component [5]. These systems rely on segmentation, which is difficult when the input document images contain noise, degradation, broken, and touching characters.

The deep neural network (DNN) model aims to carry out segmentation-free Telugu word recognition. This approach does not require connected component extraction, which is difficult due to the presence and occurrence of noise, broken characters, and touching characters in document images. Usually, the conventional Telugu OCR system has preprocessing, feature extraction, followed by training and testing stages. And the preprocessing consists of noise removal, line segmentation, word segmentation, and connected component extraction. The conventional systems involve the modeling of connected components. In this paper, we do not rely on tasks such as connected component extraction and feature extraction. This work bypasses both of the traditional functions. We intend to model Telugu characters as Unicode symbols to facilitate direct recognition of Telugu word images. The DNN framework makes it possible to eliminate many traditional preprocessing tasks in the design of Telugu OCR and supports end-to-end recognition [6].

K. R. Devarapalli (✉)
Department of Computer Science and Engineering, Mahatma Gandhi Institute of Technology, Hyderabad 500046, India
e-mail: dkoteswararao_cse@mgit.ac.in

K. R. Devarapalli · A. Negi
School of Computer and Information Sciences, University of Hyderabad, Gachibowli, Hyderabad 500046, India
e-mail: atulcs@uohyd.ernet.in

C. K. Mai et al. (eds.), *Machine Learning Technologies and Applications*, Algorithms for Intelligent Systems,
https://doi.org/10.1007/978-981-33-4046-6_11

Advances in neural networks introduced several successful applications in speech processing, computer vision, and text processing. This paper describes recent trends in the modeling of neural networks and the need to employ a distributed modeling approach to utilize available GPUs and TPU practically [7]. A hybrid DNN model uses distributed modeling [8]. We intend to model a deep neural network (DNN) for segmentation-free Telugu word recognition of Telugu OCR application in the computer vision domain. The CNN-RNN-CTC is the most successful model for segmentation-free recognition of input line and word images. It involves the convolutional neural network (CNN) and recurrent neural network (RNN) layers. The connectionist temporal classification (CTC) layer generates the output.

1.1 Distributed Training

A typical deep neural network (DNN) modeling carried out on a single CPU computer. It is time-consuming and does not utilize advanced computational resources such as GPUs and TPU. A distributed modeling approach provides an abstraction to distribute training across the nodes in a TPU, or multiple GPUs in a computer system. This approach is suitable to reduce modeling time required to train the character models.

There is a necessity to use distributed modeling in DNN training for speedy modeling. We use the standard *distribute* module of Tensorflow API to define and carry out the distributed modeling task. The distributed modeling was designed to distribute existing models and training code with no significant changes. The common ways of distributed training with parallelism are synchronous and asynchronous training approaches. The synchronous method involves training all workers over different slices of input data in a synchronous manner and aggregating gradients. Further, asynchronous training involves all workers are independently trained over the input data. The variables are updated asynchronously.

In this paper, we intend to use a mirrored strategy algorithm to achieve distributed modeling. The mirrored strategy is one of several distribution strategies available in TensorFlow core. This approach involves in-graph replication with synchronous training on the tensor processing unit (TPU) of a computer system. Essentially, it copies all of the model's variables to each processor or node. It uses all-reduce to combine the gradients from all processors and applies the combined value to all copies of the model.

We discuss the details of the distributed training of DNN for modeling Telugu characters in the following sections of this paper. Section 2 describes the related work dealing with segmentation-free word recognition. Section 3 presents the DNN structure and algorithm for distributed training. Section 4 reports experiments and results obtained, and Sect. 5 draws some conclusions.

2 Related Work

Advances in neural networks lead to many successful applications in the branches of speech processing, natural language processing, and computer vision. The widely employed neural network models are CNN, long short-term memory (LSTM), and bidirectional LSTM (BLSTM).

Some hybrid neural network models, such as CNN-LSTM-CTC and CNN-BLSTM-CTC, are successfully used for segmentation-free input data recognition. Further, a deep neural network (DNN) consists of many neural network layers that support end-to-end recognition and do not require any explicit feature extraction. Typically a DNN model specified for a computer vision problem comprises many CNN layers, which cause several hours of training. The CNNs are designed for computer vision problems such as image recognition and video processing. The convolutional operations are computationally intensive. The neural network models consisting of CNN layers consume more training time. One way to handle deep neural networks is the residual neural network (ResNet) [9], which does not take much training time and supports faster training.

Toward the recognition of Telugu character images, different classification methods proposed for the last two decades. These include the classifiers based on simple template matching, k-nearest neighbor (KNN), support vector machine (SVM), hidden Markov model (HMM), artificial neural network (ANN), and the recent deep architectures [4]. In conventional TOCR, one of the most widely employed preprocessing tasks is the connected component extraction. However, there is a recent effort [10] that assumes no prior segmentation of the input document images into connected components. The goal of a segmentation-free approach is to avoid segmentation errors and improve accuracy. In another work, the HMM classifier [11] is presented for segmentation-free Telugu word recognition. They also employed the *akshara* bigram language model. The bigram model is used to supply prior knowledge about the characters during recognition.

Telugu script Telugu is an Indian language and has its phonetic script. Telugu script has about 100 primary glyphs. The glyphs are usually divided into four types: vowels, consonants, vowel modifiers, and consonant modifiers. It has a great character set due to separate glyphs for vowel modifiers and consonant modifiers. The combinations of basic glyphs form *aksharas* (characters) are called the basic orthographic units. Telugu document images comprise lines and words. The words are composed of characters. The character may contain modifiers, which cause the word segmentation a problematic task.

3 Proposed Distributed Training of DNN

We propose a distributed training approach to model DNN classifier for the Telugu OCR system. It is intended to handle larger batches and reduce modeling time significantly.

The input and the output of this modeling are Telugu word images and character models, respectively. The block diagram in Fig. 1 depicts the distributed training approach. The significant tasks involved are CNN-LSTM-CTC neural network model specification, training models using distributed modeling approach, and word recognition. Table 1 lists the layers of the hybrid CNN-LSTM-CTC model. The model has four CNN layers, two LSTM layers, and a CTC output layer. The purpose of CNN layers is not the recognition, but implicit feature extraction. On the other hand, LSTM layers are specified to support the sequence to sequence [6] mapping in Telugu OCR. The input and output sequences of the LSTM are the feature vector sequence obtained from CNNs and character probability sequences, respectively. Moreover, the connectionist temporal classification (CTC) maps the character probability sequence into a character sequence called the predicted word.

In the proposed DNN model, each of the four CNN layers involves operations such as convolution, batch normalization, leaky Relu, and maximum pooling. Convolution uses a 3×3 kernel filter for interpreting snapshots of the word image, and the maximum pooling layer consolidates the image regions and gives a downsized version. For each word image of fixed size 180×60, CNN layers transform the word image into a sequence of 64 features. The LSTM stage of the DNN classifier is a stacked LSTM with two layers of LSTM. Each one has 128 hidden LSTM cells. The output of the stacked LSTM is a matrix. The CTC layer takes the LSTM output

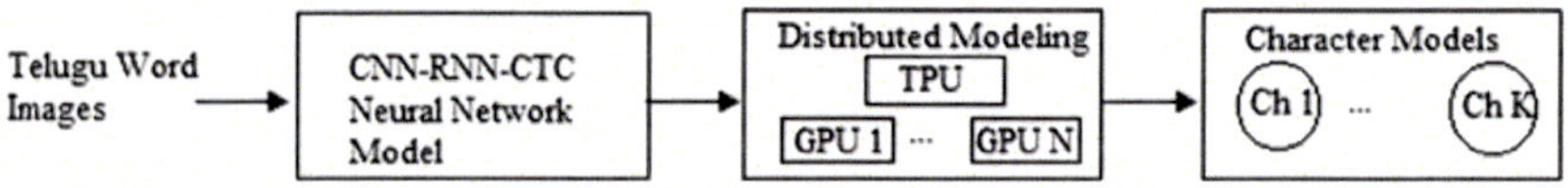

Fig. 1 Block diagram of distributed modeling approach

Table 1 The structure of CNN-LSTM-CTC model for Telugu OCR

S. No.	Layer name	Layer type	Dimension
0	Input	180×60	Batch: 32, 64
1	Hidden	Conv2D, Batch Norm, Leaky Relu, Max Pool	1×64
2	Hidden	Conv2D, Batch Norm, Leaky Relu, Max Pool	64×128
3	Hidden	Conv2D, Batch Norm, Leaky Relu, Max Pool	128×128
4	Hidden	Conv2D, Batch Norm, Leaky Relu, Max Pool	128×64
5	Hidden	LSTM	128 units
6	Hidden	LSTM	128 units
7	Output	CTC	35 labels

and ground truth to compute the loss value during training, whereas it decodes the LSTM's output into text sequence during testing.

Distributed training [12] of deep neural networks depends on two fundamental concepts: data parallelism and model parallelism. Data parallelism enables us to divide the data set equally across the processing units of the system. Each processing unit has a copy of the neural network and its local weights. The distributed training can be in synchronous and asynchronous modes. Algorithm 1 presents the mirrored strategy. It relies on the synchronous mode of training. The synchronous way of training is faster on single machines than multiple machines. It is stable and more efficient to train on a single machine with either multiple GPUs or tensor processing units (TPUs). Training on many GPUs will allow a model to scale with additional resources. The proposed distributed training is performed in the Google colaboratory platform using the TPU system.

Algorithm 1 Mirrored Strategy Algorithm

```
1: procedure MIRROREDSTRATEGY(WordImages,Labels,BatchSize)
2:     Replicate the model graph and variables on the replicas in TPU.
3:     Distribute the input evenly across the replicas.
4:     Calculate the loss and gradients at each replica for the input it received.
5:     Synchronize the gradients across all the replicas by summing them.
6:     Update the copies of variables on each replica.
7: end procedure
```

4 Experiments

We use sophisticated computer systems with GPU, and TPU, which offered in Google Colaboratory [13] environment. These computational resources are provided at no cost and helped in carrying out Telugu character modeling experiments.

Our deep neural network model is specified using Keras [14], and Tensorflow APIs. The DNN classifier for Telugu OCR was trained using a distributed modeling approach on a TPU system. The data set of binary word images is prepared from the standard corpus of Telugu document images. The words are fed to the deep neural network, which needs many instances per character. The data set is provided to the DNN model in binary form. The most frequently occurring words are chosen for training Telugu character models. The task of data set preparation involves ensuring the correctness of word images and their ground truth text. It consumes a significant amount of time to prepare a large data set. In this experiment, we use 8110-word images, consisting of 16330 occurrences of 34 Unicode labels of 52 Unique characters. Since the training is aimed at modeling characters in a segmentation-free manner, we choose word images with at least 150 instances each.

The word images are randomly divided into a training set and a test set of 7500, 610, respectively. The word images and label sequences are fed to the DNN model during training. The plot in Fig. 2 shows the validation accuracy as the function of epochs. The training and validation carried out in 20 epochs and the performance is reported.

Table 2 lists the examples of ground truth text mapping into an equivalent label sequence. We observe the highest accuracy of 97.0%. The word error rate (WER) of the test set is 5%, and the training data set needs to increase for extending this work. Our segmentation-free method is particularly useful when word images contain broken and touching characters. Thus, it offers an optimal solution to the Telugu text recognition that is required in the present context. The examples in Fig. 3 illustrate the word image recognition.

The experiments involve analyzing the impact of batch size and the modeling time taken with a change in processing units such as CPU, GPU, and TPU. We

Table 2 Ground truth text conversion into its equivalent numerical class label sequence

S. No.	Ground truth	Unicode character sequence	Label sequence
1	అతను	3077+3108+3112+3137	3+16+18+30
2	అంటూ	3077+3074+3103+3138	3+2+13+31
3	శ్యామ్	3126+3149+3119+3134+3118+3149	25+34+21+27+20+34
...	...	...	...
N	మాత్రం	3118+3134+3108+3149+3120+3074	20+27+16+34+22+2

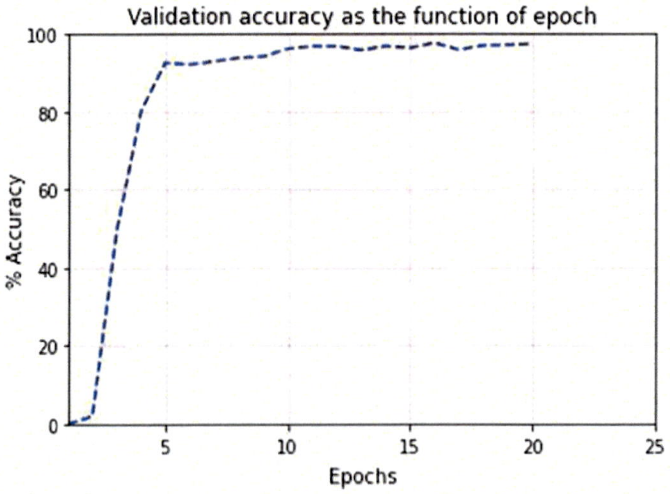

Fig. 2 The validation accuracy curve depicts the DNN model's performance on the data set of Binary word images

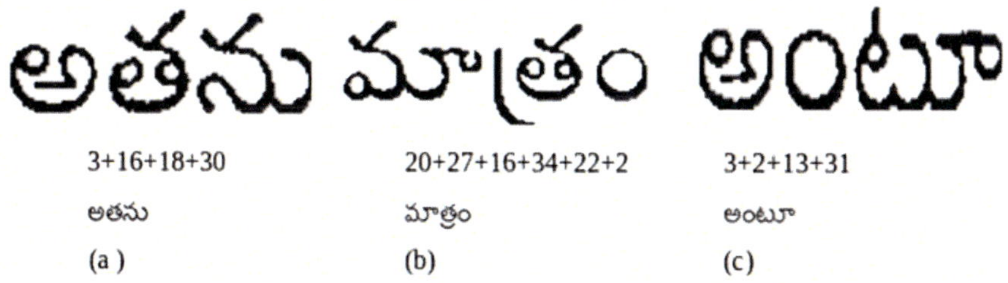

Fig. 3 Recognition of test set word images. It shows three examples

Table 3 Distributed training performance of CNN-LSTM-CTC model for Telugu word recognition

Training data set	CPU/GPU/TPU	Batch size	Elapsed time
7500 Words, 20 Epochs	1-CPU, 2.3GHz	32	3 h 12 min
	1-GPU, Tesla K80	32	38 min
	TPU	32	3 h 00 min
	TPU	64	2 h 41 min

model Telugu characters in our Telugu OCR that enables direct word recognition. The experiments are carried out in two settings: conventional modeling using CPU, GPU, and distributed modeling of characters using TPU. Our modeling approach involves handling large batches and reducing the elapsed time taken in modeling.

The computer system with a single GPU can execute the code when the batch size is small and in between 32 and 50 images. However, the memory allocation problems are observed when the batch size is increased above 50 images. Table 3 lists the processing units and elapsed times of training the CNN-LSTM-CTC model on the data set. We rely on distributed training to use TPU for handling large batches and reducing total modeling time. The distributed training uses the TPU system for working with batch sizes ranging from 32 to 128 images.

There is a significant improvement in modeling Telugu characters with a distributed training approach that supports faster training. Distributed training enables us to utilize the underlying hardware of the TPU system effectively. We mainly employ Google Colaboratory (Colab) that allows us to develop Python code for machine learning applications. Our experimental setup does not need installation of Python, Keras, and Tensorflow APIs except a client computer, browser, Internet facility, and an account in Google. The Google Colab hosts Jupyter notebook service and provides free access to computing resources such as CPU, GPUs, and TPUs. Different kinds of GPUs and TPU such as Nvidia K80s, T4s, P4s, and P100s are dynamically made available to train the models in Google Colab.

5 Conclusion

The distributed training is performed on the machine with TPU in synchronous mode, which is faster and efficient.

When the DNN model trains on a computer server with a GPU that gives the best promising performance, the TPU system has a trade-off in handling large batches as well as data sets. Our segmentation-free word recognition approach enables us to implement many similar computer vision tasks such as handwritten word recognition and line image recognition. We intend to extend this work to facilitate modeling on multiple server systems or client machines in an asynchronous manner using the Hadoop open-source framework.

References

1. Negi, A., Bhagvati, C., Krishna, B.: An OCR system for Telugu. In: ICDAR, pp. 1110–1114. IEEE Computer Society (2001)
2. Lakshmi, C.V., Patvardhan, C.: An optical character recognition system for printed Telugu text. Pattern Anal. Appl. **7**(2), 190–204 (2004)
3. Kumar, P.P., Bhagvati, C., Negi, A., Agarwal, A., Deekshatulu, B.L.: Towards improving the accuracy of Telugu OCR systems. In: ICDAR, pp. 910–914. IEEE Computer Society (2011)
4. Achanta, R., Hastie, T.: Telugu OCR framework using deep learning. CoRR (2015). arXiv:1509.05962
5. Borovikov, E.: A survey of modern optical character recognition techniques. CoRR (2014). arXiv:1412.4183
6. Graves, A., Jaitly, N.: Towards end-to-end speech recognition with recurrent neural networks. In: Proceedings of the 31st International Conference on International Conference on Machine Learning, vol. 32, pp. II–1764–II–1772. ICML'14. JMLR.org (2014)
7. Ben-Nun, T., Hoefler, T.: Demystifying parallel and distributed deep learning: an in-depth concurrency analysis. CoRR (2018). arXiv:1802.09941
8. Breuel, T.M.: High performance text recognition using a hybrid convolutional-lstm implementation. In: 2017 14th IAPR International Conference on Document Analysis and Recognition (ICDAR), vol. 1, pp. 11–16 (2017)
9. He, K., Zhang, X., Ren, S., Sun, J.: Deep residual learning for image recognition. CoRR (2015). arXiv:1512.03385
10. Mathew, M., Singh, A.K., Jawahar, C.V.: Multilingual OCR for Indic scripts. In: 2016 12th IAPR Workshop on Document Analysis Systems (DAS), pp. 186–191 (2016)
11. Rao, D.K., Negi, A.: An implicit segmentation approach for telugu text recognition based on hidden markov models. In: Thampi, S.M., et al. (eds.) Advances in Signal Processing and Intelligent Recognition Systems, pp. 633–644. Springer International Publishing, Cham (2016)
12. Chahal, K.S., Grover, M.S., Dey, K.: A hitchhiker's guide on distributed training of deep neural networks. CoRR (2018). arXiv:1810.11787
13. Google colaboratory. https://research.google.com/colaboratory/
14. Chollet, F., et al.: Keras (2015). https://github.com/fchollet/keras

Word Sense Disambiguation for Telugu Using Lesk

Sudheendra Poluru, A. Brahmananda Reddy, Rahul Manne, Lokesh Bathula, and Nikhilender B. Reddy

1 Introduction

Word Sense Disambiguation (WSD), Reddy et al. [1, 2] refer to the process of ascertaining which 'sense' of a particular word is used in a sentence based on the context of the sentence. Determining which sense of a word is used in a sentence is very important. Word Sense Disambiguation is used in doing the same. There are numerous applications to Word Sense Disambiguation. Some of them are as follow.

Machine Translation: Some words can have different translations based on different senses of the word. Since a machine cannot implicitly determine the correct meaning of a word based on context, Word Sense Disambiguation can be useful in such cases.

Information Retrieval: As the name suggests, Information Retrieval is a software program that deals with storage and effective retrieval of information from document storage space. WSD can be used to remove ambiguities from queries provided by the user to search the document base.

Text Mining: Word Sense Disambiguation can be used in the context of Text Mining to flag relevant words based on their senses.

S. Poluru · A. B. Reddy (✉) · R. Manne · L. Bathula · N. B. Reddy
Department of CSE, VNR VJIET, Hyderabad, India
e-mail: brahmanandareddy_a@vnrvjiet.in

S. Poluru
e-mail: sudheendrapoluru@gmail.com

R. Manne
e-mail: manne.rahul98@gmail.com

L. Bathula
e-mail: bathulalokesh@gmail.com

N. B. Reddy
e-mail: bnikhilender@gmail.com

C. K. Mai et al. (eds.), *Machine Learning Technologies and Applications*,
Algorithms for Intelligent Systems,
https://doi.org/10.1007/978-981-33-4046-6_12

1.1 WordNet

According to *Princeton University* [3–5], Reddy et al. [6], WordNet is a large lexical database for the English language. The different parts of speech of the English language are clustered into sets of synonyms, called synsets. These generated sets are interlinked by means of lexical relations and conceptual semantics. This particular orientation of WordNet is the reason for its popularity in natural language processing.

It represents a thesaurus since it clusters the words into groups based on their meanings.

1.2 IndoWordNet

Bhattacharyya [7] developed an IndoWordNet allowing access to multiple Indic languages. The available languages include Hindi, Bengali, Gujarati, Tamil, Telugu, etc., a total of 18 languages. Like WordNet, it also provides hyponyms and gloss for the purposes of disambiguation.

2 Literature Survey

This section describes the literature survey performed on different papers to understand the existing systems and their methodologies.

1. Sreedhar et al. [8], in their paper, spoke about the various methods which can be used to perform Word Sense Disambiguation for Telugu. The paper began by first talking about the taxonomy of Word Sense Disambiguation in NLP and explained in brief about each of them.

In the following sections of the paper, the current state of the art methods used for Word Sense Disambiguation is briefly explained. Some of the methods discussed in the paper are as follows:

The method proposed by Walker, in which a thesaurus is used. Each word is assigned a particular subject category in the thesaurus. Each subject assigned to a particular word is assumed to be a particular sense and WSD is performed based on this assumption.

The method proposed by Quillian in the mid-1960s in which he proposed the use of semantic network representation of a machine-readable dictionary. In this, a node represents the meaning, and this node is used to connect words.

Many such methods have been specified in this paper along with a brief description of each method. Lesk is one of the proposed methods which uses WordNet.

- **Merits**: The merits of this paper are that it gives an overview of the work that has been done in the field of Natural Language Processing (NLP) and more specifically in Word Sense Disambiguation, which is a subset of NLP.
- **Improvements in this project**: Even though the paper discusses various methods for Word Sense Disambiguation, it doesn't provide any implementation details about it. Lesk has been chosen to perform WSD in this project as it has not been researched extensively for Telugu, unlike other algorithms.
- **Conclusion**: Many methods were analyzed in this paper and provided a concise explanation. Since no implementation details were provided, this project aims at implementing the Lesk algorithm described in the paper.

2. Ritesh et al. [9], in their paper spoke about their Python-based Application Programming Interface (API) for Indian WordNets. This can be used as a module by getting it from pypi.org, where it is present as an open-source project.
   The paper starts off by talking about related work that already exists, such as 'The Java WordNet Library', which is a Java API to access WordNet. Next, the paper describes the API design and the different features that can be utilized in this module.
   - **Merits**: The paper provides a detailed description of how the pyiwn project was built, its motivation, procedure and utilization procedures. It also gives examples of its usage. Being a Python module, it is easier to be integrated into projects as it doesn't require additional overhead.
   - **Improvements in this project**: The paper provides details about the pyiwn project but not its implementation in Word Sense Disambiguation. This project uses the Python module described in the paper and implements it to achieve Word Sense Disambiguation for Telugu. Since it provides an API to access Telugu IndoWordNet along with other languages, this paper was chosen for this project's implementation.
   - **Conclusion**: Finally, by implementing the 'pyiwn' module from the paper, which stands for 'Python Indo Word Net', this project aims to achieve Word Sense Disambiguation for Telugu.

3 Existing System

The observations made by Sreedhar et al. [8] on different approaches of performing Word Sense Disambiguation for Telugu include solutions given by Walker, Wilk, Kozima and Furugori, Lesk, etc. After an extensive literature survey, the Lesk algorithm has been opted as the most prominent one for Telugu sentences.

Most of the existing systems for Word Sense Disambiguation are developed for the English language.

4 Proposed System

This paper tries to extend Lesk to the Telugu language as well. Lesk has been largely applied to sentences in the English language. Most of the research has been performed for English WSD but not much work has been done to implement the same for Telugu language sentences.

The proposed system aims to perform Word Sense Disambiguation for Telugu in the following ways:

- The input is taken in English and translated to Telugu using the Python package—googletrans.
- Then, required preprocessing steps are performed on the sentence:
- Removing stop words,
  Stemming—a stemmer is used to stem words in Telugu.
- Next, pyiwn—a Python package is used, which provides an API to access IndoWordNet.
- Lastly, the relevant synsets are used and provided as input to the Lesk algorithm.

4.1 Lesk Algorithm

The Lesk algorithm is a classical word sense disambiguation algorithm introduced by Michael E. Lesk. It is based on the assumption that the sense of a word is dependent upon the words present in its neighborhood. Steps in the Lesk algorithm:

- Consider a word for processing.
- Next, take the remaining words and create a list of words which consists of the gloss and examples of those words. Let's call it 'matcher_list'.
- Remove stop words from these lists to reduce irregular matches.
- Now for every synset of the word to be processed, create a similar list of words containing the gloss and examples of that synset. Let's call this list 'synset_list'.
- **Match the above two lists to find the overlap of words.**
- The synset_list which has the maximum overlap with the matcher_list is taken as the best synset and the gloss of the synset is taken as the best sense.

5 Implementation

5.1 Implementation Module

See Fig. 1.

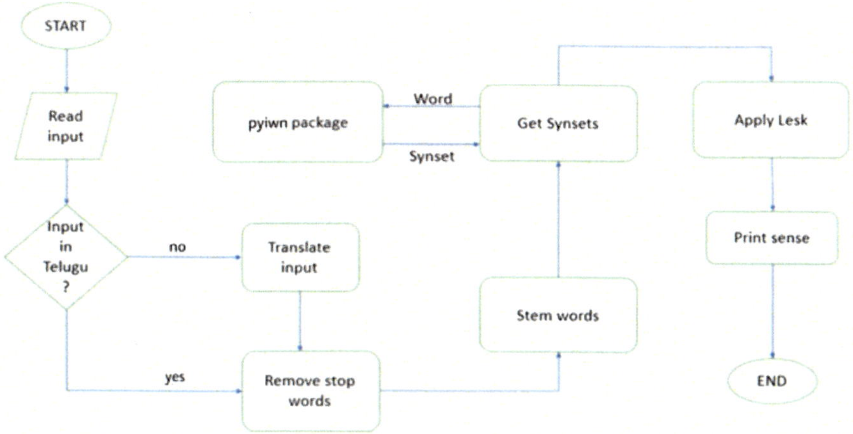

Fig. 1 Implementation module for WSD in telugu using Lesk

5.2 *Implementation*

To implement Lesk, the following tools and packages are used:

- Python 3.6.10
- Flask (for UI)
- Pyiwn—Python package to access IndoWordNet [9]
- googletrans

From Fig. 1, the steps involved in the implementation are as follows.

Step 1: Translating input
Input can be given in both English and Telugu. If the given input is in English, then it is first translated into Telugu using the Python package—googletrans. This package takes English sentences as input and translates them into desired target languages. For our use case, the target language is Telugu.

Step 2: Removing stop words
After the input is translated, it has to be processed before disambiguating the words in the sentence. The initial step in this process is the removal of stop words. Stop words are those words present in a language which add no semantic value to the sentence.

A list of Telugu stop words is generated and it is used to filter the translated sentence.

Step 3: Stemming
The next step, which is a part of preprocessing, is Stemming. It is the process of reducing a particular word to its root form. For example, the stem of the word

running is *run*. Telugu words are stemmed into their root form so that they can be given as an input to the pyiwn package.

Step 4: Applying Lesk
After all the preprocessing steps are performed, the resulting words are given as an input to the Lesk algorithm. The pseudocode for this algorithm is as follows:

```
function LESK(sentence) returns correct sense of words
result := a dictionary of word and sense
correct-sense := most frequent sense for word
max-overlap := 0
words := list of words in sentence
for each word in words
    for each sense in sense(word) do
        matcher := list of words created from gloss and hyponyms
        current-overlap := COMPUTE_OVERLAP (matcher, word)
        if current-overlap > max-overlap then
            max-overlap := current- overlap
            correct-sense := sense
    end
    result[word] := correct-sense
end return(result)
```

6 Results

Consider the input sentence—'silk saree is beautiful'.

- The translation of the sentence into Telugu is

పట్టుచీరఅందంగాఉంది

- To process it better, it is split into individual words as

['పట్టు', 'చీర', 'అందంగా', 'ఉంది']

- The next step is to filter stop words. This is performed so that the words having no semantic meaning can be excluded from consideration during sense determination.

The result after filtering stop words is

['పట్టు', 'చీర', 'అందంగా']

The word 'ఉంది'is removed as it is a stop word.

- Next, these words are stemmed. Any word not in its root form is stemmed. The result after performing stemming is

['పట్టు', 'చీర', 'అందం']

- Now all the preprocessing steps are done. Next, these processed words are supplied as input to the Lesk algorithm.
- In this, each word is considered at a time. Let us consider 'పట్టు'.
- The remaining words are ['చీర', 'అందం'].Their synsets are as follows:

[[Synset('చీర.noun.2184'),

Synset('చీర.noun.5191')],[Synset('సౌందర్యం.noun.34138')]]

- The list of words containing the meanings of the synsets of remaining words is ['చీర', 'దానిపైన', 'ముత్యాలు', 'మొదలైనవి', 'ఉన్నటువంటి', 'చీర', 'సుమారుగా', 'ఆరు', 'గజాలు', 'ఉండి', 'స్త్రీలు', 'ధరించే', 'వస్త్రం', 'స్త్రీలకు', 'ఇచ్చే', 'వస్త్రం', 'చక్కని', 'రూపం', 'వుండటం']
- Now for every sense of the word 'పట్టు',a similar list of words is created from its meaning and matched with the list of words in the previous step. One such list is— ['వస్త్రం', 'రేషమ్', 'దారంతో', 'తయారు', 'చేస్తారు', 'గొంగళిపురుగు', 'నోటి','లభించే', 'దారం.'].The word 'వస్త్రం'is matched a maximum number of times and hence this sense of the word 'పట్టు'is taken as the best sense.
- Thus for the sentence, ' పట్టుచీరఅందంగాఉంది'the output is as follows:

పట్టు =>గొంగళిపురుగునోటిద్వారాలభించేదారం.

చీర =>సుమారుగాఆరుగజాలుఉండిస్త్రీలుధరించేవస్త్రం

అందం =>చక్కనిరూపంకలిగివుండటం

Table 1 Example of WSD

English sentence	Telugu translation	Ambiguous word	Best sense
Silk saree is beautiful	పట్టుచీరఅందంగాఉంది	పట్టు	గొంగళిపురుగునోటిద్వారాలభించేదారం
Good grip on the subject	విషయం పై మంచి పట్టు ఉంది	పట్టు	ఏ పనిలోనైనా నైపుణ్యం కలిగి ఉండటం
He is playing	అతను ఆడుతున్నాడు	ఆడు(play)	క్రీడలో భాగమవటం
He is acting in a play	అతను ఒక ఆటలో నటిస్తున్నాడు	ఆట (play)	రంగస్థలంపై వేయబడినది.

6.1 *Some More Examples*

See Table 1.

6.2 *Single Word as Input*

Lesk works best when the input has more than one word. This is because Lesk determines the sense of a word based on the neighboring words of a particular word under consideration.

Consider the word ‘silk’ as input in Table 2.

It takes the last synset as the best sense when no word matches.

But, when the input contains more words,

The above sense is the right sense. Thus, Lesk works best when the context around the word is rich and relevant (Table 3).

Table 2 Example of a single word as input

English word	Telugu translation	Ambiguous word	Sense
Silk	పట్టు	పట్టు	ఏ పనిలోనైనా నైపుణ్యం కలిగి ఉండటం

Table 3 Example on importance of context

English word	Telugu translation	Ambiguous word	Sense
Silk saree	పట్టుచీర	పట్టు	గొంగళిపురుగునోటిద్వారాలభించేదారం

7 Conclusion

This paper mainly focuses on what Word Sense Disambiguation exactly is, and what methods were involved in solving them. Until now, many approaches have been made in solving them which mainly focused on the English language and other few regional languages where the approaches included both supervised and unsupervised techniques. So, this paper focuses on WSD for Telugu which is an Indian regional language spoken by the people of the states Telangana and Andhra Pradesh.

Word Sense Disambiguation for Telugu is achieved after performing necessary preprocessing steps such as translating the words, stop-word filtration, and stemming, and later implementing the Lesk algorithm on the processed words. This works for single sentences where the input is given in a sentence form. There is a scope to extend WSD for a whole paragraph or a whole document.

References

1. Reddy, A.B.: Integrated feature selection methods for text document clustering. Integrated feature selection methods for text document clustering (2015)
2. Reddy, A.B., Govardan, A.: Ontology for an education system and ontology based clustering. In: The 5th International Conference on Fuzzy and Neuro Computing (FANCCO—2015), IDRBT, Hyderabad (2015)
3. Fellbaum, C.: WordNet: an electronic lexical database. MIT Press, MA, Cambridge (1998)
4. Murty, M.R., Murthy, J.V.R., Reddy, P.P., Sapathy, S.C.: A survey of cross-domain text categorization techniques. In: 2012 1st International Conference on Recent Advances in Information Technology RAIT-2012, ISM-Dhanabad, 978-1-4577-0697-4/12 IEEE Xplorer (2012)
5. Miller, G.A.: WordNet: a lexical database for english. Commun. ACM **38**(11), 39–41 (1995)
6. Reddy, A.B., Govardan, A.:A novel approach for similarity and indexing based ontology for semantic web educational system. Int. J. Intell. Eng. Inf. (2016)
7. Bhattacharyya, P.: IndoWordNet. IIT Bombay, March 2010
8. Sreedhar, J., Raju, V.S., Babu, A.V.: A study of critical approaches in WSD for telugu language nouns: current state of the art. Int. J. Sci. Eng. Res. **5**(6) (2014), ISSN 2229-5518
9. Panjwani, R., Kanojia, D., Bhattacharyya, B.: pyiwn: a python-based API to access Indian language WordNets. IIT Bombay (2018)

Identifying Duplicate Questions in Community Question Answering Forums Using Machine Learning Approaches

Divya Vanam and Venkateswara Rao Pulipati

1 Introduction

Online education nowadays is playing an important role, and it is flexible for society to learn via the Internet. It is a process of teaching the students via the Internet. Every client is using online search engines to retrieve knowledge. A search engine provides more information but it contains duplicates. Duplicate detection is done in the database to save memory. Duplicate means the same intent is found from the memory. Due to duplicates, the database consumes more space in the memory. Duplicate detection is done by comparing two sentences or words using dimensional similarity. If a similarity is found, it is deleted. Otherwise, it is stored in the database. As in [1] by finding the Duplicate question, it gets beneficial to the Software Development Community.

Web forums are websites where clients can communicate with each other by posting messages. Web forums for Community Question Answering (CQA) are a highly famous mechanism for data searching and sharing. Reference [2] states that the questions are retrieved from the CQA forums. References [3, 4] states that CQA is one of the most important portals such as search engines. Search engines such as Stack Overflow in [5], Quora, Reddit, Yahoo! Answers, etc. They allow clients to ask or answer the questions in the hope of getting high-quality answers. We can find the relevant answers which were asked by the clients. The questions which were asked may be already present in the websites with different intent. To overcome this problem, we are finding duplicate questions and removing them. These are done by using Natural Language Processing (NLP) techniques.

D. Vanam (✉) · V. R. Pulipati
VNR Vignana Jyothi Institute of Engineering and Technology, Hyderabad, India
e-mail: divya509vd@gmail.com

V. R. Pulipati
e-mail: venkateswararao_p@vnrvjiet.in

C. K. Mai et al. (eds.), *Machine Learning Technologies and Applications*,
Algorithms for Intelligent Systems,
https://doi.org/10.1007/978-981-33-4046-6_13

Lately, an NLP application has growth in the increase of accuracy requirements. It is a communicating to an intelligent system using a Natural Language. In real-time, we use "Google Assistant", "Alexa", etc.; these are used for speech recognition. NLP helps a computer to put unstructured data in the proper format. NLP concepts are Text preprocessing, Stop Words, Tokenization, Normalization, Stemming, Bag of Words, Continuous Bag of Words, Parts-Of-Speech (POS) [6], Tagging and Classification techniques. It is used to remove noisy data. To handle this task, we extract the questions from the dataset. We construct a model by using Machine Learning techniques like Random Forest and the Gradient Boosting model.

For classifying the problem of duplicate questions, we have used random forest classification and the gradient boosting model. Random forest classifier is a decision-making tree; it extracts the sub-data from training data randomly and then aggregates the sub-data to get the best result. The gradient boosting model is used to identify the errors of the model and then add weak learning models together to make the best predictions. It minimizes the loss when a new model is added.

2 Related Works

[1] It states that they have used different supervised learning algorithms on textual features. They worked on 2 stages of "ranking classification", one is historical questions based on their similarities and the other is a novel feature based on text similarity and latent semantics. Here, DupDetector is a newly issued question which is paired with the candidate selection question; it uses the vector similarity feature and association feature, and relevance feature to predict the best result in PCQA.

[2] It explores ranking the question answering based on their likes and comments. This is a web application where user and admin has access to login into the page. They were developed by Java and MySQL as back-end and JavaScript, HTML 5, CSS 3 as front-end technologies. The Levenshtein algorithm is used to find the similarity between the two strings. The naive probability of question pairs is applied based on likes and comments and it results in the top-ranked answer.

[3] It detects the duplicate questions in CQA using multidimensional feature combination. It is based on classifying the answers as good, bad or neutrally associated with the question, and answering right or wrong based on a list of all answers to a question. They used different technologies like Bag of Words, SVM, Gradient Decision Tree, and Random Forest. LDA as a probability is finding and solving the duplicates in question answers.

[4] It finds the similarity in the text in the Natural Language Processing based on spark architecture using the k-means algorithm. Word vector distance decentralization (WVDD) is used to measure the distance, and clustering analysis is compared with Doc2Vec and Bag of words to find the similarity measurements.

[5] It states the issue of Community Question Answering for the Arabic language. They have implemented it by using the Machine Learning approaches. To leveraging the syntactic structure of sentences by using kernel trees to compute the similarity measures. This can be overcome by Continuous Bag of Words.

[6] It analyzes the duplicates using reputation-based heuristics and compares the content of duplicates both in terms of their questions and answers to determine the degree of similarity between each duplicate pair using NLP techniques. They cannot find the precise techniques for similarity.

3 Existing Work

Existing works framed the task as a supervised learning issue on the question pairs and relied on only textual features, and [7] states the problem to find the duplicate detection as a two-stage "ranking-classification" issue over question pairs. These problems can be overcome by using some techniques like Term Documentation, Bag of Words, and Term Frequency–Inverse Document Frequency.

3.1 Bag of Word

The Bag of Words [8] model is a representation used in Natural Language Processing. Bag of words is used to count the total occurrences of the most frequently used words which can also be used for text classification. In this model, the text is extracted and divided into sentences, words or tokens and the number of occurrences of the words or tokens that occurred are calculated using word2count. The frequently occurring words are collected into one unit, i.e., one bag.

For instance, "Poona likes ice cream and Mary too likes ice cream". Bag of Words is used to count the total occurrences of most frequently used words as BOW = {"Poona": 1, "Likes": 2, "Ice": 2, "Cream": 2, "And": 1, "Mary": 1, "too": 1}. Here, there are 7 words in the sentence. The words like "likes", "ice", and "cream" are a count of 2 words. They are a set/union of two words in a bag of words. The words are called terms also; we calculate the term frequency–inverse document frequency as they found in [9].

3.2 TF-IDF Vectorization

Term Frequency: Term frequency is used to calculate the frequency of words in the current dataset. Every dataset a different length. It may be possible that the words in the dataset may appear more times. The term frequency is calculated as

$$TF(t) = \frac{No.\,of\ times\ Words\,(t)\,appears\,in\,a\,sentence}{Total\,No.\,of\ \ Words\,in\,the\,sentence} \tag{1}$$

Inverse Document Frequency (IDF): It calculates how the rare word is across the dataset. IDF is a measure of how rare a term is

$$IDF(t) = loge\left(\frac{Total\,Number\,of\ sentence}{Number\,of\ Sentence\,with\,term\,(t)\,in\,it}\right) \tag{2}$$

Thus,

$$TF - IDFscore = TF * IDF \tag{3}$$

3.3 N-Gram

An n-gram model is used to calculate the continuous sequence of n items; the term or word or token is given in a sequence of the text. The "n" indicates the number of sequences of words. For example, let us take a sentence as "NLP is a great language". An n-gram is a sequence of N words: a 2-gram (or bigram) is like "NLP is", "is a", "a great", "great language" and a 3-gram (trigram) is a 3-word sequence of words "is a great" or "a great language". In this way, the words are tokenized for the n-gram model; then we fit the model according to the n-gram model on a test set, and then it is made into a new model. This model is used for further approaches.

3.4 Continuous Bag of Words

In the continuous bag of words (CBOW) algorithm, the text is represented in the form of numeric words for given neighboring words. For instance, "Raja is playing at the river bank". In this example, we are analyzing the adjacent or neighboring words, because one particular word may contain different intent. For example, the word "bank" contains a different intent as "bank" may be the meaning of a "financial place where money is stored" and it may be "the land adjacent to the river". So from the above example, the sentence is representing that the boy Raja is playing near the river bank, i.e., on the land adjacent to the river. This can be found by using the CBOW algorithm; in this, we are using the window size. If the window size is 2, then we take the words as {"Raja, is", "is, playing", "playing, at", "at, the", "the, river", "river, bank"}. By using the neighboring words, we find the intent of the words (Fig. 1).

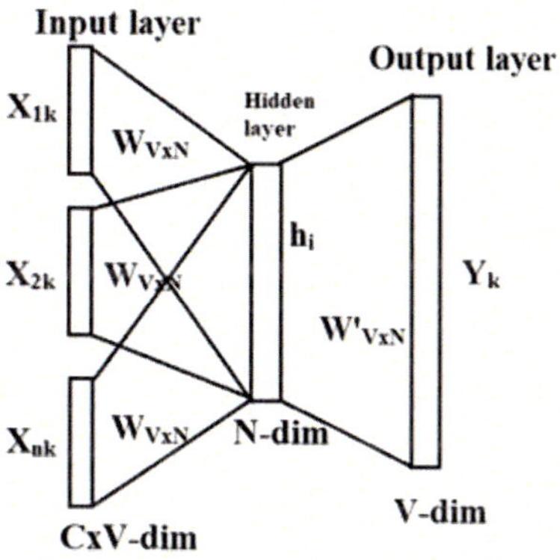

Fig. 1 Continuous bag of words model

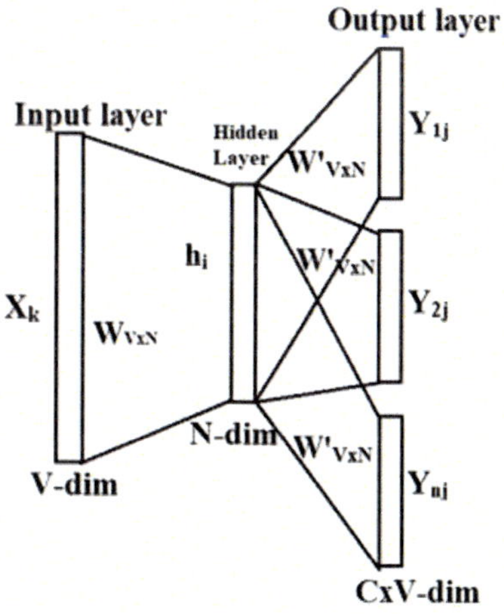

Fig. 2 Skip-gram model

3.5 *Skip-Gram Model*

Word2Vec works practically as an auto-encoder. We will prepare on one side a neural system to perform a specific task, and on the other side fix it to return to the first outcome [6, 10]. It states the different variations like Word2Vec worldview, skip-gram and CBOW. The skip-gram variation takes an objective word and tries to analyze words, while the CBOW (continuous bag of words) variation takes a lot of context words and attempts to analyze an objective word (Fig. 2).

3.6 *Feature Extraction*

During the study of this project, we have used different aspects of our model depending upon the model. Preprocessing is done before acquiring a computer precise and structured data from unstructured data [11, 12]. Techniques such as text cleaning, tokenization, stemming or lemmatization, and stop word removal are used. Text cleaning is utilized to clean or remove the noise from the unstructured data of the text which is collected from various sources. NLP applications need to divide large records of improper data into sentences or words to get the best data. Tokenization is a procedure of dividing the raw data into meaningful tokens. Stemming is a technique to find the essential word from the given word. It is an absolutely rule-based procedure; all tokens are cut down to get the root word. For example, Word "eat"—we have different implications as "eating", "ate", etc. We tend to utilize stemming to club each syntactic difference to the base of the word. Lemmatization might be a methodology of changing all the linguistic/syntactical types of the base of the word. Lemmatization utilizes the specific situation and Parts-Of-Speech (POS) tagger to decide the arched kind of the word, and different standardization rules are applied for each POS tag to get the root word. Lemmatization [12, 13] is a powerful, proficient and orderly method for consolidating linguistic varieties to the base of the word.

Stop words are frequently occurring words; these words add weightage to the sentences [12, 13]. They act as the main role or bridge in the sentence and form a meaningful semantical sentence. Classification techniques are not affected by these stop words. Example: {"and", "as", "at", "be", "but", "both", "by", "can", "in", "for" ...}.

4 Proposed Methodology

4.1 Workflow

The Text data is extracted from the dataset. On this data, Text Preprocessing techniques are applied the same as [11]. Some Preprocessing techniques are Word Embedding, Calculation TF-IDF Weights, Bag of Words (BOW), etc. By using these techniques, Feature Extraction is done. Reference [14] explains the classification algorithms like Random Forest and the Gradient Boosted model are used for text classification. With the help of prediction of the data, duplicate detection is done and comparison-based accuracy is done. We are using some Supervised Classification as [15] like Random Forest and the Gradient Boosting model (Fig. 3).

4.2 Random Forest

The Random Forest algorithm is utilized for classification and regression. It is an ensemble learning method and is developed by using different and many other decision trees as referred to in [9]. Equating the impact of several decision trees will improve the prediction method of a random forest.

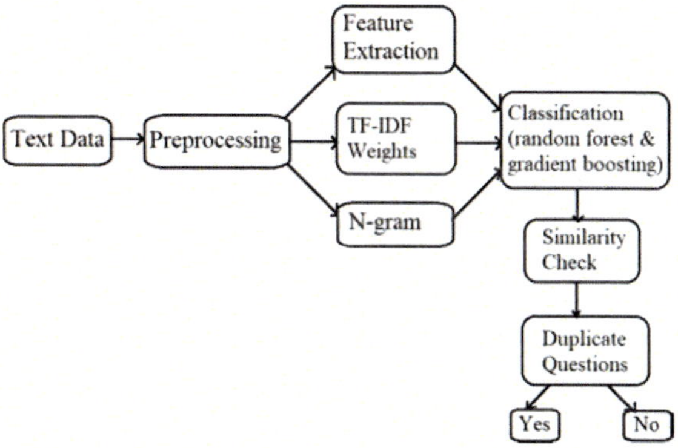

Fig. 3 Workflow of proposed system

Apart from machine learning supervised techniques like Logistic regression, Support Vector Machines (SVM), K-Nearest Neighbor (KNN), Naive Bayes, etc., [16] states that the random forest classification algorithm has many features. This may individually have more collective power and weak predictive power.

4.3 Gradient Boosting (XGBOOST)

Gradient Boosting is a machine learning classification and regression model. It is used to build models using boosting [8]. This algorithm is used to optimize the cost function over function space by choosing the weak hypothesis. This weak hypothesis is made to predict the values and the additive model is to add the weak hypothesis to the cost function. It can overfit the training dataset quickly.

Algorithm: Gradient_Boosting ():

1. Initialize model to a constant value:

$$F0(x) = \arg\min \Sigma n\ i = 1\ L(yi, \Upsilon). \tag{1}$$

2. For m = 1 to M:

   1. Compute pseudo residuals:

$$\Upsilon im = -\left[\frac{\partial L(yi, F(xi))}{\partial F(xi)}\right]_{F(x) = F\,m\text{-}1\,(x)} for\ i = 1, 2, 3 \ldots N \tag{2}$$

   2. Fit a weak Hypothesis $h_m(x)$ to pseudo residuals

   3. Compute

$$\Upsilon m = \arg\min \Sigma n\ i = 1\ L(yi, Fm - 1\ (xi) + \Upsilon hm\ (xi)). \tag{3}$$

   4. Update the model:

$$Fm(x) = Fm - 1\ (x) + \Upsilon mhm\ (x). \tag{4}$$

3. Output $F_M(x)$

4.4 Dataset

In this passage, we classify the proposed techniques on benchmark datasets. This dataset was extracted from Kaggle. The dataset consists of id, qid1, qid2, question1, question2 and is_duplicate. The question pairs are subsequently reduced to the following subtasks: Classification classifies the QA sets dependent on their likeness with the Question. Ranking calculates the rank of QA matches by reducing comparability with respect to the question.

We utilized Model Evaluation as [6, 14] scorer to access the acquired outcomes as precision, recall, F1-score and support.

5 Experiments Evaluation and Results

In this experiment I have compared the two different classification techniques like random forest and gradient boosting model by adding the new features to these classifications for resulting the best accuracy compared with existing work. From this Random forest classification is resulting the best accuracy compared with Gradient boosting model. So, it is one of the best classification techniques to find duplicate questions in CQA forums (Tables 1 and 2).

Table 1 Comparison of training score and validation score

	Training score	Validation score
BOW + XgBoost	0.82	0.76
TFIDF + XgBoost	0.89	0.76
N-gram + XgBoost	0.74	0.67
Random forest	0.89	0.83

Table 2 Comparison of precision, recall, F1-score and support

	is_duplicate	Precision	Recall	F1-score	Support
BOW + XgBoost	0	0.79	0.90	0.84	84355
	1	0.78	0.60	0.68	49080
TFIDF + XgBoost	0	0.80	0.90	0.84	84355
	1	0.77	0.61	0.68	49080
N-gram + XgBoost	0	0.73	0.91	0.81	84355
	1	0.74	0.43	0.54	49080
Random forest	0	0.83	0.91	0.86	–
	1	0.89	0.61	0.69	–

6 Conclusion and Future Enhancement

In this work Random forest algorithm is resulting the best accuracy compared with gradient boosting model. Upon experimentation of the existing work, I got a better accuracy of model evaluation results in my proposed work than the existing work. We evaluate the performance like accuracy, precision, recall and F1-score. The random forest algorithm is the best compared to the Gradient Boosting model.

Further, this work can be implemented by using another dimensionality reduction algorithm such as Principal Component Analysis (PCA) and other classification algorithms.

References

1. Abric, D., Clark, O.E., Caminiti, M., Gallaba, K., McIntosh, S.: Can duplicate questions on stack overflow benefit the software development community? In: IEEE/ACM 16th International Conference on Mining Software Repositories (MSR), Montreal, QC, Canada, pp. 230–234 (2019)
2. Cao, X., Cong, G., Cui, B., Jensen, C.S., Yuan, Q.: Approaches to exploring category information for question retrieval in community question-answer archives. ACM Trans. Inf. Syst. (TOIS) (2012)
3. Zhang, W.E., Sheng, Q.Z., Lau, J.H., Abebe, E., Ruan, W.: Duplicate detection in programming question answering communities. ACM Trans. Internet Technol. **18**(3), Article 37 (2018)
4. Cai, L., Zhou, G., Liu, K., Zhao, J.: Learning the latent topics for question retrieval in community QA. In: Proceedings of IJCNLP, pp. 273–281 (2011)
5. Silva, R.F.G., Paixao, K.V., Maia, Md.A.: Duplicate question detection in stack overflow: a reproducibility study. In: IEEE 25th International Conference on Software Analysis, Evolution and Reengineering (SANER) (2018)
6. https://medium.springboard.com/identifying-duplicate-questions-a-machine-learning-case-study-37117723844
7. Shah, S.S., Bavaskar, T.S., Ukhale, S.S., Kalyankar, A.S., Patil, R.A.: Answer ranking in community question answer (QA) system and questions recommendation. In: 4th International Conference on Computing, Communication, Control and Automation (2018)
8. https://towardsdatascience.com/finding-similar-quora-questions-with-word2vec-and-xgboost-1a19ad272c0d
9. https://towardsdatascience.com/finding-similar-quora-questions-with-bow-tfidf-and-random-forest-c54ad88d1370
10. Zhou, S., Xu, X., Liu, Y., Chang, R., Xiao, Y.: Text similarity measurement of semantic cognition based on word vector distance decentralization with clustering analysis. IEEE Access **7**, 107247–107258 (2019)
11. Fan, H., Ma, Z., Li, H., Wang, D., Liu, J.: Enhanced answer selection in CQA using multi-dimensional features combination. Tsinghua Sci. Technol. **24**(3), 346–359 (2019)
12. https://www.analyticsvidhya.com/blog/2018/02/the-different-methods-deal-text-data-predictive-python/
13. https://medium.com/data-science-101/detecting-duplicate-questions-on-quora-beating-stanfords-accuracy-6f18b3634d06
14. Zhou, T.C., Lyu, M.R., King, I.: A Classification-based approach to question routing in community question answering (2012). https://doi.org/10.1145/2187980.2188201

15. Conneau, A., Kiela, D., Schwenk, H., Barrault, L.: Bordes: supervised learning of universal sentence representations from natural language inference data. In: Proceedings of EMNLP, pp. 681–691 (2017)
16. El Adlouni, Y., Rodríguez, H., Meknassi, M., El Alaoui, S.O., En-nahnahi, N.: A multi-approach to community question answering. Expert Syst. Appl. **137**, 432–442 (2019). ISSN 0957-4174

A Comparison of Classical Machine Learning Approaches for Early Structural Damage Identification

Jay Karan Telukunta and Myneni Madhu Bala

1 Introduction

Structural Health Monitoring (SHM) is a method of detecting damage to the structures over their lifetime using sensor systems. It can be regarded as a technology, which continuously assesses the health condition of a structure. Typical SHM mechanism involves the continuous observation of the system, acquisition of response from a range of sensors, extraction of features, and evaluation of the current state of structure's health driving us to make major decisions regarding design, maintenance, and recovery of structures. Hence, monitoring is critical as it prevents human and infrastructural loss which might be incurred later.

Poor design built, improper maintenance, manual negligence are human factors triggering damage to the structures. Variation in temperature, wind speeds, moisture are environmental factors triggering damage to the structures. To reduce the manual effort involved in SHM and fetch more intelligent decisions consistently about the state of damage, the paper proposes a mechanism to segment and transform the vibrational data as input features to machine learning approaches. Hence, the paper further compares the usage of different machine learning approaches to the proposed mechanism for damage identification and diagnosis. The data was collected from a two-level structure that was installed in the laboratory for carrying out this study.

Machine learning has gained importance for the application of 'prediction' and 'classification'. Supervised and unsupervised learning are the two main classes of

J. K. Telukunta (✉) · M. M. Bala
Computer Science and Engineering, Institute of Aeronautical Engineering, Hyderabad, Telangana, India
e-mail: jaykarantelukunta@gmail.com

M. M. Bala
e-mail: baladandamudi@gmail.com

C. K. Mai et al. (eds.), *Machine Learning Technologies and Applications*,
Algorithms for Intelligent Systems,
https://doi.org/10.1007/978-981-33-4046-6_14

machine learning. These mechanisms use labeled and unlabeled datasets, respectively. A machine learning process proceeds through data collection, pre-processing of data, specifying the train and test set, selection of algorithm, training the model, and evaluating the model performance using a test set of data which is briefed as follows.

a. An **Entity** is the item of interest on which machine learning study analysis is to be made. Structures are the entities regarded in this study.
b. **Data Collection** involves the gathering of data from chosen entities. This step of a collection can be carried out using manual observations, sensors, acquisition systems, and related techniques.
c. **Data Preprocessing** ensures the quality of collected data by applying various data cleaning techniques which help in identifying required features resulting in a good predictive model.
d. A **Split of Train—Test Set** involves randomizing the dataset, followed by segregation of data into Training and Test data. These sets are used to train and test the model, respectively.
e. **Selection of the Algorithm** is a crucial step onto the data for fetching good results.
f. **Training Phase** involves learning of correlations by the model to fit the data gathered using Train set.
g. **Evaluation Phase** involves determining the performance of a trained model using the Test set.

The remainder of the paper is explained and formatted as 2-Literature Review, 3-Experimental Setup, 4-Experiment, 5-Results and Analysis, and 6-Conclusions.

2 Literature Review

Many researches have been conducted in the domain of Structural Health Monitoring. The techniques were broadly categorized into qualitative and quantitative approaches.

In earlier days, many qualitative techniques such as the sound of a hammer strikes over train wheels and railroads, visual inspections on civil structures, and many such similar methods had been adopted for monitoring structures. As time progressed, quantitative and data-driven techniques replaced most of the qualitative techniques.

Quantitative techniques are modeled based on the analysis of changes in vibration, wind, temperature, and pressure and stress loads of the structures. These changes are essential for identifying and localizing the damage to the structures. Reference [1] explains the importance of the placement of sensors on systems for proper data acquisition along with the development of the Statistical Model. It uses the X-chart and Shewhart chart for the calculated mean and standard deviation (SD). Upper and Lower limits are defined as mean + k(SD) and mean − k(SD) for healthier structure sets. Using the upper limit and lower limit, a newer response is being compared and the state of health is predicted. Natural frequencies, mode, and modal shapes and dynamic flexibility matrices are the other set of parameters that were considered in other studies for damage detection. Reference [2] uses natural frequencies for structural health monitoring and damage assessment. It explains the integrations of Global Shape Correlation and Global Amplitude Correlation functions along with the frequency points over the measurement range for damage indication. The importance of using frequency response functions as input to neural networks, genetic algorithms, and fuzzy logic are elaborated in [3]. Use of Principal Component Analysis for dimensionality reduction of larger Frequency Response Functions was explained in [4]. Other studies discuss how changes in mode shape curvature [5–7], changes in curvature damage factor [8], and changes in modal flexibility matrices help in damage detection [9].

3 Experimental Setup

A two-level test structure was created and installed in a laboratory for the experimental study. The two-dimensional schematic of the test structure is displayed in Fig. 1.

The labeled components of the test structure are explained as follows:

a. The top section of the structure is defined as the Upper Level. The bottom section of the structure is defined as the Lower Level.
b. The components used to hold Upper level, Lower level, Shaker table, and Shaker are defined as bindings. Metallic screw-like components were used in the experiments.
c. Type 4533-B CCLD accelerometer from Brüel and Kjær was used. It is designed for a wide frequency range, low noise, and low sensitivity to environmental factors.
d. The structure was mounted onto the platform called Shaker Table. This table was clamped and fixed to the Shaker.

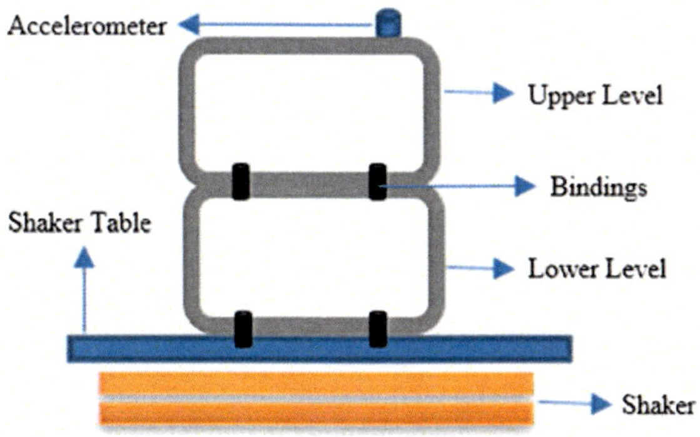

Fig. 1 Experimental test structure

e. Brüel and Kjær Shaker VA-450 equipment was used to induce vibrations and excitations to the setup.

4 Experiment

The conducted experimental study is divided into six phases as displayed in Fig. 2. The detailed explanation is elaborated as follows.

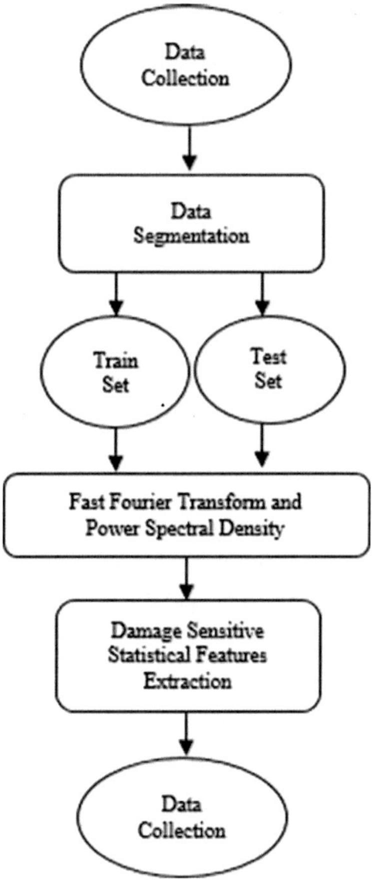

Fig. 2 Block diagram for applying machine learning in structural damage identification

Table 1 Structural damage cases

Case	Description
1	Fully healthy structure
2	Upper level is unhealthy
3	Lower level is unhealthy
4	Fully unhealthy structure

4.1 Data Collection

Vibrations are waveforms. These can be represented in the time domain, frequency domain, and time-frequency domain. Time-domain signals convey the information regarding the vibration at any given instance. There is no derived information regarding the rate at which the signal is varying. Frequency domain plays as the alternative, which analyzes the vibration to thc frequency. Therefore, vibrations were used for damage assessment in the study performed.

Four damage cases were considered in this experiment as outlined in Table 1. Time-series raw data was collected by the accelerometer sensor when the Shaker was operational for each damage case separately. A sampling rate of 20 Hz was considered.

4.2 Dataset Segmentation and Data Splitting

Raw data that was collected for each case was framed as segments. Each segment has a hundred data points. 180 set of segments of data for each damage case was framed. A total of 720 sets of segments were made for all the damage cases making the dataset balanced. The framed segments were labeled as per the respective categories.

Framed and labeled segments are then spilt into the training dataset and testing dataset. The percentage split of data segments was 80 and 20 respectively.

4.3 Fast Fourier Transform and Power Spectral Density

Segmented data was vibrational data in the time-domain. Vibrational data in other domains may fetch more intelligent decisions. Hence, signal processing transformations were used as feature engineering tasks to transform the time-domain segments into other domains before using machine learning techniques. The transformation techniques used in this study were Fast Fourier Transform and Power Spectral Density.

Fast Fourier Transform is an efficient algorithm for time-domain to frequency-domain transformation. It reduced the computational complexity from n2 to n*log

(n) [10]. Power Spectral Density describes the frequency spectrum of a signal along with the power distribution at each frequency [11]. All segments were subjected to Fast Fourier Transforms and Power Spectral Density. Corresponding transformed values are updated in the respective segments. These transformations are the feature engineering of the data.

4.4 Damage Sensitive Statistical Feature Selection

All the transformed segments were pre-processed further for feature selection. The following statistical parameters were identified and introduced as input features to machine learning models. These parameters were computed on the Transformed Segmented FFT and PSD data separately. Models will learn the correlations among these transformed segments and determine the damage in the structure. The features which were introduced on FFT and PSD segments are as follows

(1) *Mean*
Mean of the segments in FFT and PSD was used as one of the input features as it behaves as a representative value for the whole segment. It is calculated as

$$Mean = \frac{\sum Xi}{n}$$

(2) *Standard Deviation*
The deviation of the data points FFT and PSD transformed segment can be given Standard Deviation. It is calculated as

$$S = \sqrt{\frac{\sum (Xi - Mean)^2}{n}}$$

(3) *Maximum, Minimum, and Range*
The maximum, minimum, and range of each FFT and PSD transformed segment were also considered as features

$$Maximum = \max(Xi)$$
$$Minimum = \min(Xi)$$
$$Range = Maximum - Minimum$$

(4) Skewness
The distortion in the probability distribution of each transformed segment about its mean will be given by Skewness. This is also considered as one of the key features. It is calculated as

$$Skewness = \frac{1/n \sum(Xi - Mean)^3}{(1/\sum(Xi - Mean)^2)^{\frac{2}{3}}}$$

(5) Kurtosis

The tailedness of the probability distribution of each transformed segment will be given by Kurtosis. This feature is computed as

$$Kurtosis = \frac{1/n \sum(Xi - Mean)^4}{(1/n \sum(Xi - Mean)^2)^2}$$

4.5 *Applying Machine Learning*

The above damage sensitive statistical features were calculated. Training of the machine learning model using these features was then followed. Classical techniques such as Support Vector Machines, Random Forest Classifier, and K-Nearest Neighbors were used to evaluate the performance of Machine Learning in identifying and detecting the structural damage case.

SVM is a statistical learning algorithm given by [12], which is widely used for classification problems. SVM algorithm finds an optimal hyper-plane that maximizes the margin between two groups of samples. This method samples the hyper-planes which separates two or more classes. Multi-class classifiers are used to classify more than one class. SVM extends the great support of kernels to deal with nonlinearly separable data. Few SVM kernels are Polynomial Kernel, Gaussian Kernel, Radial Basis Function (RBF), Laplace RBF Kernel, Sigmoid Kernel. Random Forest is a bagging methodology that works on weak learners. This classifier is a combination of decision tree predictors. The classifier consists of a collection of tree-structured classifiers $\{h(X, \theta^k), k = 1, 2, 3, \ldots\}$, where the $\{\theta^k\}$ are independent identically distributed random vectors and each tree casts a unit vote for the most popular class at input X [13]. KNN is a supervised algorithm that classifies the new sample's category based on 'K' training samples that are nearest to the sample.

5 Results and Analysis

Extracted features from the transformed segments of the collected dataset were input to Support Vector Machine, Random Forest Classifier, and KNN. Precision, Recall, F1-Score, and Support were the metrics taken into consideration for comparing the performance. Classification reports of Support Vector Machine, Random Forest Classifier, and KNN for each damage case are displayed in Table 2, Table 3, and Table 4, respectively.

Table 2 Classification report—support vector machines

Cases	Precision	Recall	F1-Score	Support
1	0.93	1.00	0.96	26
2	1.00	1.00	1.00	26
3	1.00	1.00	1.00	26
4	1.00	0.92	0.96	26
Average	0.98	0.98	0.98	104

Table 3 Classification report—random forest classifier

Cases	Precision	Recall	F1-Score	Support
1	0.84	1.00	0.91	26
2	1.00	0.92	0.96	26
3	1.00	1.00	1.00	26
4	0.91	0.81	0.86	26
Average	0.94	0.93	0.93	104

Table 4 Classification report—K-Nearest Neighbors

Cases	Precision	Recall	F1-Score	Support
1	0.84	1.00	0.91	26
2	1.00	1.00	1.00	26
3	1.00	1.00	1.00	26
4	1.00	0.81	0.89	26
Average	0.96	0.95	0.95	104

SVM stands good in Precision, Recall, and F1-Score for all the damage cases surpassing Random Forest and KNN. The study fetched an average classification accuracy of 98% for SVM, followed by KNN with an accuracy of 95% and Random Forest with an accuracy of 93% (Fig. 3).

6 Conclusions

Health monitoring of structures is vital to prevent immeasurable loss. With the advent of evolved technology especially in the area of machine learning, it can be now implemented and applied in the realm of structures.

It is evident from the experiments conducted that the proposed method involving segmentation of data and signal processing transformations may result in proper damage detection in structural health monitoring using machine learning. Support Vector Machine performed with notable accuracy.

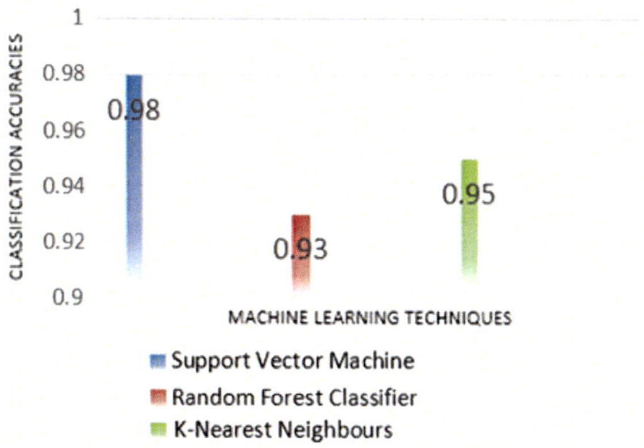

Fig. 3 Classification accuracies of machine learning techniques

Furthermore, this paper is a motivation for other researchers to take the advantage of evolved Machine Learning and Deep Learning era to contribute and carry more advanced research studies in structural health monitoring.

Acknowledgements The authors gratefully acknowledge the computational facility created in the college under DST's FIST Programme (SR/FST/College-2017/28(C)) which helped to carry out the work. The authors thank the management of IARE for their support and kind encouragement.

References

1. Fugate, M.L., Sohn, H., Farrar, C.R.: Vibration based damage detection using statistical process control. Mech. Syst. Signal Process. **15**(4), 707–721 (2001)
2. Zang, C., Friswell, M.I., Imregun, M.: Health monitoring and damage assessment using measured FRFs from multiple sensors. Key Eng. Mater. 245–246 (2003)
3. Bandara, R.P., Chan, T.H.T., Thambiratnam, D.P.: Structural damage detection method using frequency response functions. Struct. Health Monitor. **13**(4), 418–429 (2004)
4. Ni, Y.Q., Zhou, X.T., Ko, J.M.: Experimental investigation of seismic damage identification using PCA-compressed frequency response functions and neural networks. J. Sound Vib. **290**(1–2), 242–263 (2006)
5. Allemang, B.J., Drown, D.L.: A Correlation Coefficient for Modal Vector Analysis (1982)
6. Pandey, A.K., Biswas, M., Samman, M.M.: Damage detection from changes in curvature mode shapes. J. Sound Vib. **145**(2), 321–332 (1991)
7. Ratcliffe, C.P.: Damage detection using a modified Laplacian operator on mode shape data. J. Sound Vib. **204**(3), 505–517 (1997)
8. Abdel Wahab, M.M., Roeck, G.: Damage detection in bridges using modal curvatures: application to a real damage scenario. J. Sound Vib. **226**(2), 217–235 (1999)
9. Pandey, A.K., Biswas, M.: Damage detection in structures using changes in flexibility. J. Sound Vib. **169**(1), 3–17 (1994)
10. Donnelly, D.: The Fast Fourier and Hilbert-Huang transforms: a comparison. In: The Proceedings of the Multiconference on Computational Engineering in Systems Applications, 30 July 2007

11. Howard, R.M.: The power spectral density. In: Principles of Random Signal Analysis and Low Noise Design: The Power Spectral Density and Its Applications. Wiley (2002)
12. Cortes, C., Vapnik, V.N.: Support-vector networks. Mach. Learn. **20**, 273–297 (1995)
13. Breiman, L.: Random forests. Mach. Learn. **45**, 5–32 (2001)

Real Estate Sales Forecasting with SVM Classification

Arti Patle and Gend Lal Prajapati

1 Introduction

The upcoming era is interacting with big data, so engaging in data analysis and mining as well. People are using machine learning algorithms in different industries according to their needs for solving the problem. The most competitive market in the domestic is the real estate market. It has a close relationship with us. For the development of the economy real estate's role has a major part. The real estate sector in India is highly affected by overall economic conditions and performance. A huge amount of investment is infused in such industry, and due to various reasons, risk is high with a high cost of servicing of debt. In this paper, we try to find out sales and booking prediction on the basis of customer enquiry regarding the property. Demand has become the core essence of the real estate domain. Demand is generated by customers purchasing scenarios. There are so many enquiries regarding flats, shops and offices but what about sale. Sake of this sales forecasting we use machine learning. A novel type of machine learning is SVM classification, which gives the solution for the prediction of real estate sales.

Basically, a sales forecast is a prediction of future sales, it is based on historical data, status of the sales pipeline, and industry trends. In real estate, this sales forecast is used to estimate sales total in annual or quarter. Good data is the most important requirement for a good sales forecast. So, in this study customer enquiry, customer feedback, and customer information is the main input data for sales prediction in the real estate business.

A. Patle (✉) · G. L. Prajapati
Department of Computer Engineering, Institute of Engineering & Technology, Devi Ahilya University, Indore 452017, India
e-mail: artipatle@gmail.com

G. L. Prajapati
e-mail: glprajapati1@gmail.com

C. K. Mai et al. (eds.), *Machine Learning Technologies and Applications*,
Algorithms for Intelligent Systems,
https://doi.org/10.1007/978-981-33-4046-6_15

2 Basics of Support Vector Machine

Machine learning and mining is very essential for prediction and forecast in many domains like medical, credit scoring, stock market, decision support system, real estate etc. In real estate many researchers work done in price forecasting, this study is based on real estate sales forecast. In the year 1959, Machine learning is coined by Arthur Lee Samuel. In which he defines that the computer itself have ability to learn without any specified programs. Tom Mitchell in Machine Learning gives definition more formally as: A learning program is said to learn from experience E with respect to some task T and some performance measure P, if its performance on T, as measured by P, improves with experience E.

There are so many concepts in machine learning but SVM is a better option for real-life problems. One of the supervised learning methods is Support Vector Machines (SVM) [1]. SVMs are used for classification and regression tasks. Support vector machines are originated from statistical learning theory. SVM is known as the global classification model, it generates non-overlapping partitions and focuses on all attributes.

Maximum margin linear discriminants and similar to probabilistic approaches are the base of SVM. Support Vector Machine has a frontier that best categorizes the two classes (hyperplane/ line) [2]. SVM training calculates similarity from a subset of training points. In the training time, find a subset of training points from which compute the similarity of any test point. These selected training points are called support vectors [3]. On the basis of support vectors, SVM decides the soft margin for classification and making a decision.

2.1 *Mathematical Concept of SVM*

Given a training set of N data points the input value represent as x_i and the corresponding binary class label is $y_i \varepsilon \{-1, +1\}$, the SVM classifier, according to Vapnik's original formulation satisfies the following conditions for decision boundary which classify all the points [4]. For explanation, some variables require such as

w: coefficient vector of the hyperplane in feature space,

b: hyper plan threshold value,

ξ_i: slack factor for error,

c: penalty factor for error,

α_i: Lagrange multipliers,

σ: width parameter,

$\varphi(\cdot)$: nonlinear function.

$$w^T\varphi(x_i) + b \geq +1, \text{ if } y_i = +1 \tag{1}$$

$$w^T\varphi(x_i) + b \leq -1, \text{ if } y_i = -1 \tag{2}$$

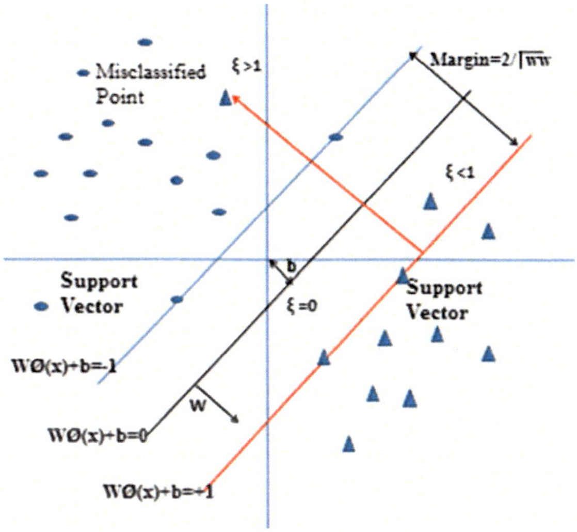

Fig. 1 SVM classification with support vector

In its decision boundary solve by the following constraint: Minimize ½ $||w||^2$.

The nonlinear function $\varphi(\cdot)$ maps the input space to a high (possibly infinite) dimensional feature space [5]. In this feature space, the above inequalities basically construct a hyperplane $w^T\varphi(x_i) + b = 0$ discriminating between both classes [6]. In primal weight space, the classifier then takes the form.

$y(x) = \text{sign}[w^T\varphi(x_i) + b]$

But, on the other hand, it is never evaluated in this form. One defines the convex optimization problem:

$$\text{Min}_{w,b,\xi}\tau(w, b, \xi) = \frac{1}{2}W_T W + C\sum\nolimits_{i=1}^{n}\xi_i$$

Subject to

$y_i[w^T\varphi(x^i) + b] \geq 1 - \xi_{i,}\ i = 1, \ldots, N$

$\xi_i \geq 0, i = 1, \ldots, N.$

The variables ξ_i are slack variables which are useful in order to allow misclassifications in the set of inequalities (Fig. 1 shows an example). The first part of the objective function tries to maximize the margin between both classes in the feature space, whereas the second part minimizes the misclassification error [7]. The positive real constant C should be considered as a tuning parameter in the algorithm [8].

2.2 *Building Blocks of SVM*

The separating hyperplane: Hyperplane which separates classes linearly and no-linearly.

The maximum-margin hyperplane: Margin which justifies the classification criteria.

The soft margin: Margin from this finalizes the optimal separation of classes.

The kernel function: The core concept of SVM from classification. Kernel function Formalize the data for classification with the help of statistics [9].

Kernel function type:

Linear kernel function: $K(x, x^{`}) = x \cdot x^{T}$.

x^{T} is the transpose of the input matrix x.

Polynomial Kernel: $K(x, x_i) = \left(1 + x \cdot x_i^{T}\right)^{d}$.

Where 'd' is the degree of the kernel function.

RBF Kernel: $K(x, x_i) = e^{-\gamma\|x - x_i\|^2}$, $\gamma > 0$.

Sigmoid Function: $K(x_i, x_j) = \tanh\left(\gamma x_i^{T} x_j + r\right)$.

Here γ, r, and d are kernel parameters.

2.3 Significance of SVM Kernel and Feature Selection.

The kernel function is a mathematical trick that allows the SVM to perform a 'two-dimensional' classification of a set of originally one-dimensional data [10]. Kernel functions are the heart of SVM; kernel is a similarity function, which captures the domain knowledge. Kernel inputs are data points and output is a group of similar data points. The advantage of kernel is: first they analyze the similarity between data points and incorporate the prior knowledge. Second is that the learning algorithm is based on the kernel function and input space [11]. It is not true that the number of operations of kernel is proportional to the number of features of dataset. The feature also affects the accuracy but at the initial level might preprocess the data according to a suitable case. The kernel parameter has own importance in classification such as:

Gamma is the Kernel coefficient for 'rbf', 'poly', and 'sigmoid'. Higher value of gamma will try to exactly fit as per training data set, i.e., generalization error and cause an over-fitting problem.

C is the penalty parameter in terms of error. It also controls the tradeoff between smooth decision boundary and classifying the training points correctly.

It is not true that the number of operations of the kernel is proportional to the number of features of dataset. Feature also affects the accuracy but at the initial level might preprocess the data according to a suitable case [12]. The SVM is an effective tool in high-dimensional spaces, which is particularly applicable to document classification and sentiment analysis where the dimensionality can be extremely large. Feature selection discards the irrelevant or redundant features from the original feature. Feature selection is a concept which mainly affects the classification [13]. The goal of feature selection is to find a minimal set of Features Fs = {fs1, fs2 ... fsd} to represent the input vector X in a lower dimensional feature space as Xs = [xs1, xs2 ... xsd].

3 Literature Review

Support Vector Machine has a strong mathematical foundation for classification. SVM classification technique is working on numerical data [14]. SVMs are often used with very complex data for applications such as recognizing handwriting, facial expression classification, and classifying images, etc.

3.1 SVM Qualities

- Very fast at classifying new data. There is no need to go through training data for new classifications.
- Can work for a mixture of categorical and numerical data.
- SVMs are "Robust to high dimensionality", meaning they can work well even with a large number of features.
- Highly accurate [15].

3.2 SVM Limitations

- Black box technique. Unlike Bayesian Classifiers and Decision Trees, SVMs don't offer easily digestible data "under the hood" [16].
- Typically requires a very large dataset. Other methods, like decision trees, can still give interesting output for small data sets; this may not be the case with an SVM.
- SVMs are not online. They will need to be updated every time new training data is to be incorporated.

Purchase in real estate is varying and being influenced by extensive factors. Many studies have great attention on how to forecast purchases accurately [17]. Earlier studies on real estate sales formulation and forecasting mainly follow the demand and supply. In forecasting spatial dynamics of the housing market using support vector machine, Chen et al. uses SVM classification for the mass appraisal of residential properties in Taipei city and come to conclude and evaluate after study that it is being useful. A large set of the variable is covered by the proposed SVM. That variable is actually identified as a price influencer. And in the intervening period of time make it possible to acknowledge the existence of geographical patterns of housing price dynamics [18]. Some researchers tried to introduce minority-sample generation approaches in their work and performance criteria for imbalanced datasets. Through this paper, we provide a framework from which we can able to predict the default rate of real estate companies. With the use of minority sample generation approach, we can avoid bias during sample selection and the prediction model also improves, applies a minority sample generation approach to create new minority samples [19].

The output is machine learning models, as well as the logistic models with a balanced dataset, had a higher G mean and F-score and a higher true positive rate (TPR) without losing the true negative rate (TNR) [20].

Prior work grade real estate project and also proposes a real estate classification model through the paper. The borrower can also make a further decision in the future. For some techniques in this experiment for real estate grad classification are Artificial Neural Network (ANN), Decision Trees (DT), Naïve Bayes (NB), K-Nearest Neighbors (KNN) and Support Vector Machines (SVM). We came to know through the study that it can use ensemble- learning-based model for real estate project classification [21]. Wang et al., focus on their work for price forecasting in real estate by SVM and PSO (particle swarm optimization). From PSO, determine the parameter for SVM and got higher prediction accuracy. Besides the supply and demand sales forecasting is an effect from so many factors like builder reputation, locality of project, prior work feedback, etc. SVM can be handle non-linear cases also. Former study on SVM proofs that, it is available to solve the problems of limited sample learning, nonlinear regression, as well as, better to overcome the "curse of dimensionality". And also determine learning and generalization through selected parameters that is why determining the parameters of SVM is essential. SVM is powerful and easy to implement as compared to ant colony algorithm, grid algorithm, genetic algorithm, particle swarm optimization (PSO). So some studies on real estate price forecasting, PSO and SVM were used. PSO-SVM models were used to prove that it has good forecasting performance [22]. Researcher's analysis of how much land is used and how much land is covered mapping is essential. Environmental processes and environmental properties both are influences through mapping. To know how much land is used there are two techniques first one is SVM and another one is MLC, through these techniques we come to know the accurate and improved method for land use classification. In the year 2006, both the methods verified by land cover map for classification and level of accuracy also measured through different parameters. And we come to conclude that the accuracy level of the SVM method is better in comparison to the MLC method [23]. According to the demographical and economic variables, researchers release that different variables tend to have different impacts on the dynamic behavior of house purchasing and the number of houses sold in different regions at different periods [24]. So in real estate, the research is carried forward for the sake of the benefit of buyer and seller.

4 Methodology

As the focus on sales prediction in real estate market dataset collection, feature evaluation, kernel selection, and result analysis are main tasks. The data for the experiment was gathered from a company that is work on real estate projects (Fig. 2).

Datasets were real estate data, it has 100 instances, 14 attributes as follows: index, name, gender, age, phone, email, company, bhk, timetobuy, projectstatus, budget, registered and last is class variable that is status. These are the variables on the basis

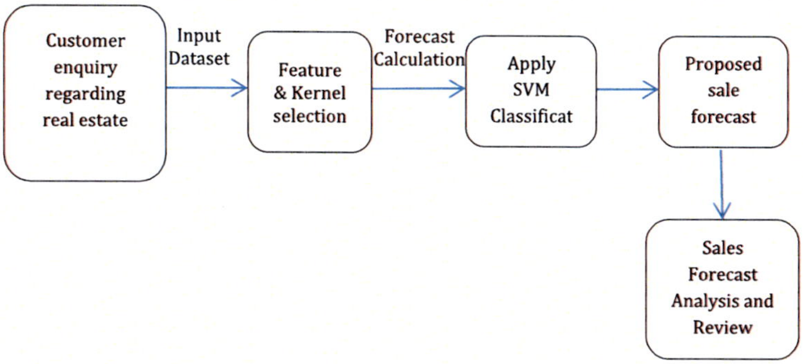

Fig. 2 Sales forecasting with SVM classification

of any real estate person predict the sales and booking scenario of any property. On the basis of customer enquiries, we predict the purchasing and sales scenario of real estate project. Sales forecasting with SVM Classification.

As shown in the diagram, the data is collected from customer enquires regarding real estate project. This data is working as an input dataset; now apply the feature selection and kernel selection on the dataset. After selection procedure, we Apply SVM classification for calculating the forecast. Classification accuracy is for training dataset from testing proposed sales forecast achieved and at the end real estate person analysis and review the sales forecast data.

4.1 *Feature Selection*

Classifier is as good as its feature is, so good feature selection is very important. The number of feature is not important but how well feature signifies the domain, this is important. Selections of relevant features for classification done by chi-square feature attribute selection method. Feature selection method reduces the dimension of dataset. Features Fs = $\{fs_1, fs_2 \ldots fs_d\}$ represent the input vector X in a lower dimensional feature space as Xs = $[x_{s1}\ x_{s2} \ldots x_{sd}]$ [25, 26]. The chi-square test is a statistical test of independence to determine the dependency of two variables. To use chi-square feature selection, we calculate χ 2 between each feature and the target and we select the desired number of features with the best $\chi 2$ scores. The intuition is that if a feature is independent of the target it is uninformative for classifying observation. It shares similarities with the coefficient of determination, R^2. Ranking of the attributes is shown in Table 1.

Chi-square feature selection: $X^2 = \sum \frac{(O-E)^2}{E}$.

where O = Observed frequencies,

E = Expected frequencies.

Table 1 Attribute rank of real estate dataset

Ranked	Attribute index no	Attribute name
80	7	company
80	2	name
80	6	email
80	12	address
80	13	registered
14.72252	9	timetobuy
9.17505	11	budget
3.35519	8	bhk
1.40351	10	projectstatus
0.00462	3	gender
0	1	index
0	5	phone
0	4	age

4.2 Kernel Selection

Kernel function is a solution or method that gives a linear classifier to solving e a non-linear problem. SVM has many kernels as discussed in research paper introduction, the combination might be any kernel because all kernels have their own potential. Here we select polynomial, RBF, and personalized kernel named Polyrobiaf. Polyrobiaf kernel is a combination of polynomial and RBF kernel. In this esxperiment, express the classification accuracy between polynomial, RBF and Polyrobiaf. The personalize Polyrobiaf kernel is defined as

$$K(x, x_i) = \left(1 + x \cdot x_i^T\right)^d + e^{-\gamma ||x - x_i||^2}$$

The parameters value of c, γ, d as c = 1.0, γ = 0.01, d = 2 are used for the experiment on real estate classification.

4.3 SVM Classification Model

Require:

Input: Training Sample, $X = [X_1, X_2, \ldots, X_k]$,

Class labels, $Y = [Y_1, Y_2, \ldots, Y_k]$.

Output: Classification accuracy with Polynomial, RBF, and Polyrobiaf kernel.

Step 1: Select the real estate dataset which has multiple instances and attributes.

Step 2: Apply Feature Selection with chi-square and find the selected feature (Feature Subset).

Table 2 Dataset classification accuracy with different kernel functions

Dataset name	# Attribute	#Instances	Correctly classification accuracy (%)		
			Polynomial	RBF	Polyrobiaf
Real estate data	14	100	49	54	55

Step 3: Select the RBF kernel, polynomial, or Polyrobiaf for classification.

Step 4: Set the parameters value of c, γ, d as $c = 1.0$, $\gamma = 0.01$, $d = 2$.

Step 5: find out the classification accuracy.

Step 6: Analysis and compare the classification accuracy for different feature subsets.

As apply the above method to real estate dataset we find out the correct classification accuracy is 49 with polynomial kernel and 54 classification accuracy with RBF kernel as described in Table 2. But with the feature selection methodology, there are so many differences. The classification accuracy with the polynomial kernel is high with different subsets of features. From these higher accuracy gives a better classification in testing data.

Thereafter subsets of new input attributes are derived and supply for the SVM model. Classification is performed with the selected feature subset. Accuracy with RBF and Polynomial kernel are shown in Table 3. According to the classification accuracy, we found that the sales forecasting of real estate have a different factor with their own significance.

Table 3 gives a practical exhibition and explanation of SVM classification. These results show that subset 2 is the best combination for real estate sale forecasting. Next, the model was applied to the testing data in order to find the results of unseen real estate data. As shown in Table 3 it is much better to use Polyrobiaf kernel for the real estate dataset. After constructing the SVM model for real estate sales forecast by training sample, sales value can be forecast for testing sample. From the sale forecasting, real estate people easily predict sales and purchases and plan accordingly (Fig. 3).

Table 3 Real estate dataset SVM classification accuracy with different feature subsets and kernels

Dataset name	Selected features subset	Classification accuracy (%) with evaluating parameters $C = 1.0$, $\gamma = 0.01$, $d = 2$		
		Polynomial	RBF	Polyrobiaf
Real estate	Subset 1 [1, 3, 4, 6, 11–14, 16]	52	51	55
	Subset 2 [1, 4, 6, 11–14, 16, 19]	86	45	87
	Subset 3 [1, 3, 4, 6, 11–14, 16, 19]	50	49	54
	Subset 4 [1, 4, 6, 11–14, 16]	52	48	52

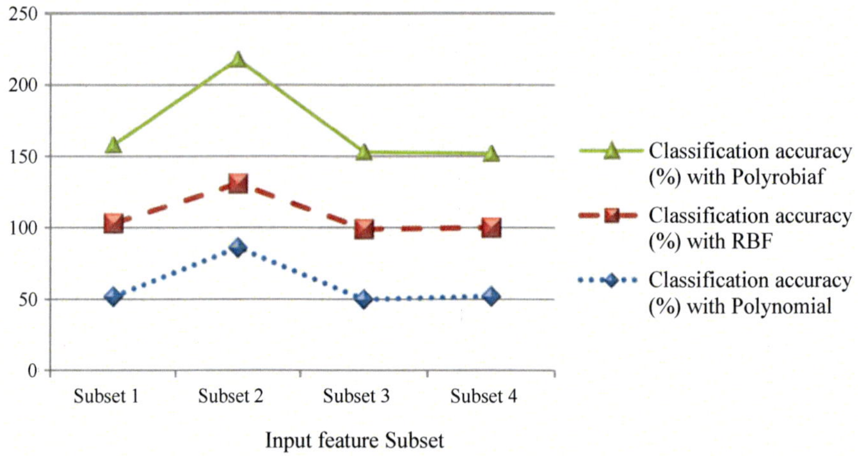

Fig. 3 Correctly classification accuracy evaluation of real estate sales forecast with a selected subset for polynomial, RBF kernel, and Polyrobiaf kernel functions in SVM

5 Conclusion

The basic need for humans is accommodation. Therefore, the real estate market is growing so fast. The category of the customer reflects the sales and booking probability of the real estate business. These sales predictions find out with machine learning on the basis of customer enquiry regarding real estate property. To the best of our knowledge, most studies base on price prediction of the real estate market. But for this paper, center of attention is real estate sales forecast model with SVM Classification. The SVM is an accurate model of machine learning for classification. Compare to different feature subsets and different kernel selections, Polyrobiaf is best. This study has formulated and optimizes the usefulness of sale forecasting in the real estate market. The key contribution in this paper is the utilization and analysis of numerous hedonic attributes during the forecasting and selection of kernel. In order to identify valuable features using the chi-square attribute selection method. According to the feature rank, reduced attributes were used for the SVM model building. Then apply polynomial, RBF, and Polyrobiaf kernel function for classification accuracy. The result finding shows that the attribute selection and selection of kernel are beneficial for sales forecasting. Finally, according to the customer enquiry features, we find out the sales forecasting of real estate for informed business decision.

References

1. Murphy, P., Aha, D.: UCI machine learning (1992). https://www.ics.uci.edu/~mlearn/MLrepository.html
2. Boujnouni, M., Jedra, M., Zahid, N.: Kernel's parameter selection for support vector domain description. Int. J. Comput. Electr. Autom. Control Inf. Eng. **7**(8) (2013)
3. Scholkopf, B., Burges, J., Smola, A.: Advances in Kernel Methods: Support Vector Machine. MIT Press (1999)
4. Debnath, R., Takahashi, H.: In efficient method for tuning kernel parameter of the support vector machine. In: IEEE International Symposium on Communications and Information Technology, vol. 2, pp. 1023–1028 (2004)
5. Prajapati, G., Patle, A.: On performing classification using SVM with radial basis and polynomial kernel functions. In:ICETET-2010 3rd International IEEE Conference, Goa, pp. 512–515 (2010)
6. Staelin, C.: Parameter selection for support vector machines. HP Laboratories Israel PL-2002–354 (R.1) (2003)
7. Gaspar, P., Carbonell, J., Oliveira, J.: On the parameter optimization of support vector machines for binary classification. J. Integr. Bioinform. **9**(3), 201 (2012)
8. Hsu, W., Yu, T.: Support vector machines parameter selection based on combined Taguchi method and Staelin method for E-mail spam filtering. Int. J. Eng. Technol. Innov. **2**(2), 113–125 (2012)
9. Chapelle, O., Vapnik, V.: Choosing multiple parameters for support vector machines. Mach. Learn. **46**, 131–159 (2002)
10. Muller, K., Mika, S., Ratsch, G., Koji, T., Bernhard, S.: An introduction to kernel-based learning algorithms. IEEE Trans. Neural Netw. 181–201 (2001)
11. Phienthrakul, T., Kijsirikul, B., Takamura, H., Okumura, M.: Sentiment classification with SVM and MKL. In: Neural Information Processing: 16th International Conference, ICONIP, pp. 583–591 (2009)
12. Schittkowski, K.: Optimal parameter selection in support vector machines. J. Ind. Manag. Optim. **1** (2005)
13. Maldonado, S., Weber, R.: A wrapper method for feature selection using support vector machines. Inf. Sci. **179**, 2208–2217 (2009)
14. Liang, X., Liu, F.: Choosing multiple parameters for SVM based on genetic algorithm. In: 6th International Conference on Signal Processing, vol. 1, pp. 117–119 (2002)
15. Zhang, X., Qiu, D., Chen, F.: Support vector machine with parameter optimization by a novel hybrid method and its application to fault diagnosis. Neurocomput. J. (2014)
16. Eitricha, T., Langb, B.: Efficient optimization of support vector machine learning parameters for unbalanced datasets. J. Comput. Appl. Math. **19**, 425–436 (2006)
17. Fan, H., Xiang, J.: Patch based visual tracking with two stages MKL. In: 8th International Conference on Image and Graphics, ICIG, pp. 20–30 (2015)
18. Prajapati, G., Patle, A.: Personalizing kernel and investigating parameters for the classification with SVM. In: IEEE Conference on Industrial Electronics and Applications ICIEA-17, Cambodia (2017)
19. Song, J., Ding, Z., Guo, C.: Research on combination kernel function of support vector machine. In: International Conference on Computer Science and Software Engineering (2008)
20. Joshua, S., RamakrihnaMurty, M., Hyma, J.: Detecting high risk property taxpayers using a new business intelligence model: a case of New York City property tax. Paideuma J. Res. **13**(3) (2020)
21. Dong, Y., Xiao, Z., Xiao, X.: Default prediction for real estate companies with imbalanced dataset. J. Inf. Process. Syst. **10**, 314–333 (2014)
22. Paireekreng, W., Choensawat, W.: An ensemble learning based model for real estate project classification. In: IEEE International Conference AHFE, pp. 3852–3859 (2015)
23. Wanga, X., Wenab, J., Zhang. Y., Wang, Y.: Real estate price forecasting based on SVM optimized by PSO. Int. J. Light Electron Opt. **125**, 1439–1443 (2014)

24. Cortes, C., Vapnik, V.: Support-vector network. Mach. Learn. 273–297 (1995)
25. Deilmai, B., Ahmad, B., Zabihi, H.: Comparison of two classification methods (MLC and SVM) to extract land use and land cover. In: IGRSM International Remote Sensing & GIS Conference and Exhibition, Malaysia (2014)
26. Chen, J., Ong, C., Zheng, L., Hsu, S.: Forecasting spatial dynamics of the housing market using support vector machine. Int. J. Strat. Prop. Manag. **3**, 273–283 (2017)

Credit Card Fraud Detection Using Spark and Machine Learning Techniques

N. V. Krishna Rao, Y. Harika Devi, N. Shalini, A. Harika, V. Divyavani, and N. Mangathayaru

1 Introduction

Based on the 2015 International Payment Report, the most widely used methodology for payment is MasterCard in 2014 compared to other methodologies like e-wallet, bank transfer. The large services provided for transactional area units are usually handled by cybercriminals and conducts dishonorable activities. There are 3 classes of MasterCard fraud specifically, standard frauds, online frauds, and businessperson connected frauds, Master Card breaches are trending alarmingly. So it is necessary to develop MasterCard fraud detection techniques.

In general, master card fraud detection is identifying the transactional area unit is real or dishonorable, because the data processing and machine learning techniques area unit immensely won't to counter cyber criminal cases, students usually embraced those approaches to check and notice MasterCard fraud activities. Data mining techniques are used to extract important information regarding MasterCard fraud detection. To discover the transactions which are fraud, we need to implement the most effective and highly efficient techniques [7, 9], which is the only way to predict the transaction data. These techniques enable us to identify various facts like authentication validation and transaction patterns.

N. V. Krishna Rao (✉) · Y. Harika Devi · N. Shalini · A. Harika · V. Divyavani
Department of CSE, Institute of Aeronautical Engineering, Dundigal, India
e-mail: nvkrishnarao3@gmail.com

N. Mangathayaru
Department of IT, VNRVJIET, Hyderabad, Telangana, India

C. K. Mai et al. (eds.), *Machine Learning Technologies and Applications*,
Algorithms for Intelligent Systems,
https://doi.org/10.1007/978-981-33-4046-6_16

Technology-driven services are originated because of coordination between banking services and Information Technology. This credit card which is simply a plastic card is issued by the bank to the user or the customer legally. Transaction of money takes place in two ways. One is a physical credit card, in which physical or face-to-face payment is done. The other is virtual credit card, in which the card is digitally issued through online purchase.

However, in the financial system, efficient and well-tested Credit Card Fraud Detection is a main requirement. Multiple data technique [2] combinations are being used by most of the researchers. Prediction of fraud before the transaction occurrence is to be notified rather than detecting the fraud after the transaction [3]. The future scope of this fraud detection will capture the picture of the fraud by adding a new module that can be sophisticated. By implementing efficient algorithms in multiple layers [1, 5], stronger protection can result.

2 Literature Survey

For maintaining associate info system, fraud detection is an important part of it. By ancient access management mechanisms, the management system will give intrusion hindrance to an extent, they are syntactically correct however transactions are semantically damaged [8]. Chung et al. say that in the info system the misuse detection is not self-addressed and proposed DEMIDS, based on the audit logs it derives user profiles. The sphere of the theory of games has been explored for issues starting from auctions to chess and its application to the domain of knowledge warfare looks promising [4]. The theory of games in IW was brought by prophet et al. To predict future attacks [6] and also the challenges and variations during this domain, one will utilize a well-developed theory of games algorithms [10] compared to ancient games like chess, like restricted.

3 System Analysis

3.1 Software Tools Used

SCALA:-

Scala programming could be a general programming language and it is object-orientated, performs well on a bigger scale. Scala language is a robust and static variety of programming language and it is influenced by the programming language called Java. One of the simplest similarities of both these languages, Java and Scala, is simply will code Java just identical means that you simply code in Scala. It's additionally doable to use heaps from Java libraries at intervals and several third-party libraries and furthermore. It has become one of the demanding technologies among developers and functioning its means in today's technology. Here some unit area topics that provide transient clarification concerning Scala.

One of the major strengths of the language Scala is its flexibility during process abstractions. Scala language has a vital element called Scala IDE(Scala Integrated Development)and this is linked with Eclipse Java tool. TScala IDE is explored with the option Eclipse. Scala can interoperate with both JRE and.NET Framework.

Scala code is simpler to check and apply, less complicated parallelization, and there will be less number of bugs in the entire program. Scala programming language is a top-down approach. Each program is attenuated into a number of chunks and each one processes in parallel, therefore, dashing methods and additionally are rising potency.

Kafka:-

There are 3 capabilities of keys in the streaming platform and they are

- Like a enterprise electronic communication system or message queue, it is also publishing and purchasing streams of records.
- During a fault-tolerant sturdy, it will store the streams of records.
- As they occur method streams of records.

The two main applications of Kafka are

- Knowledge between systems or applications is used for building time period streaming knowledge pipelines.
- To rework or react to the streams of knowledge we use building time period streaming applications.

Kafka runs as a cluster on one or additional servers that may cover multiple data centers. In classes referred to as topics, Kafka cluster will store streams of records. In Kafka each record consists of a value, a timestamp, and a key.

The API's of Kafka are: Producer API, Patron API, Streams API, Connexion API.

Communication between purchasers and servers are straightforward in Kafka and provides high performance, language-agnostic TCP protocol. With the help of the older version, the TCP protocol maintains backward compatibility. Kafka offers Java consumers, however, purchasers' area unit is offered in several languages (Fig. 1).

Cassandra

Once you want measurability and high availability without compromising on performance the right selection is Apache Cassandra. Fault-tolerance and measurability on cloud infrastructure or trade good hardware build the right platform for mission-critical knowledge. Cassandra's support for lower latency and replicating multiple data centers.

Apache Cassandra could be a free and ASCII text file. It is a distributed database and stores the records in columnar format. It is a NoSQL management system in order to handle a large amount of knowledge from several trade servers and provides high convenience without any single failure. It offers clusters for a spanning number of data centers, with less replication, low latency operations, asynchronous mater for all purchasers.

Fig. 1 Kafka processing

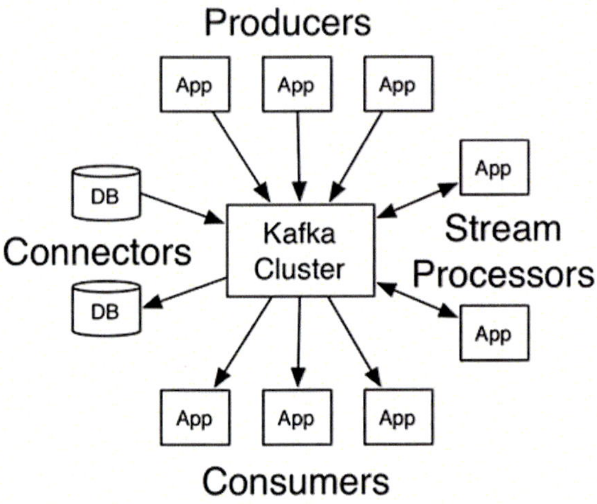

Spark:-
One of the fast cluster computing technology is Apache Spark. It is used for quick computation designing. It supports Hadoop map cut back and MapReduce model is extended and uses it efficiently for additional kinds of computations, Apache Spark could be a lightning-fast cluster computing technology, designed for quick computation. it's supported Hadoop Map cut back and it extends the MapReduce model to efficiently use it for additional kinds of computations, which incorporates interactive queries and stream process. In-memory cluster computing is an important feature of Spark that increases the associate degree application process speed. It is used to hide a good vary of work loads like streaming, batch applications, interactive queries and iterative algorithms. During individual systems it will reduce the management burden for maintaining separate tools by excluding supporting of these workloads.

Spark streaming:

An extension of core Spark API is Spark streaming which allows fault-tolerant stream, ascendible, and high-turnout live knowledge streams are processed (Fig. 2).

Fig. 2 Spark streaming

4 System Design

4.1 System Model

See (Fig. 3).

Machine learning is not expressly programmed to a system and it is the ability of a system to mechanically learn and improve from expertise. It focuses on PC programs that access information and uses it for themselves (Fig. 4).

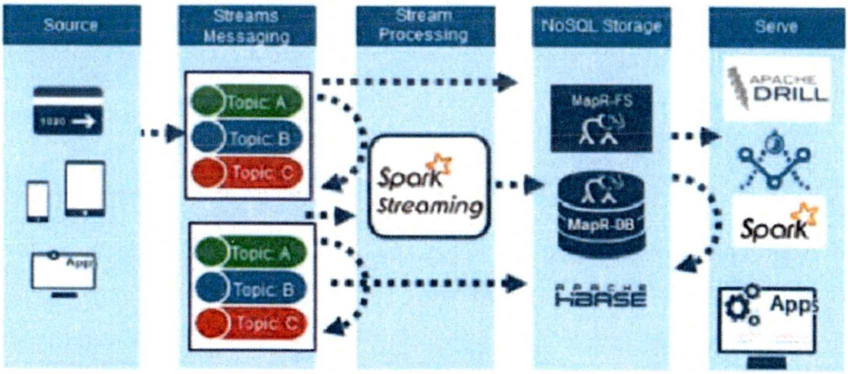

Fig. 3 Classifier system model

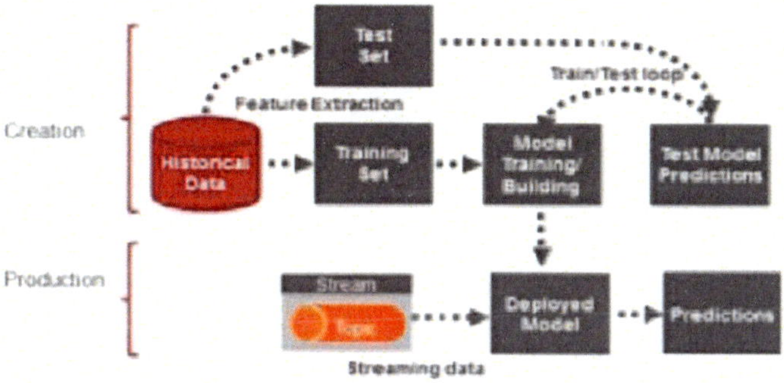

Fig. 4 Classifier steps

4.2 Algorithms and Their Usage

Logistic Regression:

A supervised classification technique is known as a logistic regression that returns binary variables that has two values zero or one. We say the model is a regression model wherever the quantity of a variable is categorical and examine the connection among many freelance variables. There were multiple varieties of LR models like multiple, binary, binomial.

To estimate binary response supported one or additional predicators binary logistical model is derived. The logistical regression equation is shown in Fig. 5

The above graph represents the distinction between mean and logistic regression. Logistic regression represents a curve whereas regression toward the mean represents a line (Fig. 6).

Support Vector Machine model:

See (Fig. 7).

For every popular machine learning rule for regression, classification there will be one SVM. The information used for classification and regression is analyzed by the SVM. It has two steps, foremost to get a model and to use this model to predict info of a testing information set and to coach an information set.

Decision Tree:

Decision tree uses graph or model calls to predicate the decisions that are ultimate and it uses conditional management statement. To work on expected, worst and best values on various situations, call tree helps by simplifying to interpret and grasp and permits to add recent potential situations (Fig. 8).

$$\mathrm{Entropy(S)} = \sum_{i=1}^{c} -\mathrm{p_i}\log_2 \mathrm{p_i}$$

$$p = \frac{e^{\alpha+\beta_n X}}{1 + e^{\alpha+\beta_n X}}$$

Fig. 5 Logistical regression equation

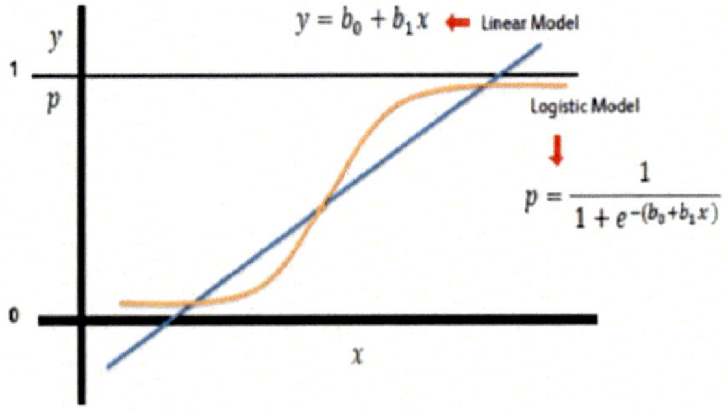

Fig. 6 Logistic curve

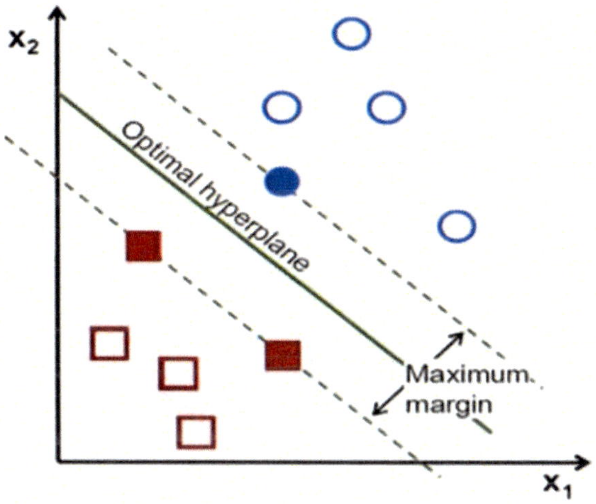

Fig. 7 Support vector machine model graph

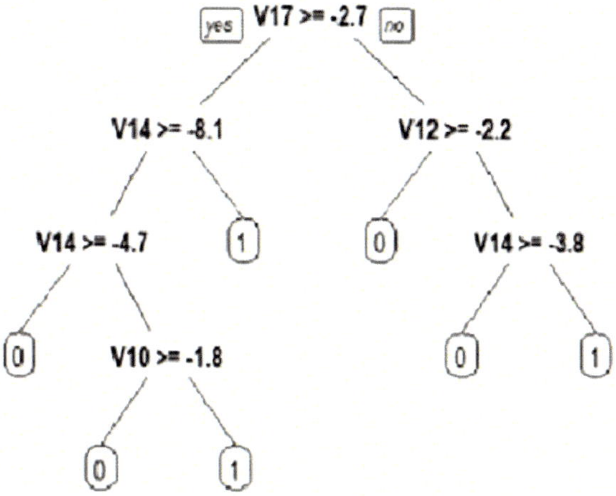

Fig. 8 Decision tree

$$Gain(S, A) = Entropy(S) - \sum_{v \subset Values(A)} \frac{|S_v|}{|S|} Entropy(S_v)$$

Random Forest:

Classification and regression is AN rule for Random Forest. It is a call tree classifier set. It corrects the overfitting to their coaching set thus the random forest has an advantage over call tree. Random forest provides an estimation for generalization error and is immune to overfitting.

$$p(c|x) = \frac{p(x|c)p(c)}{p(x)}$$

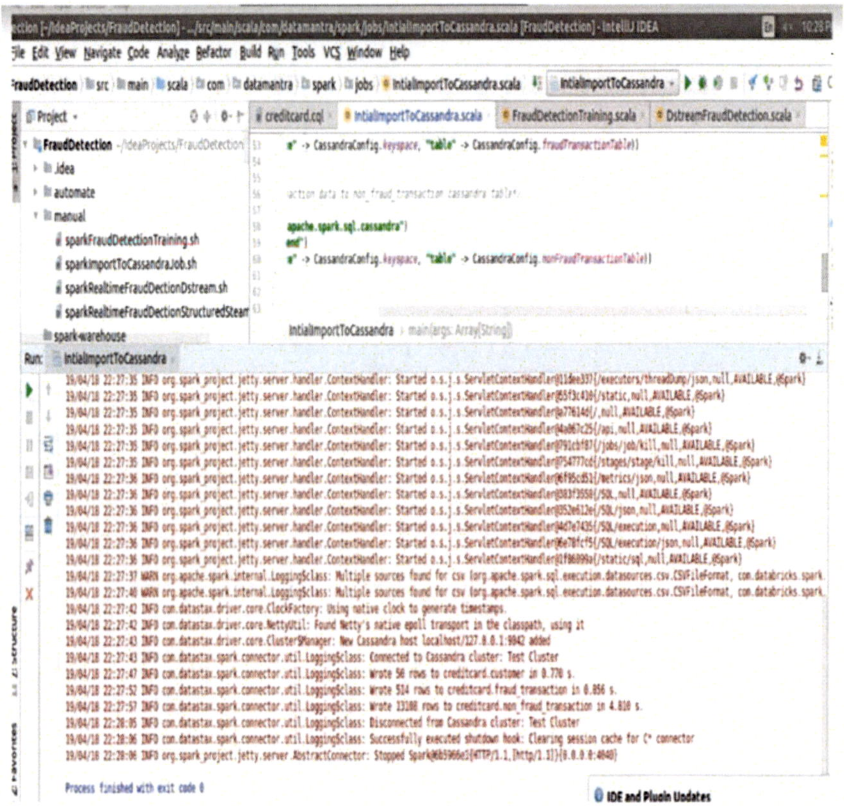

Fig. 9 Initial import to Cassandra

$$p(c|x) = p(x_1|c) \times p(x_2|c) \times \cdots \cdot \times p(x_n|c) \times p(c)$$

5 Results

See (Figs. 9, 10, 11).

6 Conclusion

Credit card fraud is an associate degree act of criminal dishonesty. MasterCard recent findings are reviewed from this text. Various kinds of frauds like counterfeit, application, bankruptcy, behavioral, larceny frauds are known in this paper and mentioned

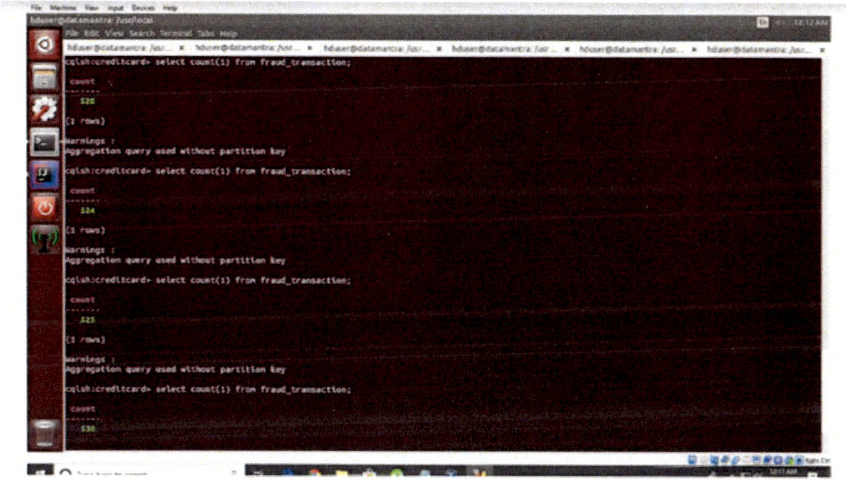

Fig. 10 Detection of fraud transaction

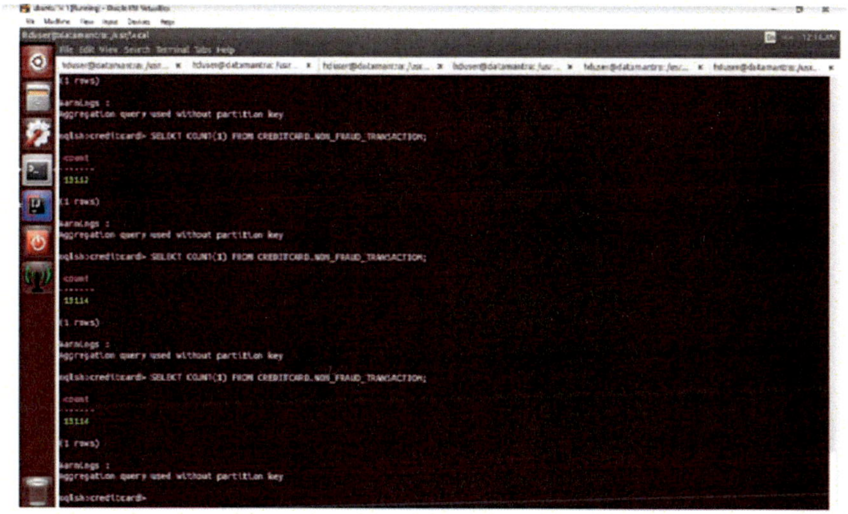

Fig. 11 Detection of non-fraud transaction

measures to find them. Those measures are cluster techniques, pair-wise matching, neural networks, genetic algorithms, and call trees.

All fallacious cases are decided from banks and MasterCard according to associate degree moral perspective. Dimensions of the skilled fraudster are unlikely to control by the amateurish fraudster so the prices to the bank are also wasteful to their detection. It can be featured with an associate degree moral quandary. ought to they fight to find such fallacious cases or ought to act in shareowner interests and avoid wasteful prices.

Implementation of a'suspicious' book on true dataset and its analysis is the next step in analysis program. Most of the tasks are used for creating grading models to predict fallacious behavior, By taking fields of behavior under consideration that are related to various kinds of MasterCard fraud are known during this paper.

References

1. Aleskerov, E., Freisleben, B., Rao, B.: Cardwatch: a neural network- based database mining system for credit card fraud detection. In: Processdings of the IEEE/IAFE on CIFE, pp. 220–226 (1997)
2. Krishna Rao, N.V., Mangathayaru, N., Sreenivasa Rao, M.: Banking data clustering and classification in DMDW-a review. JARDC (2017)
3. Sarmilabanu, A., Mohamed Surputheen, M., Fast and efficient credit card fraud detection on real-time data using spark. IJARSET **3**(9), 2668–2672 (2016)
4. Chiu, C., Tsai, C.: A web services-based collaborative scheme for credit card fraud detection. In: Proceedings of 2004 IEEE International Conference on e-Technology, e-Commerce and e-Service (2004)
5. Brause, R., Langsdorf, T., Hepp, M.: Neural data mining for credit card fraud detection. In: 1999 Proceedings of 11th IEEE ICTAI. IEEE Xplore (2002).
6. Varre Perantalu, K., Kiran, B.: Credit card fraud detection using predictive modeling. IJIRT **3**(9), (2017)
7. Bolton, R., Hand, D.: Statistical fraud detection: a review. Stat. Sci. **17**, 235–249 (2002)
8. Chan, P., Stolfo, S., Fan, D., Lee, W.: A prodromidis, credit card fraud detection using meta learning: issues and initial results. In: Working Notes of AAAI Workshop on AI Approaches to Fraud Detection and Risk Management (1997)
9. Soltaniziba, S., Balafar, M.A.: The study of fraud detection in financial and credit institutions with real data. Comput. Sci. Eng. **5**(2), 30–36 (2015)
10. Dorronsoro, J.R., Ginel, F., Sgnchez, C., Cruz, C.S.: Neural fraud detection in credit card operations. IEEE Trans. Neural Netw. **8**, 827–834 (1997)

A Real-Life Decision-Making Problem via a Fuzzy Number Matrix

Rakesh Kumar Tripathi and Showkat Ahmad Bhat

1 Introduction

Fuzzy logic allows decision-making under incomplete or uncertain information. Making a decision is one of the most activities human beings facing every day. Fuzzy logic has provided a powerful tool for decision-making systems. Ranking and ordering of different fuzzy sets can be obtained with the help of a comparison matrix. The optimizing decision defined as the maximum value of the corresponding membership function. D. Dubois and H. Prade defined the fuzzy number as a fuzzy subset of real line [1]. Hexagonal, octagonal fuzzy numbers have been introduced to clear the vagueness [2–4]. Fuzzy matrices play an important role in engineering and scientific development. Namarta, Umesh Chandra gupta, and Neha Ishesh Thakur solve the fuzzy game theory problem [5]. N. Sarala, I. Janatul Firthouse, and R. Rajeshwari showed which one of the grandchildren resembles their grandmother, using the triangular fuzzy number and the corresponding triangular fuzzy number matrix [6]. Masteaasa [7] pointed out in his study the spouse selection process for women has various limitations and there was no specific mechanism to define it. There are different ways to choose a spouse and there were few opportunities available for the selection of spouse due to limited access to geographical area and information. In the literature of Jejeebhoy [8], parental arranged marriages are marriages in which the family members did the selection of a spouse of female and a male. However, the selection of a life partner done by the candidate itself, the marriages organized in such a way, has been defined as love marriages [9]. Ankur Chauhan and Parveen Kumar [10] present a procedure to assist the females in their decision-making for

R. K. Tripathi · S. A. Bhat (✉)
DR. A.P.J. Abdul, Kalam University, Indore, India
e-mail: sabhatmaths@gmail.com

R. K. Tripathi
e-mail: drrakeshkumartripathi@gmail.com

C. K. Mai et al. (eds.), *Machine Learning Technologies and Applications*,
Algorithms for Intelligent Systems,
https://doi.org/10.1007/978-981-33-4046-6_17

selecting an appropriate life partner. The hybrid method of analytic hierarchy process (AHP) and technique for order preference and similarity to ideal solution (TOPSIS) have been applied to calculate the importance of each criterion which is identified through literature review and field survey and rank the various alternatives (candidate profiles), respectively. Buss and Barnes [11] studied that females and males have specific preferences such as intelligence, height, or extraversion for their mate selection. Joshi and Kumar [12] points out in studies that the selection of a partner for marriage is supposed to be the most important decision-making in India. Therefore, there is a need of some mathematical decision-making model which can provide the best candidate among the given profiles. Marriages are not just a wedding of two persons but it also connects the families of the two persons and strengthens their roots in the community. Nowadays, a candidate can choose a life partner via social networking sites such as Humsafar marriage counsels, matrimonial websites, newspapers, etc., these websites usually provide a long list of candidates to select as a suitable spouse. Choosing from various alternatives becomes a tedious task and very difficult to decide. Therefore, to solve this problem the model in the present may be fruitful. The present model is for arranged marriages and not for the love marriages. In this paper, we use a decagonal fuzzy number matrix for real decision-making problem that is for choosing a suitable spouse.

2 Preliminaries

Definition 2.1 *[13]*

Decagonal fuzzy number: A fuzzy Number $\widetilde{D} = (a, b, c, d, e, f, g, h, i, j)$ is said to be decagonal fuzzy number if its membership function is defined as

$$m_{\tilde{D}}(x) = \begin{cases} \frac{(x-a)}{4(b-a)} & a \le x \le b \\ \frac{1}{4} + \frac{(x-b)}{4(c-b)} & b \le x \le c \\ \frac{1}{2} + \frac{(x-c)}{4(d-c)} & c \le x \le d \\ \frac{3}{4} + \frac{(x-d)}{4(e-d)} & d \le x \le e \\ 1 & e \le x \le f \\ 1 - \frac{(x-f)}{4(g-f)} & f \le x \le g \\ \frac{3}{4} - \frac{(x-g)}{4(h-g)} & g \le x \le h \\ \frac{1}{2} - \frac{(x-h)}{4(i-h)} & h \le x \le i \\ \frac{(j-x)}{4(j-i)} & i \le j \le j \\ 0 & otherwise \end{cases}$$

Definition 2.2 Decagonal fuzzy number matrix: The elements of decagonal fuzzy number matrix are defined as $A = \left(a_{ij}\right)_{m\times n}$, where $a_{ij} = \left(a_{ij}L, a_{ij}S, a_{ij}S, a_{ij}S, a_{ij}S, a_{ij}T, a_{ij}T, a_{ij}T, a_{ij}T, a_{ij}U\right)$ is the ijth element of decagonal fuzzy number matrix of A. Then

$$0 \le a_{ij}L \le a_{ij}S \le a_{ij}S \le a_{ij}S \le a_{ij}S \le a_{ij}T \le a_{ij}T \le a_{ij}T \le a_{ij}T \le a_{ij}U \le 30 \quad (1)$$

where $a_{ij}L$ is the lower bound, $a_{ij}S, a_{ij}T$ are the moderate value and $a_{ij}U$ is the upper bound.

Definition 2.3 Corresponding membership Function: The membership function of $a_{ij} = \left(a_{ij}L, a_{ij}S, a_{ij}S, a_{ij}S, a_{ij}S, a_{ij}T, a_{ij}T, a_{ij}T, a_{ij}T, a_{ij}U\right)$ is defined as

$$\frac{a_{ij}L}{30}, \frac{a_{ij}S}{30}, \frac{a_{ij}S}{30}, \frac{a_{ij}S}{30}, \frac{a_{ij}S}{30}, \frac{a_{ij}T}{30}, \frac{a_{ij}T}{30}, \frac{a_{ij}T}{30}, \frac{a_{ij}T}{30}, \frac{a_{ij}U}{30}$$

If $0 \le a_{ij}L \le a_{ij}S \le a_{ij}S \le a_{ij}S \le a_{ij}S \le a_{ij}T \le a_{ij}T \le a_{ij}T \le a_{ij}T \le a_{ij}U \le 30$ Where

$$0 \le \frac{a_{ij}L}{30} \le \frac{a_{ij}S}{30} \le \frac{a_{ij}S}{30} \le \frac{a_{ij}S}{30} \le \frac{a_{ij}S}{30} \le \frac{a_{ij}T}{30} \le \frac{a_{ij}T}{30} \le \frac{a_{ij}T}{30} \le \frac{a_{ij}T}{30} \le \frac{a_{ij}U}{30} \le 1 \quad (2)$$

is called a decagonal fuzzy number matrix into its membership function.

2.1 Arithmetic Operations on Decagonal Fuzzy Number Matrix

Let $A = \left(a_{ij}\right)_{n\times n}$ and $B = \left(b_{ij}\right)_{n\times n}$ are two decagonal fuzzy number matrices of the same order $n \times n$.

Definition 2.1.1 Subtraction: (A − B) = $\left(a_{ij} - b_{ij}\right)_{m\times n}$, where $a_{ij} - b_{ij} = (a_{ij}L - b_{ij}U, a_{ij}S - b_{ij}T, a_{ij}S - b_{ij}T, a_{ij}S - b_{ij}T, a_{ij}S - b_{ij}T, a_{ij}T - b_{ij}S, a_{ij}T - b_{ij}S, a_{ij}T - b_{ij}S, a_{ij}T - b_{ij}S, a_{ij}U - b_{ij}L)$ is ijth element of A − B

Definition 2.1.2 Addition: (A + B) = $\left(a_{ij} + b_{ij}\right)_{m\times n}$, where $a_{ij}+b_{ij} = (a_{ij}L+b_{ij}L, a_{ij}S + b_{ij}S, a_{ij}S + b_{ij}S, a_{ij}S + b_{ij}S, a_{ij}S + b_{ij}S, a_{ij}T + b_{ij}T, a_{ij}T + b_{ij}T$, ijth element of A + B

Definition 2.1.3 Maximum operation on decagonal fuzzy number: Let $A = \left(a_{ij}\right)_{n\times n}$, where $a_{ij} = \left(a_{ij}L, a_{ij}S, a_{ij}S, a_{ij}S, a_{ij}S, a_{ij}T, a_{ij}T, a_{ij}T, a_{ij}T, a_{ij}U\right)$ is the ijth element of decagonal fuzzy number matrix of A and $C = \left(c_{ij}\right)_{m\times n}$, where $c_{ij} = \left(c_{ij}L, c_{ij}S, c_{ij}S, c_{ij}S, c_{ij}S, c_{ij}T, c_{ij}T, c_{ij}T, c_{ij}T, c_{ij}U\right)$ is the ijth element of decagonal fuzzy number matrix of B. Then the maximum operation for the two decagonal fuzzy number matrices is given by max(A, C) = sup(a_{ij}, c_{ij}), where

$\sup(a_{ij}, c_{ij}) = \sup(a_{ij}L, c_{ij}L), \sup(a_{ij}S, c_{ij}S), \sup(a_{ij}T, c_{ij}T), \sup(a_{ij}U, c_{ij}U)$ is ijth element of max(A, C).

Definition 2.1.4 Arithmetic mean for decagonal fuzzy number (AM): Let A = (p, q, r, s, t, u, v, w, x, y) be a decagonal fuzzy number. Then AM (A) = $\frac{p+q+r+s+t+u+v+w+x+y}{10}$.

3 Decision-Making Under Decagonal Fuzzy Number

Definition 3.1 Relativity function: Let u and v be variables defined on a universal set X. The relativity function is denoted by $f(u/v)$ and is defined as

$$f(u/v) = \frac{\mathrm{m}_v(u) - \mathrm{m}_u(v)}{\max\{\mathrm{m}_v(u), \mathrm{m}_u(v)\}} \tag{3}$$

where $m_v(u)$ is membership function of u with respect to v for decagonal fuzzy number and $m_u(v)$ is the membership function of v with respect to u for decagonal fuzzy number. Here $m_v(u) - m_u(v)$ is calculated using subtraction operation on decagonal fuzzy number matrix, by using the Definition 2.1.1 and $\max\{m_v(u), \mathrm{m}_u(v)\}$ is calculated using maximum operation on decagonal fuzzy number that is by using the Definition 2.1.3.

Definition 3.2 Comparison Matrix: Let $A = \{x_1, x_2, x_3, ...x_{i-1}, x_i x_{i+1}, \ldots .x_n\}$ be a set of n variables defined on X. Then form a matrix of relativity values $f\left(x_i/x_j\right)$, where $x_i's$ for $i = 1ton$, are n values defined on a universe X. The matrix $\mathrm{C} = (C_{ij})$ is a square matrix of order n is called the comparison matrix. Or) C-matrix with

$$C_{ij} = AM\big(f\left(x_i/x_j\right)\big) = \frac{AM\left(m_{x_j}(x_i) - m_{x_i}\left(x_j\right)\right)}{AM\left(\max\left\{m_{x_j}(x_i), m_{x_i}\left(x_j\right)\right\}\right)}$$

where AM denote the Arithmetic Mean and it is calculated using the arithmetic mean for decagonal fuzzy number (Definition 2.1.4) the comparison matrix is used to make the ranking of different fuzzy sets, The elements of C-matrix belongs to $[-1\ 1]$. The smaller value in the ith row of the comparison matrix, that is $C_i = \min\{f(x_i/x), i = 1ton\}$ will have the lowest weight for ranking purpose, so for the optimal or best solution we have to choose the maximum of $C_i's$. that is we take $\max\{C_i, i = 1\,to\,n\}$ in this way we can make ranking the variables $x_1, x_2, x_3, x_4, \ldots x_n$ by ordering the membership values.

Linguistic variables used: The decision-maker uses the linguistic variables for giving the preference of each criterion of each of the three alternatives. "No predilection", "very low predilection", "low predilection", "medium predilection", "high

Table 1 fuzzy numbers and corresponding fuzzy linguistic variables

Linguistic variables	Linguistic values
No predilection (NP)	(0, 0, 0, 0, 0, 0, 0.03, 0.06, 0.09, 0.12)
Very low predilection (VLP)	(0, 0.03, 0.06, 0.09, 0.012, 0.15, 0.18, 0.21, 0.24, 0.27)
Low predilection (LP)	(0.15, 0.18, 0.21, 0.24, 0.27, 0.3, 0.33, 0.36, 0.39, 0.42)
Medium predilection (MP)	(0.3, 0.33, 0.36, 0.39, 0.42, 0.45, 0.48, 0.51, 0.54, 0.57)
High predilection (HP)	(0.54, 0.57, 0.6, 0.63, 0.69, 0.72, 0.75, 0.78, 0.81, 0.84)
Very high predilection (VHP)	(0.84, 0.87, 0.9, 0.93, 0.96, 1, 1, 1, 1, 1)

predilection" and "very high predilection" are the fuzzy terms used by the decision-maker. These linguistic terms can be expressed in decagonal fuzzy number. For the selection of the appropriate alternative decagonal fuzzy numbers are used. The decagonal fuzzy linguistic scale is shown in Table 1.

Procedure for a real-life decision-making problem:

Step 1: The fuzzy linguistic terms assigned by the decision-maker, for each criterion or we can say for each perimeter are evaluated and then translated into fuzzy numbers and the same is represented in the matrix.

Step 2: Convert the given fuzzy number matrix into the decagonal fuzzy membership matrix using the definition of decagonal fuzzy membership matrix.

Step 3: Calculate all the relativity values $f(x_i/x_j)$ by 3.

Step 4: form the comparison matrix from the values $AMf(x_i/x_j)$ by using the definition of a comparison matrix.

Step 5: Find the minimum value from each row and then the maximum value among the minimums is the required solution.

A real-life decision-making problem: Let A be a set of choices of a person to select a spouse, let 'x_1' (Ruby), let 'x_2' (Shuby) and 'x_3' (Ifsha). Let us find out the spouse that would be appropriate for a person by comparing three options (x_1, x_2 and x_3), for which their parameters are health, family background, education, religion, ethnicity, looks, age, wealth, job, and chastity.

Step 1: Matrix A represents the scores in the form of a decagonal fuzzy number matrix.

A =

$$\begin{array}{c} \\ x_1 \\ x_2 \\ x_3 \end{array}\begin{bmatrix} x_1 & x_2 & x_3 \\ (9, 12, 12, 12, 21, 21, 30, 30, 30, 30) & (0, 3, 6, 9, 9, 9, 15, 15, 21, 21) & (0, 12, 12, 12, 15, 18, 21, 24, 27, 30) \\ (6, 6, 6, 6, 12, 12, 18, 21, 27, 30) & (18, 18, 18, 21, 21, 21, 24, 27, 30, 30) & (0, 3, 6, 9, 12, 15, 18, 21, 24, 30) \\ (6, 9, 12, 15, 18, 21, 24, 27, 30, 30) & (9, 12, 15, 18, 21, 21, 24, 24, 27, 30) & (0, 0, 9, 12, 15, 18, 21, 24, 27, 27) \end{bmatrix}$$

Step 2: $(\mathrm{A})_{mem} =$

$$\begin{matrix} x_1 \\ x_2 \\ x_3 \end{matrix}\begin{bmatrix} x_1 & x_2 & x_3 \\ (0.3, 0.4, 0.4, 0.4, 0.7, 0.7, 1, 1, 1, 1) & (0, 0.1, 0.2, 0.3, 0.3, 0.3, 0.5, 0.5, 0.7, 0.7) & (0, 0.4, 0.4, 0.4, 0.5, 0.6, 0.7, 0.8, 0.9, 1) \\ (0.2, 0.2, 0.2, 0.2, 0.4, 0.4, 0.6, 0.7, 0.9, 1) & (0.6, 0.6, 0.6, 0.7, 0.7, 0.7, 0.8, 0.9, 1, 1) & (0, 0.1, 0.2, 0.3, 0.4, 0.5, 0.6, 0.7, 0.8, 1) \\ (0.2, 0.3, 0.4, 0.5, 0.6, 0.7, 0.8, 0.9, 1, 1) & (0.3, 0.4, 0.5, 0.6, 0.7, 0.7, 0.8, 0.8, 0.9, 1) & (0, 0, 0.3, 0.4, 0.5, 0.6, 0.7, 0.8, 0.9, 0.9) \end{bmatrix}$$

$$m_{x_1}(x_1) = (0.3, 0.4, 0.4, 0.4, 0.7, 0.7, 1, 1, 1, 1),$$

$$m_{x_1}(x_2) = (0.2, 0.2, 0.2, 0.2, 0.4, 0.4, 0.6, 0.7, 0.9, 1),$$

$$m_{x_1}(x_3) = (0.2, 0.3, 0.4, 0.5, 0.6, 0.7, 0.8, 0.9, 1, 1),$$

$$m_{x_2}(x_1) = (0, 0.1, 0.2, 0.3, 0.3, 0.3, 0.5, 0.5, 0.7, 0.7),$$

$$m_{x_2}(x_2) = (0.6, 0.6, 0.6, 0.7, 0.7, 0.7, 0.8, 0.9, 1, 1),$$

$$m_{x_2}(x_3) = (0.3, 0.4, 0.5, 0.6, 0.7, 0.7, 0.8, 0.8, 0.9, 1),$$

$$m_{x_3}(x_1) = (0, 0.4, 0.4, 0.4, 0.5, 0.6, 0.7, 0.8, 0.9, 1),$$

$$m_{x_3}(x_2) = (0, 0.1, 0.2, 0.3, 0.4, 0.5, 0.6, 0.7, 0.8, 1),$$

$$m_{x_3}(x_3) = (0, 0, 0.3, 0.4, 0.5, 0.6, 0.7, 0.8, 0.9, 0.9),$$

Step 3:

$$f(x_1/x_1) = \frac{(0.3, 0.4, 0.4, 0.4, 0.7, 0.7, 1, 1, 1, 1) - (0.3, 0.4, 0.4, 0.4, 0.7, 0.7, 1, 1, 1, 1)}{\max\{(0.3, 0.4, 0.4, 0.4, 0.7, 0.7, 1, 1, 1, 1), (0.3, 0.4, 0.4, 0.4, 0.7, 0.7, 1, 1, 1, 1)\}}$$

$$= \frac{(-0.7, -0.6, -0.6, -0.6, 0, 0, 0.6, 0.6, 0.6, 0.7,)}{(0.3, 0.4, 0.4, 0.4, 0.7, 0.7, 1, 1, 1, 1)}$$

$$AMf(x_1) = \frac{AM\{m_{x_1}(x_1) - m_{x_1}(x_1)\}}{AM\{\max\{m_{x_1}(x_1), m_{x_1}(x_1)\}\}} = \frac{0}{0.69} = 0$$

$$f(x_1/x_2) = \frac{(0, 0.0.2, 0.3, 0.3, 0.3, 0.5, 0.5, 0.7, 0.7) - (0.2, 0.2, 0.2, 0.2, 0.4, 0.4, 0.6, 0.7, 0.9, 1)}{\max\{(0, 0.1, 0.2, 0.3, 0.3, 0.3, 0.5, 0.5, 0.7, 0.7), (0.2, 0.2, 0.2, 0.2, 0.4, 0.4, 0.6, 0.7, 0.9, 1)\}}$$

$$= \frac{(-1, -0.8, -0.5, -0.3, -0.1, -0.1, 0.3, 0.3, 0.5, 0.5)}{(0.2, 0.2, 0.2, o.3, 0.4, 0.4, 0.6, 0.7, 0.9, 1)}$$

$$AMf(x_1/x_2) = \frac{AM\{m_{x_2}(x_1) - m_{x_1}(x_2)\}}{AM\{\max\{m_{x_2}(x_1), m_{x_1}(x_2)\}\}} = \frac{-0.12}{0.49} = -0.244$$

$$f(x_1/x_3) = \frac{(0,0.4,0.4,0.4,0.5,0.6,0.7,0.8,0.9,1) - (0.2,0.3,0.4,0.5,0.6,0.7,0.8,0.9,1,1)}{\max\{(0,0.4,0.4,0.4,0.5,0.6,0.7,0.8,0.9,1),(0.2,0.3,0.4,0.5,0.6,0.7,0.8,0.9,1,1)\}}$$
$$= \frac{(-1,-0.6,-0.5,-0.4,-0.2,0,0.2,0.4,0.6,0.8)}{(0.2,0.4,0.4,0.5,,0.6,0.7,0.8,0.9,1,1)}.$$

$$AMf(x_1/x_3) = \frac{AM\{m_{x_3}(x_1) - m_{x_1}(x_3)\}}{AM\{\max\{m_{x_3}(x_1), m_{x_1}(x_3)\}\}} = \frac{-0.07}{0.65} = -0.107.$$

$$f(x_2/x_1) = \frac{(0.2, 0.2, 0.2, 0.2, 0.4, 0.4, 0.6, 0.7, 0.9, 1) - (0, 0.1, 0.2, 0.3, 0.3, 0.3, 0.5, 0.5, 0.7, 0.7,)}{\max\{(0.2, 0.2, 0.2, 0.2, 0.4, 0.4, 0.6, 0.7, 0.9, 1), (0, 0.1, 0.2, 0.3, 0.3, 0.3, 0.5, 0.5, 0.7, 0.7,)\}}$$
$$= \frac{(-0.5, -0.5, -0.3, -0.3, 0.1, 0.1, 0.3, 0.5, 0.8, 1)}{(0.2, 0.2, 0.2, 0.3, 0.4, 0.4, 0.6, 0.7, 0.9, 1)}$$

$$AMf(x_2/x_1) = \frac{0.12}{0.49} = 0.244$$

$$f(x_2/x_2) = \frac{m_{x_2}(x_2) - m_{x_2}(x_2)}{\max\{m_{x_2}(x_2), m_{x_2}(x_2)\}} = 0, \qquad AMf(x_2/x_2) = 0$$

$$f(x_2/x_3) = \frac{(0, 0.1, 0.2, 0.3, 0.4, 0.5, 0.6, 0.7, 0.8, 1) - (0.3, 0.4, 0.5, 0.6, 0.7, 0.7, 0.8, 0.8, 0.9, 1)}{\max\{(0, 0.1, 0.2, 0.3, 0.4, 0.5, 0.6, 0.7, 0.8, 1), (0.3, 0.4, 0.5, 0.6, 0.7, 0.7, 0.8, 0.8, 0.9, 1)\}}$$
$$= \frac{(-1 - 0.8, -0.6, -0.5, -0.3, -0.2, 0, 0.2, 0.4, 0.7)}{(0.3, 0.4, 0.5, 0.6, 0.7, 0.7, 0.8, 0.8, 0.9, 1)}$$

$$AMf(x_2/x_3) = \frac{AM\{m_{x_3}(x_2) - m_{x_2}(x_3)\}}{AM\{\max\{m_{x_3}(x_2), m_{x_2}(x_3)\}\}} = \frac{-0.21}{0.67} = -0.313$$

$$f(x_3/x_1) = \frac{(0.2, 0.3, 0.4, 0.5, 0.6, 0.7, 0.8, 0.9, 1, 1) - (0, 0.4, 0.4, 0.4, 0.5, 0.6, 0.70.80.9, 1)}{\max\{(0.2, 0.3, 0.4, 0.5, 0.6, 0.7, 0.8, 0.9, 1, 1), (0, 0.4, 0.4, 0.4, 0.5, 0.6, 0.70.80.9, 1)\}}$$
$$= \frac{(-0.8, -0.6, -0.4, -0.2, 0, 0.2, 0.4, 0.5, 0.6, 1)}{(0.2, 0.4, 0.4, 0.5, 0.6, 0.7, 0.8, 0.9, 1, 1)}$$

$$AMf(x_3/x_1) = \frac{AM\{m_{x_1}(x_3) - m_{x_3}(x_1)\}}{AM\{\max\{m_{x_1}(x_3), m_{x_3}(x_1)\}\}} = \frac{0.7}{6.5} = 0.107$$

$$f(x_3/x_2) = \frac{(0.3, 0.4, 0.5, 0.6, 0.7, 0.7, 0.8, 0.8, 0.9, 1) - (0, 0.1, 0.2, 0.3, 0.4, 0.5, 0.6, 0.7, 0.8, 1)}{\max\{(0.3, 0.4, 0.5, 0.6, 0.7, 0.7, 0.8, 0.8, 0.9, 1), (0, 0.1, 0.2, 0.3, 0.4, 0.5, 0.6, 0.7, 0.8, 1)\}}$$
$$= \frac{(-0.7, -0.4, -0.2, 0, 0.2, 0.3, 0.5, 0.6, 0.8, 1)}{(0.3, 0.4, 0.5, 0.6, 0.7, 0.7, 0.8, 0.8, 0.9, 1)}$$

$$AMf(x_3/x_2) = \frac{AM\{m_{x_2}(x_3) - m_{x_3}(x_2)\}}{AM\{\max\{m_{x_2}(x_3), m_{x_3}(x_2)\}\}} = \frac{0.21}{0.67} = 0.313$$

$$f(x_3/x_3) = \frac{(0, 0, 0.3, 0.4, 0.5, 0.6, 0.7, 0.8, 0.9, 0.9) - (0, 0, 0.3, 0.4, 0.5, 0.6, 0.7, 0.8, 0.9, 0.9)}{\max\{(0, 0, 0.3, 0.4, 0.5, 0.6, 0.7, 0.8, 0.9, 0.9), (0, 0, 0.3, 0.4, 0.5, 0.6, 0.7, 0.8, 0.9, 0.9)\}}$$
$$= \frac{(-0.9, -0.9, -0.5, -0.3, -0.1, 0.1, 0.3, 0.5, 0.9, 0.9)}{(0, 0, 0.3, 0.4, 0.5, 0.6, 0.7, 0.8, 0.9, 0.9)}$$

$$AMf(x_3/x_3) = \frac{AM\{m_{x_3}(x_3) - m_{x_3}(x_3)\}}{AM\{\max\{m_{x_3}(x_3), m_{x_3}(x_3)\}\}} = \frac{0}{0.51} = 0$$

Step 4: The comparison matrix $C = (C_{ij}) = AMf(x_i/x_j)$. is given by

$$A = \begin{matrix} \\ x_1 \\ x_2 \\ x_3 \end{matrix} \begin{bmatrix} x_1 & x_2 & x_3 \\ 0 & -0.244 & -0.107 \\ 0.244 & 0 & -0.313 \\ 0.107 & 0.313 & 0 \end{bmatrix}$$

Step 5:
$c_1 =$ minimum of ith row
$c_1 = -0.244$, $c_2 = -0.313$, $c_3 = 0$,
Hence the maximum value is $c_3 = 0$,

For this problem the ranking is x_3, x_1 and x_2. So a person chooses x_3 (Ifsha) as his/her appropriate spouse by the proposed model.

4 Results and Discussions

The weights obtained for each criterion using fuzzy theory show the significance of a particular criterion in decision-making. Religion has been noticed as the most significant factor for the candidate considered in this study. However, the candidates describe that it becomes easier to convince parents if they choose a life partner from the same religion. The ethnicity of a candidate is another significant criterion, which is considered by candidates for the selection of their better half. This shows that cultural or language differences play an important role in the selection process of a spouse. Educational qualification, age, and looks of a candidate can be noticed as the other important criteria for being selected to be an appropriate one. When a candidate has more than one option, it is difficult for him to decide which one is appropriate. Therefore, the ranking of the candidates is need by weighing each important criterion. That is why this work has been done. This work is about the ranking of candidates, using the fuzzy decision-making technique. To prove this decision-making technique experimentally, whether this technique works on ground level on not, it is a time-consuming study. Therefore, that may be our next paper in near future.

5 Conclusion

Fuzzy decision-making is most important as it is not only used in scientific, social, economic and industrial fields, but also used in medical diagnosis as well. There exist several approaches within fuzzy decision-making. In this paper, we have used the ranking order via a decagonal fuzzy number matrix to solve the decision-making problem. Here one has to choose a spouse when he/she has to make a decision, which one he/she should select among the three. Finally, it may be concluded that the present study would assist the candidates in the selection of a suitable spouse.

References

1. Dubois, D., Prade, H.: Operations on fuzzy numbers. Int. J. Syst. Sci. **9**(6), 613–626 (1978)
2. Victor Devadoss, A., Felix, A.: A new Fuzzy DEMATEL method in an uncertain linguistic environment. Adv. Fuzzy Sets Syst. **16**(2), 93–123 (2013)
3. Rajarajeshwari, P., Sudha, A.S., Karthika, R.: A new operation on hexagonal fuzzy number. Int. J. Fuzzy Logic Syst. **3**(3), 15–26 (2013)
4. Malini, S.U., Kennedy, F.C.: An approach for solving fuzzy transportation using octagonal fuzzy numbers. Appl. Math. Sci. **54**, 2661–2673 (2013)
5. Namarta, U.C.G., Thakur, N.I.: Solution of fuzzy game problem by using decagonal fuzzy numbers. Int. J. Sci. Eng. Res. **8**(12). ISSN 2229–5518
6. Sarala, N., Janatul Firthouse, I., Rajeshwari, R.: Decision making problems of membership matrix and comparison matrix under fuzzy environment. Int. J. Adv. Trends Eng. Sci. Technol. **2**(2)
7. Tarakeswara Rao, B., Ramakrishna Murty M., et al.: A comparative study on effective approaches for unsupervised statistical machine translation. In: International Conference and Published the Proceedings in AISC Springer Conference, vol. 1076, pp. 895–905 (2020)
8. Jejeebhoy, S.J., Santhya, K.G., Acharya, R., Prakash, R.: Marriage-related decision making and young woman's marital relations and agency. Asian Publ. Stud. **9**(1), 28–49. https://doi.org/10.1080/17441730.2012.736699
9. Tsutsui, J.: The traditional phase of mate selection in East Asian contraries. Int. Sociol. **28**(1), 67–83. https://doi.org/10.1177/0268580913484775
10. Chauhan, A., Kumar, P.: Selection of a spouse for females using hybrid-criteria decision model in India. Int. J. Model. Oper. Manage. **10**(10)
11. Buss, D.M., Barnes, M.: Preferences In human mate selection. J. Pers. Soc. Psychol. **50**(3), 559–570
12. Joshi, K., Kumar, S.: Matchmaking using fuzzy analytical hierarchy process, compatibility measure and stable matching for online matrimony in India. J. Multi-criteria Decis. Anal. **19**(1–2), 57–66
13. Virginraj, A., Hemavati, S.: Solving the decision making problem using decagonal fuzzy number and fuzzy matrix. Int. J. Math. Appl. **4**(4), 155–162 (2016). ISSN 2347-1557

Extractive Summarization Using Frequency Driven Approach

V. Mohan Kalyan, Chukka Santhaiah, M. Naga Sri Nikhil, J. Jithendra, Y. Deepthi, and N. V. Krishna Rao

1 Introduction

In today's world, where we have vast amounts of data to discern, an efficient and effective understanding of information is an essential requirement. The idea of shortening this vast information into smaller texts is called Text Summarization. These smaller texts are intended to be coherent and fluent, covering the gist of the entire text. Text Summarization is broadly grouped into two techniques: Extractive summarization and Abstractive summarization. In extractive summarization, the idea is to identify the important sections based on the sentence weightage by using the word frequency count. The term word frequency refers to the number of times each word is repeated. Sentences with high weightage are considered as important and relevant and are added to the summary. On the other side, abstractive summarization is different as it creates new phrases of text and thereby generates the summary by holding the gist of the text. But this method is still under research and development due to its

V. Mohan Kalyan · C. Santhaiah · M. Naga Sri Nikhil (✉) · J. Jithendra · Y. Deepthi · N. V. Krishna Rao
Department of CSE, Institute of Aeronautical Engineering, Dundigal, Hyderabad, Telangana, India
e-mail: mnagasrinikhil123@gmail.com

V. Mohan Kalyan
e-mail: mohankalyan01@gmail.com

C. Santhaiah
e-mail: chukka.santh@gmail.com

J. Jithendra
e-mail: jithendrakumar819@gmail.com

Y. Deepthi
e-mail: deepthisagar7@gmail.com

N. V. Krishna Rao
e-mail: nvkrishnarao3@gmail.com

C. K. Mai et al. (eds.), *Machine Learning Technologies and Applications*,
Algorithms for Intelligent Systems,
https://doi.org/10.1007/978-981-33-4046-6_18

complexity and in many cases, extractive summarization yields better results when compared to abstractive summarization as the latter deals with semantics, inferences, and other natural language generation which are complicated and complex to model.

Text summarization has various applications in different fields. It is used in Q&A bots, search marketing, internal document workflow, and media monitoring, etc., It generates an abstract of the entire document or given text which helps humans and even other machines to understand the gist of given text without processing the whole document. One important factor to be considered while generating the summary is to maintain the coherence and fluency between the logic and grammar of the sentences. In a real-time implementation, an extractive technique might miss the logical connection between the sentences and produce grammatically incorrect results. But many techniques are discovered to mitigate the problem and for producing an efficient summary. On the whole, text summarization is all about increasing the readability and decreasing the reading time by extracting the important and relevant sentences.

In this paper, our goal is to achieve maximum readability with minimum reading time using the frequency driven approach. In this approach, the term frequency is calculated for each word and sentence weightage is determined using the term frequency thereby generating a summary with sentences having maximum weightage.

This paper is organized as follows. Section 2 describes the literature study related to different works on extractive text summarization. Section 3 explains the architecture of the model. Section 4 illustrates the implementation process of the model. Section 5 depicts the results and measures the accuracy of the model and Sect. 6 contains the conclusion and future work.

2 Related Work

This section discusses the other works related to text summarisation including the word and sentence level features. The methods used for the extractive text summarisation are divided into 2 categories [1] 1. Supervised Learning and 2. Unsupervised learning. LSA(Latent Semantic Analysis) [1] is considered as an unsupervised learning technique in text summarization. This method of unsupervised learning captures the frequently occurred words in different sentences [2]. As the words which are repeatedly present have a weightage in the whole text. These repeated words are used to calculate the word frequency [3] from which an important sentence or paraphrase is identified.

Term-Frequency inverse document frequency (TF-IDF) [4] is another unsupervised learning method that uses numeric statistics to assign the weightage for each of the sentences present in the document. The problem for this method mentioned in [4] is that it assigns more weightage to sentences that have more words, i.e., longer sentences. Another approach would be considering the keywords which are

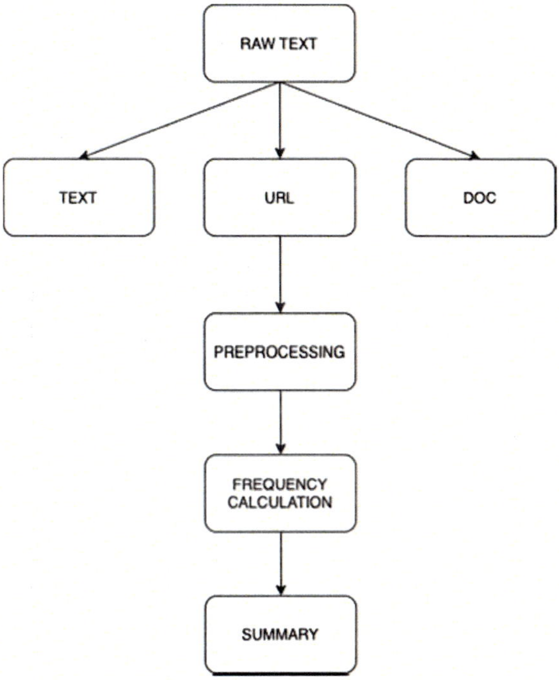

Fig. 1 Architecture

obtained after removing the stop words from the entire text [5] and the weightage for the sentences in that text are assigned through the co-occurrences of the words in different sentences within the sliding window [6].

The authors of [7] introduced two summaries for every text as heading summary and main summary where they assign the important sentences to these two summaries using the sentence ranking and term frequency approach. They concluded that the heading summaries were more efficient and lucid than main summaries as they have considered 1/3 of the sentences for each heading and achieved better result and Rule-based technique where features are calculated using the tf-idf which uses term frequency and sentence position for feature extraction works for single document [8].

The process of extractive text summarization is divided into 3 phases: analysis, extraction, and generation in [9]. They have described the analysis as a phase containing lemmatization and assigning word weightage [1], extraction as using the rank [10] for considering the important sentences [7], and generation for reranking the important sentences and forming a complete summary.

3 Architecture

The architecture of the algorithm is explained in the Fig. 1. At first, the raw text is considered in the form of 3 sources—direct text feed, URL, document. If the URL is specified, then the text information present in that website is extracted and given as input to the model. As the same continues for the document, the information from the document is read and provided as input to the model. Preprocessing also referred to as text cleaning is performed on the text extracted from the first step. Then frequency driven techniques are used for implementing the model. Then the most important and relevant text is generated and tagged as "summary".

4 Implementation

The entire process of generating the summary can be divided into 4 steps. The first step involves the extraction of text from either of the sources like documents, URL, and even direct text feed is allowed for summarization. The second step is concerned with the preprocessing (Text Cleaning) of the text extracted from the first step. The third step is about assigning a frequency to the words and sentences present in the text by which we determine the importance of the sentences. The final step in this process is to extract the important sentences as a summary.

4.1 Packages Used

- nltk
- re
- urllib
- bs4
- tkinter (GUI).

4.2 Extracting the Text

- Input to the model is extracted from websites using the packages bs4 and urllib which are used to open a specified URL and extract the information from that URL.
- A direct document of ".txt" type can be fed as input to the model. Text in the documents is read using the read and open methods of the python.
- We can provide a direct text to the model which will further be applied for text processing.

- For GUI, various packages and tools are available, out of which tkinter is the one we choose to apply.

4.3 Text Preprocessing

- By using the package "re": Regular expressions, the text is cleaned by removing the contents (Alphanumeric characters) present in brackets, a parenthesis which is unwanted information.
- Stop words like and, the, an, in, is, are, etc., are removed by using "stop words" from the nltk package.

4.4 Frequency Method

- The entire text is tokenized into sentences and words using the nltk.tokenize methods.
- Sent_tokenize is used to tokenize the entire text into sentences and word tokenize is used to tokenize these sentences into words.
- A word dictionary is created to maintain the word frequency of each word present in the text.
- Based on each word frequency, the sentence weightage is calculated.

$$wordfreq_i = 1 + 1 + 1 + \ldots j_{\text{occurances}} \tag{1}$$

$$\text{sentence weight}_i = \sum_{j=1}^{m} wordfreq_j \tag{2}$$

4.5 Generating Summary

- A threshold value is assigned by considering, half the value of max(sentence_frequency).
- All the sentences which have a weightage less than or equal to the threshold are considered into the resulting summary.

$$\text{threshold} = \frac{\max(\text{sentence weight}_1, \text{sentence weight}_2, \ldots \text{sentence weight\_}n)}{2} \tag{3}$$

5 Results

- The experiment was conducted out on Intel i5 Processor, 8 GB Ram, and 64-bit Operating System. We have used the text module of the model and got the summary for the given text. The implemented method is the frequency approach through which we assign the weightage to the sentences and extract them based on importance.
- For evaluating the model performance we have considered a built-in module of python called "difflib" in which "Sequence Matcher" is used for comparing the extracted summary with the human perceived summary. The method provides a value between 0 and 1 which indicates the accuracy of the model. "0" refers to the zero percent relevance between the model generated and human perceived summaries. Whereas "1" refers to the maximum relevance between model generated and human perceived summaries.

5.1 *Test Cases*

Model is verified with the input given and the word count, sentence count, average reading time is calculated for the given input and generated summary (Figs. 2 and 3).

5.2 *Accuracy Check 1*

The model is tested with an article from Wikipedia and the generated summary is compared with the human-made summary. The accuracy in this case was 78% as the sequential matcher produced 0.78 (between 0 and 1). This refers that the generated summary is 78% relevant to the human-made summary (Fig. 4).

5.3 *Accuracy Check 2*

The model is tested with another article from a blog and the generated summary is compared with the human-made summary. Sequential matcher returned the value 0.88. So, the accuracy in this case was 88%. This refers that the generated summary and human-made summary are 88% relevant (Fig. 5).

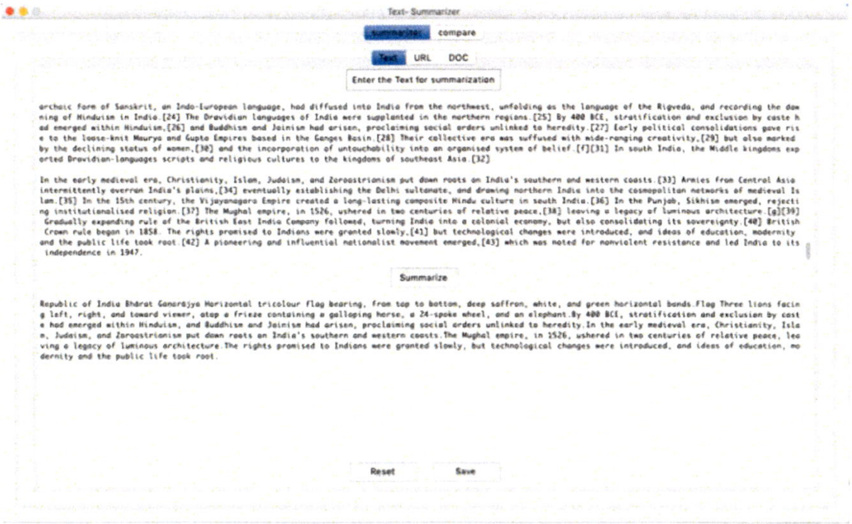

Fig. 2 Text input and generated summary

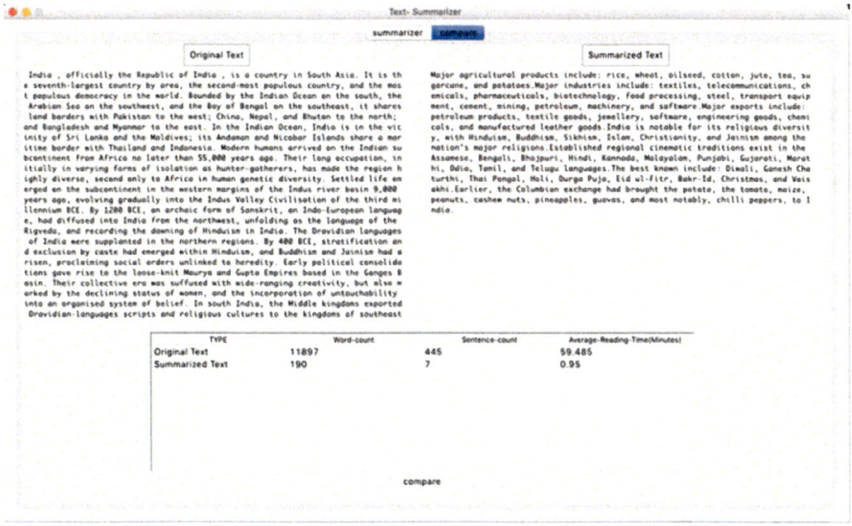

Fig. 3 Comparison of original text and generated summary

6 Conclusion and Future Work

In the aspect of quantitative analysis, we were to ascertain that the output of the algorithm is quite efficient. On an average 30%–40% of the entire text is considered as the summary. In this view, when evaluated, the generated summary contains the

```
actual_summary = "India , officially the Republic of India , is a country in South Asia.It

expected_summary = "The india , officially the Republic of India , is a seventh-largest cou

print(difflib.SequenceMatcher(
    None, expected_summary, actual_summary).ratio())

0.7820683533067022
```

Fig. 4 Accuracy check 1

```
actual_summary_1 = "Cricket is a bat-and-ball game played between two teams of eleven pl

expected_summary_1 = "Cricket is a bat-and-ball game played between two teams of eleven

print(difflib.SequenceMatcher(
    None, expected_summary_1, actual_summary_1).ratio())

0.8845381526104418
```

Fig. 5 Accuracy check 2

important and relevant sentences present in the whole text. In this model, we can measure the number of words, sentences, and average reading time for the original summary and generated summary.

The present technique works efficiently on new articles, text documents, direct text, URLs, and other text sources. However, if applied to conversations and fictional content (stories), it might not work as expected. Future enhancements on the model would be improving the perseverance of contextual meaning by performing structural analysis and contextual understanding using POS tags and other semantics.

References

1. Moratanch, N., Chitrakala, S.: A survey on extractive text summarization. In: 2017 International Conference on Computer, Communication and Signal Processing (ICCCSP) (2017)
2. Annapurna, V., Ramakrishna Murty, M. et al.: Comparative analysis of frequent pattern mining for large data using FP-tree and CP-tree methods. In: International Conference FICTA-17 at KIIT University, Bhubaneswar, Springer, AISC series (2017)
3. Madhuri, J. N., Ganesh Kumar, R.: Extractive text summarization using sentence ranking. In: 2019 International Conference on Data Science and Communication (IconDSC), pp. 1–3. Bangalore, India (2019). https://doi.org/10.1109/IconDSC.2019.8817040
4. Andhale, N., Bewoor, L. A.: An overview of text summarization techniques. In: 2016 International Conference on Computing Communication Control and automation (ICCUBEA), pp. 1-7. Pune (2016). https://doi.org/10.1109/ICCUBEA.2016.7860024
5. Patil, A.P., Dalmia, S., Ansari, S.A.A., Aul, T., Bhatnagar, V.: Automatic text summarizer. In: 2014 International Conference on Advances in Computing, Communications and Informatics (ICACCI), pp. 1530–1534. New Delhi (2014). https://doi.org/10.1109/ICACCI.2014.6968629
6. Ganiger, S., Rajashekharaiah, K.M.M.: Comparative study on keyword extraction algorithms for single extractive document. In: 2018 Second International Conference on Intelligent Computing and Control Systems (ICICCS), IEEE (2018)

7. Krishnaveni, P., Balasundaram, S.R.: Automatic text summarization by local scoring and ranking for improving coherence. In: 2017 International Conference on Computing Methodologies and Communication (ICCMC) (2017)
8. Naik, S.S., Gaonkar, M.N.: Extractive text summarization by feature-based sentence extraction using rule-based concept. In: 2017 2nd IEEE International Conference on Recent Trends in Electronics Information & Communication Technology (RTEICT), India, May 19–20, 2017
9. Mishra, A.R., Panchal, V.K., Kumar, P.: Extractive text summarization—an effective way to extract information from text. In: International Conference on contemporary Computing and Informatics (IC3I). IEEE (2019)
10. Afsharizadeh, M., Ebrahimpour-Komleh, H., Bagheri, A.: Query-oriented text summarization using sentence extraction technique. In: 2018 4th International Conference on Web Research (ICWR) (2018)
11. Rahimi, S.R., Mozhdehi, A.T.: An overview of extractive text summarization. In: IEEE 4th International Conference on Knowledge-Based Engineering and Innovation (KBEI) Dec. 22, 2017 (Iran University of Science and Technology), Tehran, Iran

An Efficient Deep Learning Based Approach for Malware Classification

Madhurima Rana and Swathi Edem

1 Introduction

In the field of information security and computer science malicious software or malware is harmful threat when it has been executed on a computing system or device [1]. Execution of malware can lead to minor to calamitous damages, like tampering, stealing, encrypting compromise crucial data [2]. For providing protection security, experts and researchers are working on robust and new tools and methods to identify and detect malware [3].

Now, there are several ways to serve this same purpose but this proposed methodology uses deep Learning Algorithms to help classify or generalize several families of Malwares present, and thus identify them with the highest degree of efficiency [4]. Deep learning techniques require more time to compute for training and retrain the models [5]. Traditional machine learning algorithms used to be fast but not that much accurate [6], on the other hand, dynamic deep learning algorithms are time taking procedure but they promise more accurate results [7]. Malware detection and analysis has two phases, first phase is malware discovery then identifiation and the next phase is classification [8]. The methodologies are used here to identify or classify each and every threat sample. To perform classification on malware sample feature selection is an important phase [9]. It can be done in two phases static and dynamic. The basic concept behind feature selection is to select the promising features from the pool of data.

M. Rana (✉) · S. Edem
Chaitanya Bharthi Institute of Technology, Hyderabad 500075, India
e-mail: madhurima.rana@gmail.com

C. K. Mai et al. (eds.), *Machine Learning Technologies and Applications*,
Algorithms for Intelligent Systems,
https://doi.org/10.1007/978-981-33-4046-6_19

With the rapid development of internet and cloud usages a huge volume of data set has been generating in every passing days. It's very difficult to deal with a huge volume of data and it is not mandatory that in the dataset all the features are important and promise for the result. So, to overcome this kind of scenario in the very first step, data should be preprocessed and then relevant features should be selected [10].

In our work, we have used CNN as a feature selection method. CNN has different layers apart from input-output layer to get the precise data. Classification of malware is based on statistical methodologies. It is a supervised machine learning technique where input data is labeled classes and models are building based. After selecting the relevant features, classifier is used to classify. Here, we have used KNN as a classifier.

Our paper has been organized into different sections. In the first section, we have introduced the topic in nutshell. Section 2 consists of literature survey where malware detection related work has been explored briefly. In Sect. 3 working process model, steps associated and methodologies has been described. In the next section, i.e., Sect. 4, we have discussed experimental results. In the last section, we have concluded our work.

2 The State of Art Processing Techniques

Fred et al. [11] have proposed an effective and efficient model which detects newly released malware by mathematical generalization. In their paper they have used a dataset which contains malware images taken from malware binaries, then deep learning technique has been used to classify each malware family. Here they have used deep learning-based Support Vector Machine to make an intelligent system and they have reached approx 86% accuracy. In this paper, authors used three deep learning methodologies like CNN-SVM, GRU-SVM, and MLP-SVM, they applied those three dl techniques on Malimg Dataset, [11] and the accuracies were following 77.22, 84.92, and 80.46%.

Cakir et al. [7] used deep learning based feature extraction method and then gradient boosting algorithm for classifying malware and they achieved success rate as 96% with limited sample data. GBM is basically a regression and classification method. In this paper, authors used two procedures to evaluate the GBM classifier logarithmic loss and accuracy. For accuracy measure, k-fold cross validation has been used. Here, they have used the Microsoft malware classification dataset.

Burnap et al. [12] has proposed a technique for APT(Advanced Persistent Threats), a sophisticated threat whose main feature is its encryption. In this paper, self organizing feature maps has been used as a classifier and along with SOFM they have compared other classifiers like Random forest, Bayesian Network, Multi Layer

Perceptron, Support vector machine. For that particular dataset, the accuracy has increased for SOFM between 7.24 and 24.68%. For accuracy measures, k-fold cross validation method has been used where the number of fold is 10.

Natraj et al. [13] has proposed an effective and simple technique to classify malware images using image processing. To get the feature vector, they have used Gabor filter which is based on a sinusoidal plane for certain frequency and Gaussian envelope for modulation which has been used for texture classification and segmentation. In this paper, they have taken GIST dataset which consists of 8 malware families and 1713 malware images. Here k-nearest neighbor is used as a classifier where $k = 3$ has been used and this classifier is successful to obtain 99.93% accuracy. K(10)-fold cross validation used for the accuracy measure of the classifier. Then they extended their work to 25 malware families, 9458 images and by using the same classifier and same accuracy measure they got approx 97% accuracy on classification.

3 Technology Used

In this section, the methodologies used have been described.

3.1 Dataset Description

Dataset is a major part of any experiment and when it comes to supervised learning, arranging of dataset plays a major role in that. In our paper, we have used Malimg dataset [11]. In this dataset, 25 malware classes (families) are there and their number of samples vary with each other. Malware samples could be kept in different subfolders which consist of binaries or digital images that make label and feature computation easy. Family names are Allaple.L (1591), Allaple.A (2949), Yuner.A (800), Lolyda.AA (1213), Lolyda.AA (2184), Lolyda.AA (3 123), C2Lop.P (146), C2Lop.gen!G (200), Instant access (431), Swizzor.gen!I 1(32), Swizzor.gen!E (128), VB.AT (408), Fakerean(381), Alueron.gen!J (198), Malex.gen!J (136), Lolyda.AT(159), Adialer.C(125), Wintrim.BX (97), Dialplatform.B (177), Dontovo.A (162), Obfuscator.AD (142), Agent.FYI(116), Autorun.K(106), Rbot!gen (158), Skintrim.N (80).

3.2 Classification

Convolution Neural Network (CNN) A CNN consists of one or more convolution layers, sometimes with a subsampling step tracked by one or more well connected layers like multilayer neural network [14]. It has been designed in such a way that it can take a 2D image as an input and process further. They have very few parameters than any other well connected network and are very easy to train. It has an input layer, output layer, and one or more hidden layers. Basically, it is used to take images in width * height * depth format and in each layer covert one volume to another by some set of functions [8]. In hidden layers there are number of layers like covolution layer, activation function layer, pool layer, fully connected layer. As CNN works on volume of image by passing it through different layers. Different function used to work on the background of each layers. In our simulaation we have used these CNN feature to select the relevant feature fron the pool of data [14].

K-Nearest Neighbor (KNN) K-nearest neighbor is a statistics-based machine learning methodology for classification and regression [15]. In KNN classification, input is k closest training sample and output is class membership. In KNN, k is a positive integer generally small [7]. The steps associated with knn are the following:

1. Loading the data.
2. Initialization of K value.
3. To get predicted class, iteration from 1 to the total number of data points of training data.

   (i) Distance calculation between training data(each row) and test data. We can use different distance metrics like Euclidean distance, Manhattan distance, Chebyshev, cosine, etc.
   (ii) Calculated distances will be sorted in order (ascending) according to distance values.
   (iii) Top k rows can be achieved from the sorted array.
   (iv) Most frequent class could be obtained of these rows
   (v) Predicted class will be obtained.

In our simulation, we have used KNN for classification where k values are considered as 5 and 7 and distance metrics used are Euclidean distance, Minkowski distance, and Manhattan distance.

Evaluation Classification can be measured by accuracy. There are different performance metrics to measure the efficiency of the algorithm like accuracy, recall, precision, and F-measure [16]. Accuracy refers how much close the value to the standard value. Precision and Recall are the predicted value of very imbalanced classes. If we take harmonic average, we can get an F-measure. The formulas are the following:
$\text{Accuracy} = \frac{\text{TP} + \text{TN}}{\text{TP} + \text{TN} + \text{FP} + \text{FN}}$ $\text{Precision} = \frac{\text{TP}}{\text{TP} + \text{FP}}$ $\text{Recall} = \frac{\text{TP}}{\text{TP} + \text{FN}}$

$$\text{F-measure} = 2 * \frac{\text{Precision} * \text{Recall}}{\text{Precision} + \text{Recall}}$$

where

TP True positive
TN True Negative
FP False Positive
FN False Negative

Confusion matrix has created according to the calculated values.

4 Experimental Results and Simulation

All the experiments of this proposed methodology are conducted in a desktop computer with Intel Core (T5) i5 processor, 8 GB DDR4 RAM, and Intel® UHD Graphics 630 graphic card. In this work, we used Malimg dataset [11] for experimentation. There are 25 different malware classes in the experiment dataset. We employed a deep learning model namely CNN for the feature extraction. In our experiment, we have used 80% of the data to train 20 epochs. The CNN's last layer output is fed as the input to the KNN classifier. For this study, the following CNN architecture has used (Fig. 1).

For the first Conv2d() function, 5 × 5 kernel is used. For the second and third Conv2d() functions, 3 × 3 kernel is used. The pool size for MaxPooling2D is 2 × 2. The accuracy of the Convolution Neural Network for 20 epochs is 97.85%.

In this work, we tuned the parameter of the KNN algorithm to get maximum accuracy and less error rate. The distances metrics are considered in this experiment are

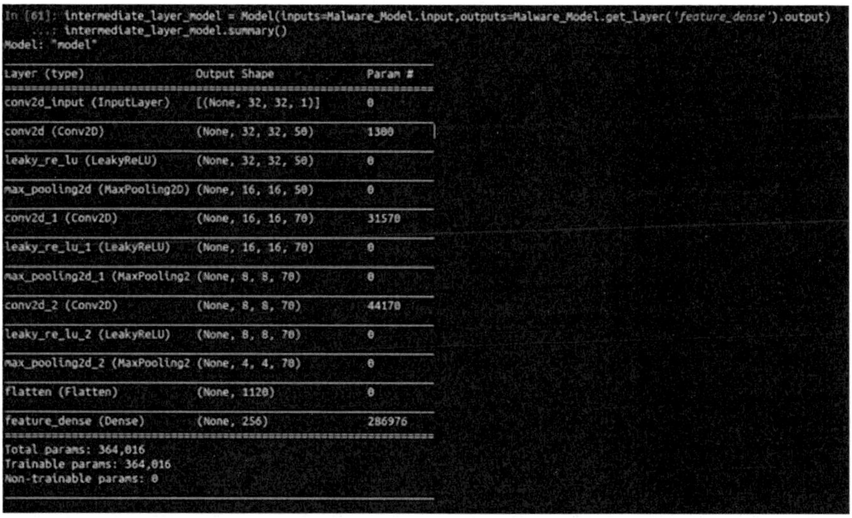

```
In [61]: intermediate_layer_model = Model(inputs=Malware_Model.input,outputs=Malware_Model.get_layer('feature_dense').output)
    ...: intermediate_layer_model.summary()
Model: "model"
_________________________________________________________________
Layer (type)                 Output Shape              Param #
=================================================================
conv2d_input (InputLayer)    [(None, 32, 32, 1)]       0
_________________________________________________________________
conv2d (Conv2D)              (None, 32, 32, 50)        1300
_________________________________________________________________
leaky_re_lu (LeakyReLU)      (None, 32, 32, 50)        0
_________________________________________________________________
max_pooling2d (MaxPooling2D) (None, 16, 16, 50)        0
_________________________________________________________________
conv2d_1 (Conv2D)            (None, 16, 16, 70)        31570
_________________________________________________________________
leaky_re_lu_1 (LeakyReLU)    (None, 16, 16, 70)        0
_________________________________________________________________
max_pooling2d_1 (MaxPooling2 (None, 8, 8, 70)          0
_________________________________________________________________
conv2d_2 (Conv2D)            (None, 8, 8, 70)          44170
_________________________________________________________________
leaky_re_lu_2 (LeakyReLU)    (None, 8, 8, 70)          0
_________________________________________________________________
max_pooling2d_2 (MaxPooling2 (None, 4, 4, 70)          0
_________________________________________________________________
flatten (Flatten)            (None, 1120)              0
_________________________________________________________________
feature_dense (Dense)        (None, 256)               286976
=================================================================
Total params: 364,016
Trainable params: 364,016
Non-trainable params: 0
_________________________________________________________________
```

Fig. 1 Description of multiple layers of CNN

Table 1 Experimental result on CNN-KNN model

Distance metric	Test data size	K value	Accuracy
Euclidean	0.1	5	0.983940042826
	0.1	7	0.9807280513
	0.2	5	0.9850107066
	0.2	7	0.986616702355
Minkowski	0.1	5	0.983940042826
	0.1	7	0.9807280513918629
	0.2	5	0.98501070663
	0.2	7	0.98661670235
Manhattan	0.1	5	0.9828693790149893
	0.1	7	0.9817987152034261
	0.2	5	0.9850107066381156
	0.2	7	0.9855460385438972

Euclidean distance, Minkowski distance, and Manhattan distance. For each metric, the k value is tuned and also different test data sizes are considered. It is observed that as the k value increases, the over fitting takes place and accuracy starts dropping. So k value is taken between 5 and 7 for this experiment. The results obtained in this work are presented in the following table (Table 1).

The highest accuracy obtained is 98.66% when k value is 7 and test data size is 20%. So, for this combination, the following is the confusion matrix.

Figure 2 Plotted using matplotlib shows the testing performance of CNN-KNN model in multinomial classification on malware families. For this, the following is the report for precision, f1-score, and recall (Fig. 3).

The reported test accuracy is 98.66% states that CNN-KNN has the strongest predictive performance than the models reported in [3]. The presented model has a complex design. The 3 convolution layers, 3 max pool layers, and 1 fully connected dense layer are presented in the design. Its 7-layer design allows it to represent increasingly complex mappings between labels and features. KNN is used for classifying the given sample to one of the 25 malware classes.

5 Conclusion

In this paper, we used Malimg dataset [11] which is used to classify malware family and consists of malware images. Here we have employed deep learning-based method to classify. It has been observed that if different deep learning-based methods like GRU-SVM, CNN-SVM, MLP-SVM [7, 11] are applied on the same dataset, then among all, GRU-SVM can give maximum classification accuracy, i.e., 84.92%. In our proposed work, improved DL-based technique, i.e., CNN-KNN has given better

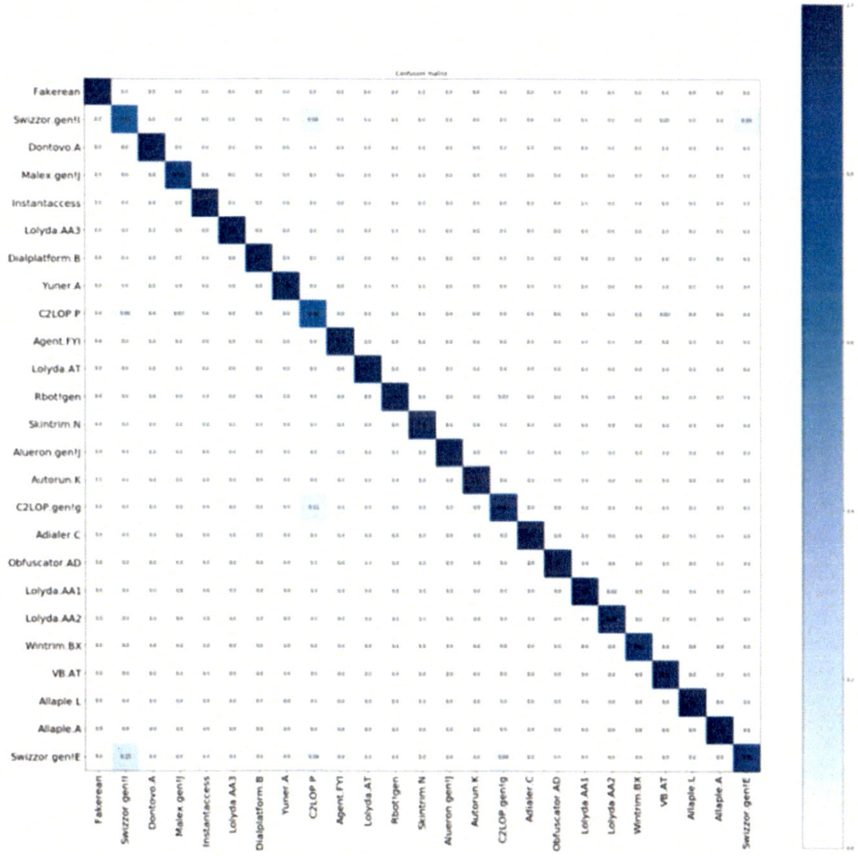

Fig. 2 Confusion Matrix for CNN-KNN testing results, showing its predictive accuracy for each malware family described in table

accuracy, i.e., 98.66% as compared to any other Deep Learning-based model. In our proposed model, better nonlinearities and hidden layers have been added along with optimized dropout which leads to better performance on the classification of malware. Such an approach may give good exposure to the information which may help to build an intelligent anti-malware system.

Fig. 3 Evaluation metrics for the KNN when distance metric is Euclidean, k = 7 and data size is 20%

	precision	recall	f1-score	support
0.0	1.00	1.00	1.00	82
1.0	0.76	0.73	0.74	22
2.0	1.00	1.00	1.00	36
3.0	1.00	0.92	0.96	26
4.0	1.00	1.00	1.00	91
5.0	1.00	1.00	1.00	20
6.0	1.00	1.00	1.00	35
7.0	1.00	1.00	1.00	157
8.0	0.87	0.79	0.83	34
9.0	1.00	1.00	1.00	23
10.0	1.00	1.00	1.00	33
11.0	0.97	1.00	0.98	31
12.0	1.00	1.00	1.00	18
13.0	1.00	1.00	1.00	30
14.0	1.00	1.00	1.00	24
15.0	0.90	0.95	0.92	38
16.0	1.00	1.00	1.00	27
17.0	1.00	1.00	1.00	26
18.0	0.98	1.00	0.99	49
19.0	1.00	0.96	0.98	26
20.0	1.00	0.92	0.96	25
21.0	1.00	0.97	0.98	96
22.0	1.00	1.00	1.00	331
23.0	0.99	1.00	1.00	562
24.0	0.80	0.92	0.86	26
accuracy			0.99	1868
macro avg	0.97	0.97	0.97	1868
weighted avg	0.99	0.99	0.99	1868

References

1. A malware detection model based on behavior analysis and KNN algorithm. Comput. Sci. Appl. **07**, 491–498 (2017). https://doi.org/10.12677/csa.2017.76060
2. Liu, W., Ren, P., Liu, K., Duan, H.-X.: Behavior-based malware analysis and detection. IEEE (2012)
3. Edem, S.: A study on the malware analysis with machine learning methods. IJRAR (2019)
4. Kim, K., Aminanto, M.E., Tanuwidjaja, H.C.: A survey on malware detection from deep learning. In: Springer Briefs on Cyber Security Systems and Networks. Springer, Singapore (2018)
5. Christodorescu, M., Jha, S.: Static analysis of executables to detect malicious patterns. In: Proceedings of the 12th Conference on USENIX. Security Symposium, Vol. 12 (SSYM'03). USENIX Association, Berkeley, CA, USA, pp. 12–12 (2003)
6. Nataraj, L., Karthikeyan, S., Jacob, G., Manjunath, B.S.: Malware images: visualization and automatic classification. In: Proceedings of the 8th International Symposium on Visualization for Cyber Security, pp. 1–7 (2011)
7. Cakir, B., Dogdu, E.: Malware classification using deep learning methods. In: Proceedings of ACMSE Conference, pp. 1–5 (2018)
8. RamakrishnaMurty, M., Murthy, J.V.R, Prasad Reddy, P.V.G.D.: Text document classification based on a least square support vector machines with singular value decomposition. Int. J. Comput. Appl. (IJCA) **27**(7), 21–26 (2012)
9. Pascanu, R., Stokes, J.W., Sanossian, H., Marinescu, M., Thomas, A.: Malware classification with recurrent networks. In: Acoustics. IEEE, pp. 1916–1920 (2015)

10. Sethi, K., Chaudhary, S.K., Tripathy, B.K., Bera, P.: A novel malware analysis framework for malware detection and classification using machine learning approach. In: Proceedings of the 19th International Conference on Distributed Computing and Networking, pp. 1–4 (2018)
11. Agarap, A.F.: Towards building an intelligent anti-malware system: a deep learning approach using support vector machine (SVM) for malware classification. arXiv preprint arXiv:1801.00318 (2017)
12. Krizhevsky, A., Sutskever, I., Hinton, G.E.: Imagenet classification with deep convolutional neural networks. In: Advances in Neural Information Processing Systems, pp. 1097–1105 (2012)
13. Burnap, P., French, R., Turner, F., Jones, K.: Malware classification using self organising feature maps and machine activity data. Comput. Secur. **73**, 399–410 (2018)
14. Gavriluţ, D., Cimpoeşu, M., Anton, D., Ciortuz, L.: Malware detection using machine learning. In: 2009 International Multiconference on Computer Science and Information Technology, pp. 735–741. IEEE (2009)
15. Zhou, H.: Malware detection with neural network using combined features. In: China Cyber Security Annual Conference 2018, pp. 96–106. Springer, Singapore
16. Idika, N., Mathur, A.P.: A survey of malware detection techniques. Technical report, Purdue University, February 2007
17. Popov, I.: Malware detection using machine learning based on word2vec embeddings of machine code instructions. In: 2017 Siberian Symposium on Data Science and Engineering (SSDSE). IEEE, pp. 1–4 (2017)

A Hybrid Deep Learning Approach for Detecting Zero-Day Malware Attacks

Shaik Moin Sharukh

1 Introduction

The twentieth century has witnessed a sheer dominance of the information society posing a major security concern. With the revolution lead by the information society, the production of IoT devices got amplified along with the increasing number of users. These billions of IoT devices generate an enormous amount of data, paving a path to data breaches. Since the mainstream users stand unaware of the default security settings in their devices, the cyber-criminals utilize the vulnerabilities to attack the devices with various malware and steal confidential data to obtain financial gains [1]. Malware is simply a code engendered by cyber-criminals to launch cyber-attacks and gain unauthorized access to various devices in a network. It has numerous variants like Trojan, worm, ransomware, command and control bot, adware, virus, and spyware [2]. Malware detection remains an unremitting process until the malware authors stop developing novel evasion techniques.

1.1 Research Background

The inception of anti-virus software was happened in 1987, to detect the existence of the first malware. Signature-based detection remained to be the foremost technique used in the anti-virus software to understand the behavior of the malware files. The signature-based detection techniques are evidenced to be ineffective to detect novel malware signatures, as they failed to bypass malware evasion techniques like stegosploit, code obfuscation, and code encryption [3]. To reverse engineer the novel

S. M. Sharukh (✉)
VNR Vignana Jyothi Institute of Engineering and Technology, Hyderabad, India
e-mail: moinsharukh001@gmail.com

C. K. Mai et al. (eds.), *Machine Learning Technologies and Applications*,
Algorithms for Intelligent Systems,
https://doi.org/10.1007/978-981-33-4046-6_20

malware signatures, the signature-based detection technique requires deep domain level knowledge which is time-consuming.

To countermeasure the malware evasion techniques, security researchers introduced machine learning algorithms to detect and categorize malware into their respective families. The Machine Learning Algorithms (MLAs) employ the domain level engineering and feature selection approaches to generate a separating plane between malware files and benign files [4, 5]. The features employed by MLAs are attained from dynamic and static analysis. If a code is inspected during execution, it falls into the class of dynamic analysis and if the same code is examined without execution it falls into the class of static analysis. When compared, dynamic analysis outperformed static analysis in distinguishing benign and malware signatures but has issues in time complexity.

For a machine learning algorithm to be successful, it requires rigorous training under various patterns of malware. Moreover, MLAs evidenced fading of outputs when enormous data is dumped, whereas deep learning seizures novel patterns and generates a relation with the longstanding patterns to attain better performance and results [6]. The lack of efficiency and accuracy in MLAs inspired the current research paper to explore the deep learning algorithms and propose an efficient architecture for malware detection.

1.2 Deep Learning Architectures

Artificial intelligence acts as the fountainhead for deep learning and machine learning architectures, as it has analogous functionalities that of a human brain. In this epoch of technology, the IoT devices generate huge amount of data and machine learning algorithms require domain level knowledge to preprocess the data and track down the malware. The deep learning architectures, such as Recurrent Neural Network (RNN) and convolution neural network, possess the competence to understand and process data in large amounts unlike machine learning algorithms [6].

In this paper, we are using deep learning algorithms such as CNN for geospatial data and Long Short-Term Memory (LSTM) to detect, classify, and categorize malwares into their respective malware families.

2 Implementation Methodology

In this paper, we implement several deep learning methodologies such as deep static analysis, deep image processing technique and proposed architecture for detecting malicious malware binaries.

2.1 Detection of Malware Binaries Using Deep Learning Methodology

The efficiency of specific machine learning algorithms such as Support Vector Machine (SVM) algorithm [13], Random forest, Decision Tree, Naive Bayes (NB), Logistic Regression, K-Nearest Neighbors (KNN) are evaluated along with two deep learning algorithms. Figure 1 depicts the working mechanism of the deep learning architecture used for splitting benign and malware binaries [9].

Initially, the dataset comprising both benign and malware binaries is uploaded where both raw bytes and domain level features are extracted from the dataset. The domain level features such as file name, file size, hashes, MD5 checksum, etc., are forwarded to the traditional MLAs for malware classification and detection. In a parallel method, the raw bytes are subjected to the deep learning methodologies for detecting, classifying, and categorizing malware binaries [5]. The voting methodology adopts both algorithms and finally classifies the binaries into two classes, namely benign and malware. The detected malwares are categorized into respective malware families using deep image processing technique dependent on deep learning methodology.

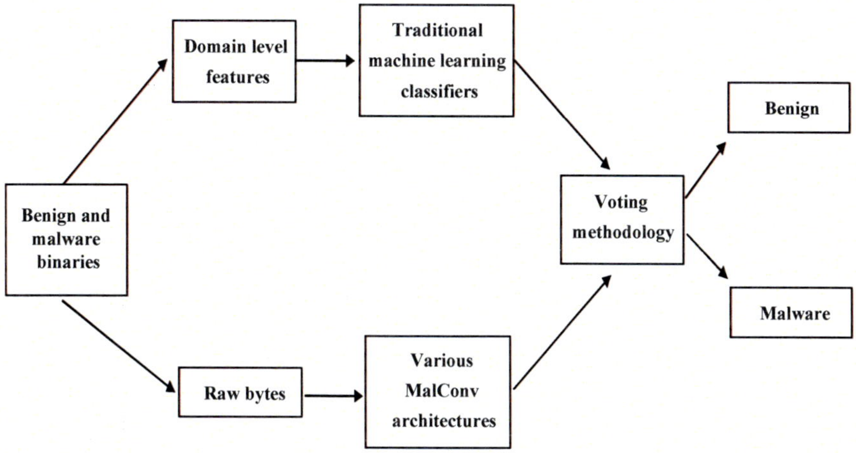

Fig. 1 Deep learning methodology for splitting benevolent and malicious malware binaries

2.2 *Categorization of Detected Malware Using Deep Image Processing Technique Dependent on Deep Learning Methodology*

The CNN alongside LSTM [16] forms a hybrid pipeline to categorize malwares based on image processing technique [7]. In this paper, we use visualization for categorizing malware into their respective families by evading the feature engineering segment. Unlike static analysis, the image processing technique uses raw bytes of information which makes it faster and added to that it can completely evade the execution phase [8] and [9]. The proposed image processing technique is compatible with malwares derived from various operating systems like windows, Linux, Android, etc. Figure 2 portrays the deep image processing technique [9].

The malware binaries in the dataset exhibit three colors, namely black, white, and gray [17]. The black, white colors in the grayscale image portray 0 and 255, respectively. Malware binaries with transitional shades of gray fall in between [0–255]. If the image comprises only black color, it means that the grayscale image holds only 0's in it and if the image is subjected to white color it portrays that the majority part of the image comprises the number 255 [7].

Description of Dataset

In this paper, we used maling dataset which comprises 9,339 malware signatures categorized in 25 different families. For training and testing the dataset, we have

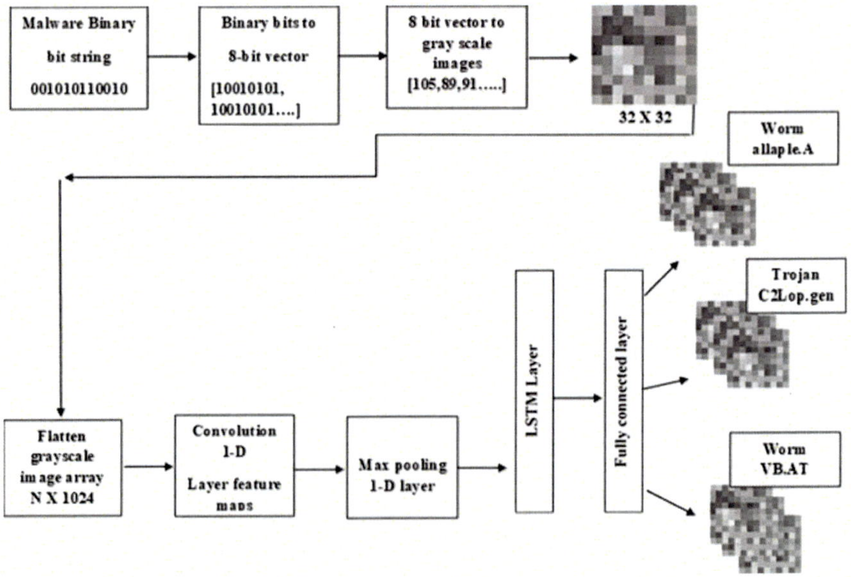

Fig. 2 Deep learning architecture based on image processing

separated the dataset into two sections [11]. The primary segment comprises 80% of malware data, used for training the dataset and the subsequent segment comprises the remaining 20% of the malware data for testing the malware binaries [9]. The two parts of the dataset comprise malware signatures. Initially, the data was existing in the format of malware binaries later these were converted into matrix format, i.e., 8-bit unsigned integer. As discussed above, these matrices are visualized as a picture or gray-scale image. After visualizing the picture in a 2D matrix, it is transformed into a 1D vector form which results in the formation of a 1024 sized array [9].

3 Proposed Architecture—DLMDN

An outline of our proposed architecture Deep Learning Malware Detection Network (DLMDN) is depicted in Fig. 3. The proposed framework has a sequential procedure partitioned into five steps, to detect the malware [9, 10].

Initially, the collected data from the maling dataset is partitioned using. Exe parser [3] and in the subsequent step the malware samples are subjected to pre-processing.

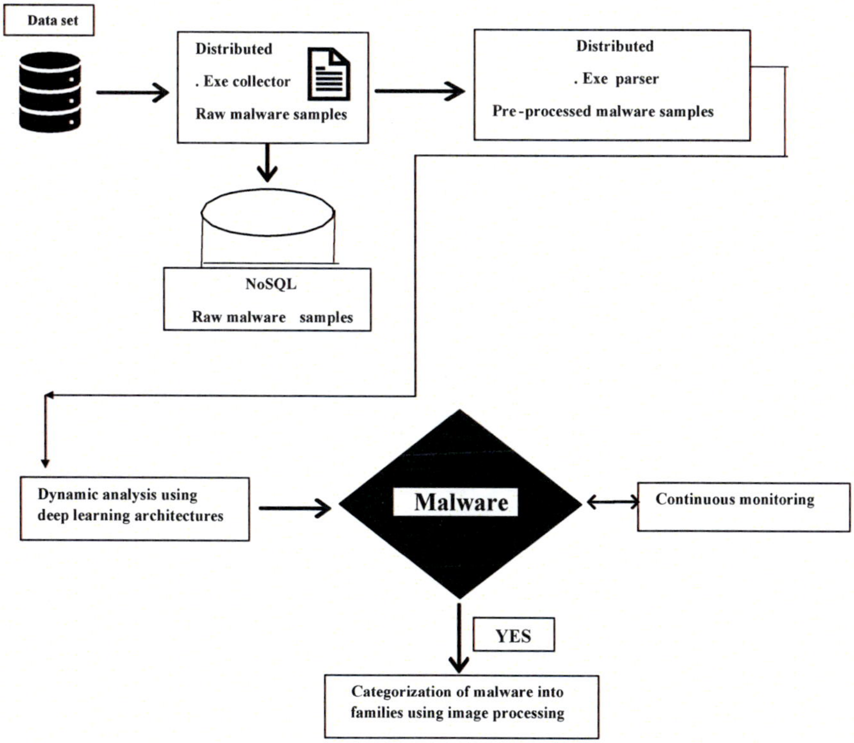

Fig. 3 Proposed architecture—DLMDN

To distinguish between the benevolent and malicious malware binaries, the pre-processed data is progressed to a voting methodology based on machine learning and deep learning algorithms [12].

As shown in the above figure, the detected malware is continuously monitored. Using the deep image processing technique, the detected malwares obtained in the binary format are transformed into matrix format, i.e., 8-bit unsigned integer [14]. These unsigned 8-bit integers are visualized as greyscale images in a 2D matrix which is converted into 1-D vector format resulting in a 1024 sized array. Upon thorough training, these malwares are categorized into their respective malware families [15].

4 Results

The results obtained in this paper showcases the brightside of the deep learning architectures by outperforming the machine learning algorithms. As shown in Fig. 4, the Convolutional Neural Network (CNN) algorithm has surpassed many MLA's algorithms such as Support Vector Machine (SVM), Naive Bayes, Decision tree, etc.

The Convolutional Neural Network (CNN) has gained the uppermost accuracy rate among all algorithms in detecting malware samples. The superlative algorithm for detecting novel malware signatures is determined by assessing major factors like accuracy rate of detection, precision, recalling factor, and F-score. Table 1 depicts the obtained results for malware detection.

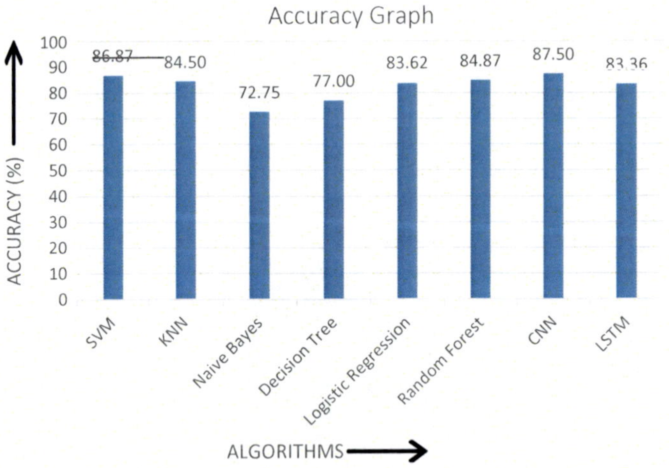

Fig. 4 Accuracy graph of MLA and deep learning algorithms

Table 1 Detailed Test results

Algorithms	Accuracy (%)	Precision (%)	Recall (%)	F-score (%)
SVM	86.87	86.77	86.80	86.79
KNN	84.50	84.49	84.55	84.52
Naive Bayes	72.75	72.70	72.79	72.77
Decision tree	77.00	77.10	76.35	77.26
Logistic Regression	83.62	83.65	83.70	83.68
Random Forest	84.87	84.90	84.87	84.86
CNN	87.50	87.49	87.51	87.50
LSTM	83.36	83.38	83.35	83.34

5 Conclusion

We conclude the paper by proposing a novel framework called DLMDN for evaluating both the MLAs and deep learning methodologies. The proposed framework has a sequential procedure partitioned into five steps, to detect the malwares. It consists of collecting raw malware samples, parsing the data, pre-processing, detecting, and categorizing the malware into respective malware families. The categorization of malwares is performed by image processing technique. This paper has proved the supremacy of deep learning methodologies over MLAs in terms of accuracy for detection of novel malwares, precision rate, recalling factor, and F-score.

References

1. Tang, M., Alazab, M., Luo, Y.: Big data for cyber security: Vulnerability disclosure trends and dependencies. IEEE Trans. Big Data, to bepublished
2. Alazab, M., Venkatraman, S., Watters, P., Alazab, M., Alazab, A.: Cybercrime: The case of obfuscated malware. In: Georgiadis, C.K., Jahankhani, H., Pimenidis, E., Bashroush, R., Al-Nemrat, A. (eds.) Global Security, Safety and Sustainability & e-Democracy (Lecture Notes of the Institute for Computer Sciences, Social Informatics and Telecommunications Engineering), vol. 99. Springer, Berlin, Germany (2012)
3. Raff, E., Barker, J., Sylvester, J., Brandon, R., Catanzaro, B., Nicholas, C.: Malware detection by eating a whole .exe" [online] available https://arxiv.org/abs/1710.09435. (2017)
4. Rhode, M., Burnap, P., Jones, K.: Early-stage malware prediction using recurrent neural networks. Comput. Secur. **77**, 578594 (2018)
5. Shibahara, T., Yagi, T., Akiyama, M., Chiba, D., Yada, T.: Efficient dynamic malware analysis based on network behavior using deep learning. In: Proceedings of the IEEE Global Commun. Conf (GLOBECOM), pp. 1–7 (2016)
6. LeCun, Y., Bengio, Y., Hinton, G.: Deep learning. Nature **521**(7553), 436–444 (2015)
7. Ni, S., Qian, Q., Zhang, R.: Malware identification using visualization images and deep learning. Comput. Secur. **77**, 871–885 (2018)
8. Sun, G., Qian, Q.: Deep learning and visualization for identifying malware families. IEEE Trans. Depend. Secure Comput. to bepublished

9. Vinaykumar, R., Alazab, M., Soman, K.P., Poornachandran, P., Venkatraman, S.: Robust intelligent malware detection using deep learning. IEEE Access. **7** (2019)
10. Raff, E., Sylvester, J., Nicholas, C.: Learning the PE header, malware detection with minimal domain knowledge. In: Proceedings of the 10th ACM Workshop Artif. Intell. Secur (pp. 121–132). ACM, New York, NY, USA (2017)
11. Conti, G., Dean, E., Sinda, M., Sangster, B.: Visual reverseengineering of binary and data files. In: Visualization for Computer (pp. 1–17). Springer, Berlin, Germany (2008)
12. Alazab, M.: Profiling and classifying the behavior of malicious codes. J. Syst. Softw. **100**, 91–102 (2015)
13. Huda, S., Abawajy, J., Alazab, M., Abdollalihian, M., Islam, R., Yearwood, J.: Hybrids of support vector machine wrapper and filter based framework for malware detection. Future Gener. Comput. Syst. **55**, 376–390 (2016)
14. Krcál, M., Švec, O., Bálek, M., Jašek, O.: Deep convolutional malware classifiers can learn from raw executables and labels only. [Online]. Available: https://open-review.net/forum?id=HkHrmM1PM (2018)
15. Saxe, J., Berlin, K.: Deep neural network based malware detection using two dimensional binary program features. In: Proceedings of the 10th International Conference on malicious Unwanted Software (Malware) (pp. 11–20) (2015)
16. Pascanu, R., Stokes, J.W., Sanossian, H., Marinescu, M., Thomas, A.: Malware classification with recurrent networks. In: Proceedings of the IEEE International Conference Acoustician Speech Signal Process (ICASSP), (pp.1916–1920) (2015)
17. Nataraj, L., Karthikeyan, S., Jacob, G., Manjunath, B.S.:Malware images: Visualization and automatic classification. In: Proceedings of the 8th International Symposium on Vision Cyber Security (p. 4). ACM, New York, NY, USA (2011)

Glaucoma Detection Based on Deep Neural Networks

Madhav Kode, Ruthvika Reddy Loka, Laasya Garapati, Yashna Lahari Gutta, and Ravikanth Motupalli

1 Introduction

The human eye is considered a chief part among organs as it works as the source of the sensation for sight. The eye's retina is the most important and critical part of the human vision system. Glaucoma is a serious disease that damages the retina thereby causing blindness. As per the survey present in https://www.ncbi.nlm.nih.gov/pubmed/24974815, Glaucoma is predicted to affect a population of around 76.0 million by 2020 extending to 111.8 million by the year 2040. Until it reaches an advanced stage, Glaucoma usually does not exhibit any kind of signs or symptoms. By the time it is detected, the damage becomes significant and irreversible, with the optic nerve damage resulting in vision reduction caused due to around 40% loss of axons. However, it may be possible to delay the vision impairment caused by Glaucoma if it gets diagnosed sufficiently early. The causes of Glaucoma are for the most part linked with the development of Intra-Ocular Pressure (IOP) in the eyes that stems from blockage in intraocular fluid's drainage flow. The increased IOP damages the optic nerve which carries visual sensory information to the brain from the eye. Damage caused to optic nerve fiber weakens the visualization ability and object

M. Kode (✉) · R. R. Loka · L. Garapati · Y. L. Gutta · R. Motupalli
Department of CSE, VNR VJIET, Hyderabad, Telangana, India
e-mail: madhavkode007@gmail.com

R. R. Loka
e-mail: ruthvikareddy2020@gmail.com

L. Garapati
e-mail: srilaasyagarapati2710@gmail.com

Y. L. Gutta
e-mail: yashnagutta27@gmail.com

R. Motupalli
e-mail: ravikanth_m@vnrvjiet.in

C. K. Mai et al. (eds.), *Machine Learning Technologies and Applications*,
Algorithms for Intelligent Systems,
https://doi.org/10.1007/978-981-33-4046-6_21

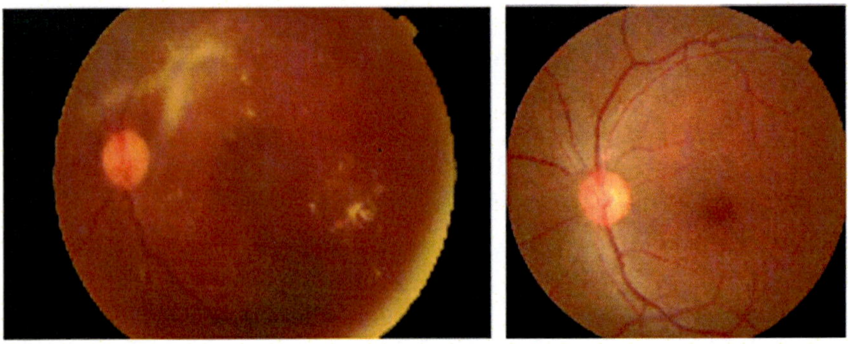

Fig. 1 Fundus image **a** Glaucoma **b** Non-Glaucoma

recognition that may lead to blindness. Figure 1 shows Glaucoma, non-Glaucoma in (a) and (b).

Numerous research scholars have employed a variety of methods to identify Glaucoma disease, but among all, the convolution neural networks method has given better results. So, in the system that we propose, Convolution Neural Networks (CNNs) are used to build the model which predicts Glaucoma from just an image given as input. Replacing the usage of handcrafted features, a CNN can be utilized as it learns a hierarchy of features, which can well be used for the purposes of image classification. As the hierarchy approach is available to learn more complex features, as well as translation and distortion features in higher layers, the accuracy of the CNN-based image classification method can be higher. Based on this assumption, we explore the use of the CNN-based method for Glaucoma test in this work.

The core intent of the proposed approach is to classify the images and get the scale of severity to what extent Glaucoma disease has affected the eye. To procure such composite classification of images, a robust and enormous training is essential. Hence our approach is to apply deep convolutional neural networks to perform training and testing. Apprehending the deep features of the disease effectively based on deep CNN is our main interest.

The later part of this paper is categorized as follows: Sect. 2 gives out an overview of previous work done in relative fields, Sect. 3 describes the training methods used in the proposed course of action along with the architecture of the CNN, Sect. 4 presents the results achieved out of our experiments, Sect. 5 ends the paper with discussion on the results obtained and the scope of advancements that can be brought in future.

2 Literature Survey

Septiarini et al. [1] have considered Retinal Nerve Fiber Layer (RNFL) loss as a main parameter for Glaucoma detection, which is not very common. The fundus image is processed from the gray level matrix of co-occurrence for feature extraction, such that the Optic Nerve Head (ONH) region is removed and the rest of the RNFL is divided into sub-sectors based on ISNT regions.

Support Vector Machines (SVMs) have been used by Sarma et al. [2], which have been memory efficient. SVMs are also observed to be highly effective when dealing with high dimensional spaces. Phase information seems to be amiss, as it is not preserved by the Radon Transform. Moreover, some information contained in the image is lost by used projections. RT further increases mathematical complexity and introduces error. They have achieved a classification accuracy of 98.8 and 95%.

Zhao et al. have implemented a combination of unsupervised learning for the purpose of feature representation in [3] and performed supervised learning for CDR regression. Feature extraction and representation are achieved by using MFPPNet which employs three dense connectivity blocks along with pyramid pooling. Dataset used for testing has 934 images from 443 clinical subjects, validation performed on ORIGA dataset with an AUC of 0.90.

In [4], Khairina, MKom et al. have taken statistical features for classification like the mean, entropy, and 3rd moment. Smoothness, uniformity, and standard deviation were also considered. They have extracted these features and fed them to KNN classifier to perform classification. The dataset they used has 84 images with an accuracy of 95.24%.

Pavithra et al. [5] have proposed a system that can be easily implemented on hardware kits which may in turn be connected directly to the optical instruments and prediction can be done along with diagnosis. They process the image using histogram equalization, finding the ROI and finally estimating the Cup-to-disc ratio, based on which prediction takes place.

The authors Atheesan et al. have achieved optic disc segmentation using observations of vasculature present in the retinal area, through red channel analysis in [6]. Clustering is applied first using Simple Linear Iterative Clustering (SLIC) followed by k-Means Clustering algorithm, along with Gabor filter for edge detection. Prediction is done using CDR. The dataset employed is of 100 images with an F-score value maintained at 96%.

An et al. [7] have employed parametric inferences of area swept by the curve characterizing receiver operations (AUC) with a cross-validation of 10 folds on a random-forest classifier. This RF was derived from color fundus images, RNFL thickness maps, GCC macular maps showing thickness, disc maps showing deviation in RNFL, and GCC macular deviation maps. They have adopted VGG19 CNN architecture consisting of 19 layers.

Adaptive Neuro-Fuzzy Inference, hemorrhage detection have been used in [8, 9], respectively. In [10], vessel structure segmentation of colored retinal images is employed.

Unlike existing methods, in which features were handcrafted from the optic disk, in our method extraction of the features is automatically done by CNN from raw images, which are eventually fed to the classifier. Then, the classifier performs classification on the images into respective labels (No Glaucoma, Mild, Moderate, Severe, Proliferative Glaucoma). Analysis of the ophthalmic images manually is a time-consuming procedure. Also, the accuracy varies with the variation in expertise and skill of the professionals. Our method offers way better responses to the drawbacks involved in manual processes of Glaucoma perception. Most established detection techniques need either selection of features or very accurate measurements of geometric ONH structures, which is CDR. The phases of detection, analysis, diagnosis, treatment time, and prevention of associated risks can be made efficient using automation. Convolutional Neural Networks (CNNs) and other Deep Learning (DL) systems are proved to be useful.

3 Objectives

- To propose a system which checks and detects for possible occurrences of Glaucoma whenever a fundus image is generated by a technician without the intervention of a skilled person like a doctor.
- To preprocess the available images so as to extract and focus on the ROI.
- To diagnose correctly on the image whether it is Glaucoma or not and if yes, measuring the severity of the effect using a deep learning model (Fig. 2).

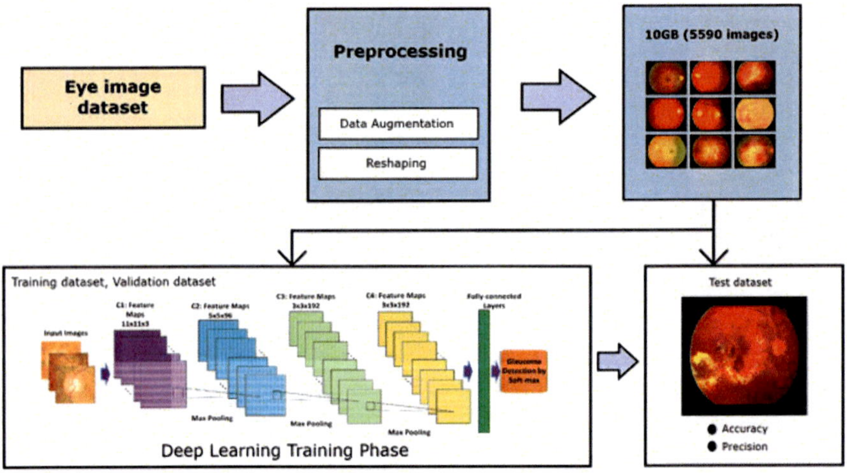

Fig. 2 The complete architecture of the system used

4 Proposed System

4.1 Dataset

The high-resolution Funduscopic images of the retina taken from the data science community website named Kaggle (https://www.kaggle.com/) are considered under the datasets used for the process. Since our project is in the bio-medical domain and requires accurate results, we have chosen a huge dataset. Each image is rated for the severity of the disease on a scale of 0–4, as portrayed in Fig. 3. (0 >> No Glaucoma, 1 >> Mild, 2 >> Moderate, 3 >> Severe, 4 >> roliferative Glaucoma). There are 5,590 images in the dataset. All the images are divided into train, test, and validation sets. High-quality training dataset is useful for training the classifier model but not all the images in our data are labeled. For this reason, we have considered a total of 3,362 images which are labeled. A normal computer wouldn't be able to accommodate such huge data due to low computation power. Hence, we have used Google Colab as a platform.

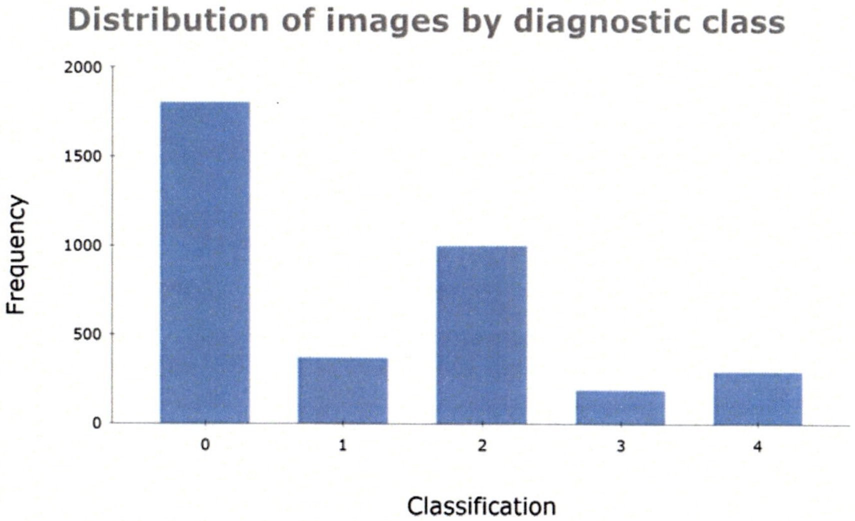

Fig. 3 Distribution of images by diagnostic class

4.2 Preprocessing

Patients of varying age groups, gender, nationality are considered for fundus photography performed at extremely distinct levels of lighting and images are collected to form the dataset. Considering such varied data leads to the change in pixel dimensions within the images which might create futile variations. In order to negate this, image pre-processing methods like cropping the image, color normalization, making the dimensions, data augmentation, etc., were performed on the images. The image data is converted to a hierarchical data format to facilitate preprocessing, followed by data augmentation, and then trained accordingly. Images were increased in real time for improvisation in the capability of network localization and also supporting reduced overfitting. Techniques like horizontal flip, vertical flip, zoom in of 20% were performed to make the data uniform since the number of images for each type was not distributed equitably in the dataset. As observed in Fig. 2, data augmentation makes the model more robust to slight variations, and hence prevents the model from Overfitting. The dataset has been formatted to 128 * 128 pixels which enabled the retention of intricate features that are to be identified.

4.3 Training and Testing

Initial training was applied on the CNN until it gained remarkable progress. This was a time-saving method executed to achieve a comparatively quick classification result with no delay. After 2 epochs of training on the images, the network was then trained for a further 5 epochs. We used a residual learning framework to ease the training of networks that are substantially deeper. Neural networks suffer from severe over-fitting. To solve this issue, we stopped training the neural network early before it over fitted the training dataset and finally improved the generalization of deep neural networks.

The proposed CNN architecture is shown in Fig. 4. This model consists of an input layer with input size 128 * 128, and it is followed by convolution layers with different window sizes followed by an activation function and max-pooling layers of pool size 3×3. Convolution 2D are filters, which specifies the number of filters to use, kernel size is length and breadth of the kernel used, padding specifies whether to use an extra layer of zeros around the image or not. Here, we have used two activation functions 1. ReLu, 2. Softmax. These functions enable to decide whether a neuron should be activated or not. These are also used to add non-linearity to the output of neurons. Max pooling is used to reduce the spatial volume of input image by taking the largest element from the rectified feature map. Batch Normalization is used to normalize the activations of each layer by transforming the inputs to be mean 0 and unit variance. It helps in regularizing the model. Then, we apply the Flatten function to convert a 2D array to 1D array. To eliminate overfitting, we drop 50% neurons using the Dropout function. Dense layers and Softmax regression are used in the

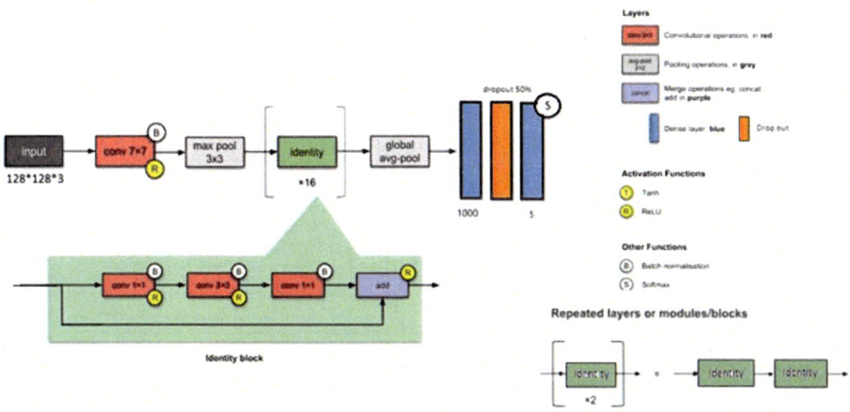

Fig. 4 CNN architecture

classification stage. Softmax activation function is used to generate probabilities of each class using the output neurons. We used Adam optimizer.

4.4 *Transfer Learning*

It's a complex task to train a CNN from the beginning. It requires a huge measure of labeled information—which can become a challenging problem. Besides, the computational resources required are large in scale. However, a substitute to train a CNN without any preparation comprises fine-tuning an existing CNN that has been trained utilizing a comparatively larger labeled dataset from a different application (e.g., ImageNet). Transfer Learning is the process of overcoming the isolated learning worldview and using knowledge earned for one use case to some other related ones. It is, in essence, using a model that was designed for some purpose and implementing it for a different purpose. Transfer Learning likewise applies to adjusting a pre-trained ANN to perform a new function. The use of pre-trained deep CNNs and subsequent fine-tuning of the weights of the network applying the new labeled images could lead to even better performance metrics and a potential reduction in training resources in terms of time, memory and computational operations, as depicted in Fig. 5.

5 Result

This project classifies the Glaucoma disease and shows the severity of the disease. We have trained our model on a total of 3,662 images. The images in the training set are divided as No Glaucoma—1805 images, Mild—370 images, Moderate—999 images,

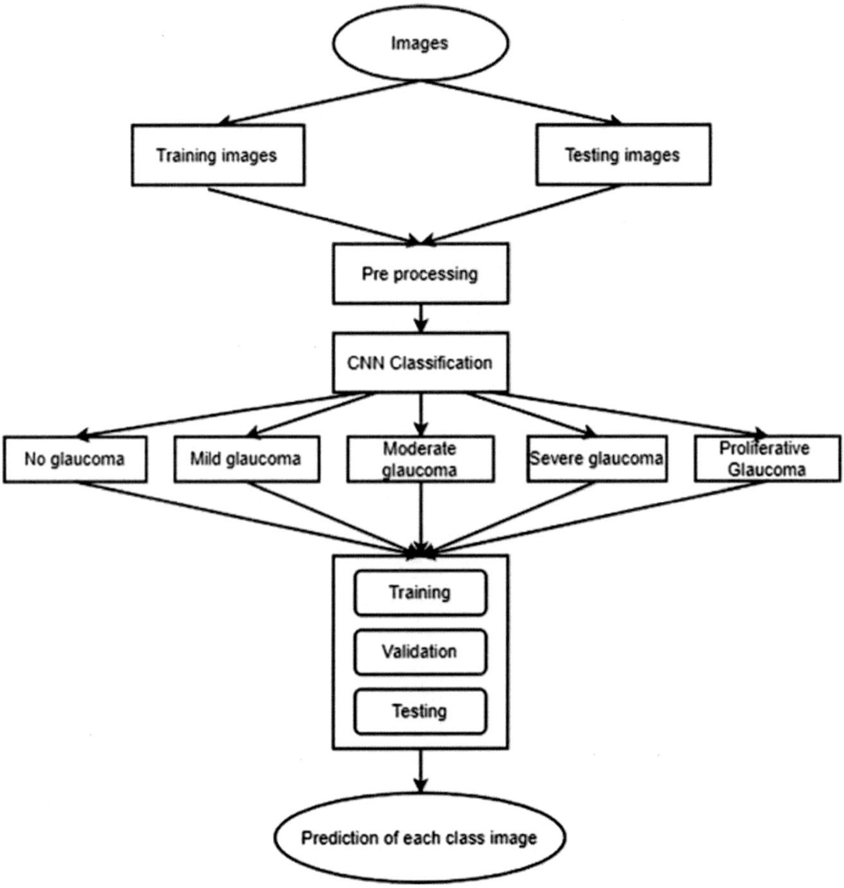

Fig. 5 Flow chart of the proposed system

Severe—193 images, Proliferative—295 images. The accuracy was calculated using the following equation.

Accuracy = (Number of accurate Predictions/Total number of Predictions).

By checking the precision and accuracy we have got the below scores for each:

Precision score—82.76839723142821%
Accuracy score—89.57219251336899%

In the interest to analyze the implementation of our CNN classifier, we have adopted the measure of accuracy. We have used 1928 images for Glaucoma detection in which we have achieved training accuracy of **93.41%** and validation accuracy of **90.96%**. Figure 6 shows the graph of model accuracy against train data and validation data.

Efficiency parameter of a classifier and its capability are described by the confusion matrix. This matrix has information about the predicted and actual classifications

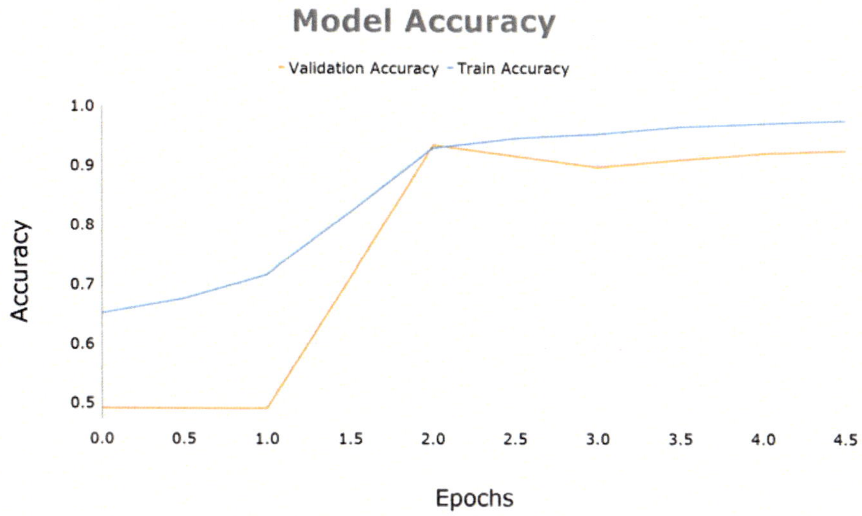

Fig. 6 Model accuracy against train data and validation data

performed by a classification system. Our proposed classification model's confusion matrix consists of five rows and columns. The performance measure parameter of the classification model in the process of classifying Glaucoma has been depicted in the matrix. The confusion matrix obtained as the output is shown in Fig. 7. The rows of the matrix in the order of top to bottom (rows 1–5) correspond, respectively, to actual classes, in the order of severity: no, mild, moderate, severe, and proliferative. The columns of the matrix from left to right (columns 1–5) correspond to predicted classes in the same order of increasing severity: no, mild, moderate, severe, and proliferative.

From a set of 187 non-Glaucoma images, 173 were predicted correctly as "No Glaucoma" thereby achieving an accuracy of 92.51%.

From a set of 37 mild-Glaucoma images, 30 were predicted correctly as "Mild Glaucoma" thereby achieving an accuracy of 81.08%.

From a set of 100 moderate-Glaucoma images, 88 were predicted correctly as "Moderate Glaucoma" thereby achieving an accuracy of 88%.

From a set of 20 severe-Glaucoma images, 16 were predicted correctly as "Severe Glaucoma" thereby achieving an accuracy of 80%.

From a set of 30 proliferative-Glaucoma images, 28 were predicted correctly as "Proliferative Glaucoma" thereby achieving an accuracy of 93.33%.

6 Conclusion

To conclude, automation aids in prediction, prevention, and early diagnosis of the risks associated with the disease. We presented a deep neural network framework to detect Glaucoma and also stages of its severity. In this paper, a structure of deep

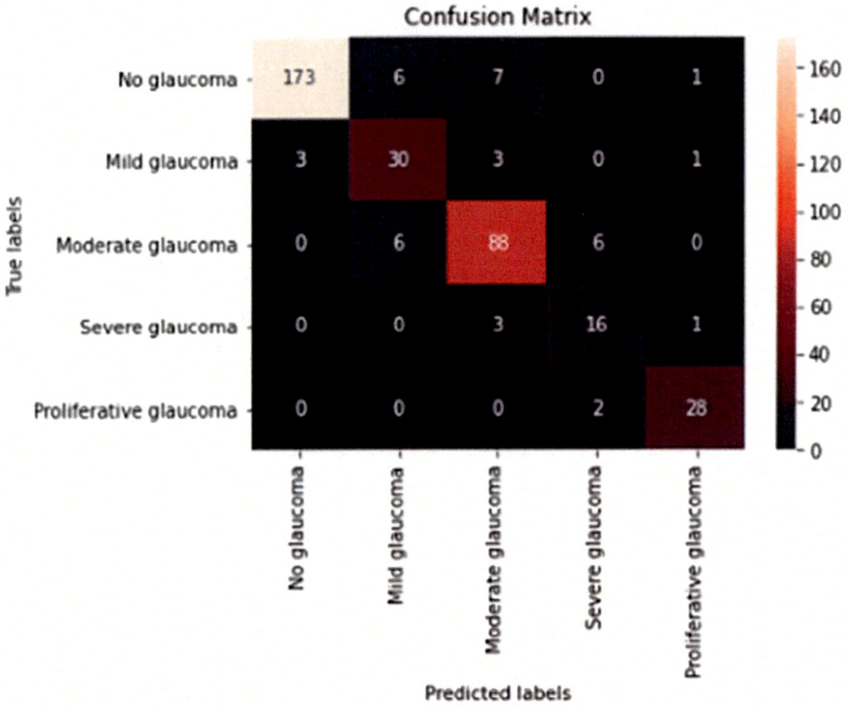

Fig. 7 Confusion matrix

learning for Glaucoma disease detection relies on significant CNN, which can get the discriminative features that better depict the hidden models related to Glaucoma. An accuracy percentage of **93.41%** is achieved through this technique which shows its efficiency to contribute in the process. Future scope of this research work is to improve the accuracy with enhanced cost effectiveness and also to implement a procedure of checking every fundus image generated in the hospital for Glaucoma so that a patient can take necessary steps at the earliest.

References

1. Septiarini, A., et al.: Automated detection of retinal nerve fiber layer by texture-based analysis for Glaucoma evaluation. Healthc. Inf. Res. **24**(4), 335–345 (2018). https://doi.org/10.4258/hir.2018.24.4.335
2. Sharma, R., Sircas, P., et al.: Automated Glaucoma detection using center slice of higher order statistics. J. Mech. Med. Biol. **19**(01), 1940011 (2019). https://doi.org/10.1142/S0219519419400116
3. Zhao, R., Chen, X., Xiyao, L., Zailiang, C., Guo, F. Li, S.: Direct cup-to-disc ratio estimation for Glaucoma screening via semi-supervised learning. IEEE J. Biomed. Health Inf.

4. Septiarini, A., Khairina, D.M., Kridalaksana, A.H., Hamdani, H.: Automatic Glaucoma detection method applying a statistical approach to fundus images. Healthc. Inf. Res. **24**(1), 53–60 (2018). https://doi.org/10.4258/hir.2018.24.1.53
5. Pavithra, G., Anushree, G., Manjunath, T.C., Lamani, D.: Glaucoma detection using IP techniques. In: 2017 International Conference on Energy, Communication, Data Analytics and Soft Computing (ICECDS), Chennai, pp. 3840–3843 (2017)
6. Atheesan, S., Yashothara, S.: Automatic Glaucoma detection by using funduscopic images. In: 2016 International Conference on Wireless Communications, Signal Processing and Networking (WiSPNET), Chennai, pp. 813–817 (2016)
7. An, G., Omodaka, K., Hashimoto, K., Tsuda, S., Shiga, Y., Takada, N., Kikawa, T., Yokota, H., Akiba, M.: Glaucoma diagnosis with machine learning based on optical coherence tomography and color fundus images. J. Healthc. Eng. (2019)
8. Diptu, N.A., et al.: Early detection of glaucoma using fuzzy logic in Bangladesh context. In: 2018 International Conference on Intelligent Systems (IS), Funchal—Madeira, Portugal, pp. 87–93 (2018)
9. Sengar, N., Dutta, M.K., Burget, R., Ranjoha, M.: Automated detection of suspected Glaucoma in digital fundus images. In: 2017 40th International Conference on Telecommunications and Signal Processing (TSP), Barcelona, pp. 749–752 (2017)
10. Odstrcilik, J., Jan, J., Kolar, R., Gazarek, J.: Improvement of vessel segmentation by matched filtering in colour retinal images. In: 2009 Springer IFMBE Proceedings of World Congress on Medical Physics and Biomedical Engineering, Munich, Germany, pp. 327–330 (2009)

Early Detection of Sepsis on Clinical Data Using Multi-layer Perceptron

N. Venkata Sailaja, Meghana Yelamarthi, Yendluri Hari Chandana, Prathyusha Karadi, and Sreshta Yedla

1 Introduction

Sepsis is a hazardous condition that happens when the body's reaction to contamination causes tissue harm, organ failure, or even demise of the person. Generally, the body releases natural synthetics into the circulation system in order to counterbalance the infection which is inside. Sepsis occurs when the body's response to these chemicals is out of balance, and this can damage many organ systems. Sepsis is caused by infection and can happen to anyone. It is most common and dangerous for senior citizens, pregnant ladies, kids below one-year old, persons suffering from chronic conditions, such as diabetes, kidney disease, lung disease, or even cancer, as they have weak immune systems. This disease is a major health concern for the public in terms of morbidity, health care expenses, and mortality. Detecting at early stages, with antibiotic treatment the outcomes can be improved. Though many professional care societies have proposed new methods in recognizing sepsis, the central requirement for early identification and treatment remains neglected. It can be treated if it can be recognized at early stages. Several examinations have demonstrated that

N. V. Sailaja (✉) · M. Yelamarthi · Y. H. Chandana · P. Karadi · S. Yedla
Department of CSE, VNR Vignana Jyothi Institute of Engineering and Technology, Hyderabad, India
e-mail: sailajanv@vnrvjiet.in

M. Yelamarthi
e-mail: ymeghana2464@gmail.com

Y. H. Chandana
e-mail: chandana6620@gmail.com

P. Karadi
e-mail: prathyushakaradi.29@gmail.com

S. Yedla
e-mail: sreshtareddyyedla@gmail.com

C. K. Mai et al. (eds.), *Machine Learning Technologies and Applications*,
Algorithms for Intelligent Systems,
https://doi.org/10.1007/978-981-33-4046-6_22

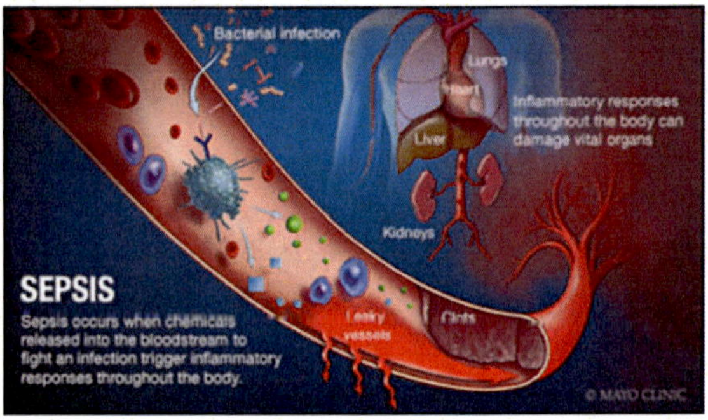

Fig. 1 Image of sepsis

delays in finding and treatment of sepsis can prompt high death rates. Our main motto is to detect sepsis as soon as the patient visits the emergency department for the treatment [16] (Fig. 1).

1.1 Causes of Sepsis

Any infection can trigger sepsis, but the following types of infections are more likely to cause sepsis:

- Pneumonia: It is a type of infection that causes inflammation in the air sacs called alveoli in the lungs, the bacteria or virus or fungi fills this with fluid which makes breathing difficult.
- Abdominal infection: It surrounds a number of infectious processes, including peritonitis, cholecystitis, diverticulitis, pancreatitis, and cholangitis.
- Kidney Infection: It generally results from an infection in the urinary tract that spreads to 1 or both the kidneys, and this can be chronic or sudden.
- Bloodstream Infection: It is an infection that occurs when bacteria are in the circulatory system. It generally describes bacteremia or sepsis.

1.2 Symptoms of Sepsis

The major symptoms of this include that a person having fever, i.e., temperature above 38 °C or even the temperature below normal temperature, i.e., 36 °C, with a heart rate greater than 90 beats/min, breathing higher than 20 breaths/min or it could be confirmed or chances of infection.

1.3 Statistical Information

Population polls found low population perception of sepsis and its impact ranging from 14% in Brazil to 40% in Australia. Sadly, public awareness data is not accessible for India. A 2016 study reported that almost 30% of patients have been admitted to ICUs in India who came out to have sepsis; it also showed that one out of three of those patients died. Research has identified a heavy sepsis burden in pregnant mothers and the neonate. A study conducted in 2017 by leading journal Lancet recorded that communicable diseases (infections) led to a significant proportion of deaths in India.

2 Literature Review

Sepsis is an infection—chronic inflammatory disorder—typically results from the spread of a localized nidus to the systemic circulation, and both with very high deaths and morbidity levels correlated [1].

Early and reliable projections of sepsis will enable further agitation and tailored treatment while antimicrobial stewardship is preserved. Established detection approaches are badly implemented and need laboratory test tests, sometimes overtime [11].

Better health outcomes have been related to MLA. That's the primary randomized random experimental managed frame in septic reconnaissance to demonstrate observable comparisons and clinical deaths [10].

Recently, automated testing has proved to save lives. Tackling and researching is also seen in ICU patients [5].

Under these conditions, beneficial local inflammatory processes, intervened by specific white platelets. The course by which a patient advances either to death or medical clinic release is notable [4].

On using The Dynamic Bayesian Networks, a time-probabilistic method of predicting a network utilizing patient data admitted into an emergency room, assessed the precision of diagnosis of sepsis in the first six hours after entry [1].

In the light of initial studies in comparison to HC and septic sequence sets, there were 42-gender markers that spoke of important intrinsic and diversified resistance capacities, cell cycling, differentiation of wireless connectivity, additional cell remodeling, and immune modulation pathways. A LogitBoost algorithm has been used to construct a symptomatic learning guideline for predicting septic series outcomes. The accuracy was around 86% [2].

All factors which are significant for prediction (i.e., predictor variables) utilized for the examination show the distinction between both sepsis and non-sepsis patients. In patients, without sepsis, the mean temperature for hypothermia was considerably less. The risk of sepsis was 2,126 for patients with 38 °C or higher temperatures [3].

A model, which was built on combining both boosting and bagging tree models (lightgbm, xgboost, and random forest) has been built to predict on the basis of patient hourly data reports the best performance achieved was 79.2% [6].

In this analysis of proof of definition, AI proposes a close-by huge solution to overcome current CDRs as well as standard strategies for estimating ED patients with septic disease mortality in the hospital. The viability of this methodology should be tentatively evaluated and whether further research should turn this into better clinical results for high-risk patients with sepsis. The approaches created to support, for example, another model for detailed crash tests that can be robotized and applied for certain clinical results of a plot and submitted to EHRs for local clinical predictions [8].

Three specific approaches to describe the obligations of this document: Improved execution by utilizing field detail extraction, Check the capabilities to extract deep neural systems by correlating with reference highlights, and Enhanced execution with LSTM, a neural network architecture feed for neural networks that are able to know patterns [7].

When using LSTM to diagnose the septic shock early, patients are identified by identical highlights and aim meanings up to 20 h earlier than the Cox relative hazard model, with equal affectability and explicitly. This result is significant as early detection and treatment of the septic condition is necessary to improve the stamina capacity of the patient [9].

The LSSVM proposed was tested using a fivefold cross-validation technique to execute with 2 separate kernels: the cubic-polynomial and the Gaussian radial base (RBF). The analysis revealed that LSSVM with RBF kert was a successful classifier to classify the development of sepsis syndrome with an accuracy of classification of 93.32% [12]. A thorough early learning calculation on multi-focus Danish information that focuses on time precision. The findings range from AUROC 0.856 (3 h prior to the onset of sepsis) to AUROC 0.756 (24 h prior to sepsis) [13].

A meta-analysis of quantitative research is conducted to test the display of the septic learning pattern. This produced a pool of 0.89, a responsiveness of 0.81; a speciality of 0.72 in the area that acknowledged the functioning bend (SAUROC) [14]. In this paper, they have compared most of the machine learning algorithms and declared that CNN-LSTM neural networks performed the best [15].

3 Proposed Methodology

The primary intention of this research is to design and develop a technique for early detection of sepsis using Multi-Layer Perceptron. The proposed technique involves three major steps, such as preprocessing, feature importance, and classification. Initially, the data will be preprocessed using the resampling technique. The feature importance is done using Xgboost Algorithm. The proposed classifier, named Multi-Layer Perceptron Classifier, gives the detection results on the sepsis. The technique we proposed is shown as the architecture—the block view of the modules of

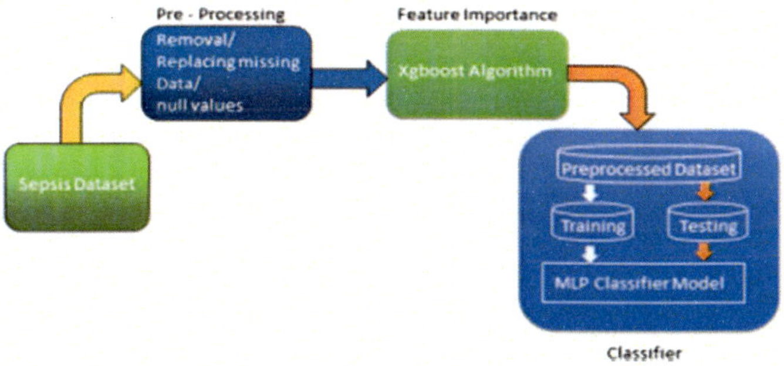

Fig. 2 Proposed Technique—Block diagram

the proposed system (in Fig. 2). The implementation of the proposed technique will be in PYTHON. The system is evaluated in terms of Accuracy and Log Loss in order to show the performance of the technique. This measured performance of the technique which is proposed in this paper will be compared with that of the existing work.

3.1 Dataset

Dataset is garnered from patients in ICU from three separate hospitals. A total of 40,336 patients' clinical data from two definite hospitals while 22,761 patients' clinical data from three definite hospitals were segregated as obscure test sets. Each patients' clinical data contained likely 40 measurements of vital signs, laboratory, and demographics data. Each file has data separated with pipes in which each row represents a 1 h's worth of data [17].

Extremely Imbalance data: The records are extremely imbalanced (More than 97.8% are having 0 sepsis label and 2.2% having sepsis) with the minority class being Sepsis (shown in Fig. 2).

Missing Data: The dataset contains the percentage of data which is missing high (shown in Fig. 3). This is handled by ignoring the features with more than 80% of missing data.

3.1.1 Features [17]

Respiratory rate, Temperature, Mean arterial Pressure, etc., are Vital Signs.

Platelet Count, Glucose, Calcium, etc., are Laboratory Values.

Age, Gender, Time in ICU Hospital Admit time, etc., are considered as Demographics.

0 (Non-sepsis) and 1 (Sepsis) are the Labels for identification.

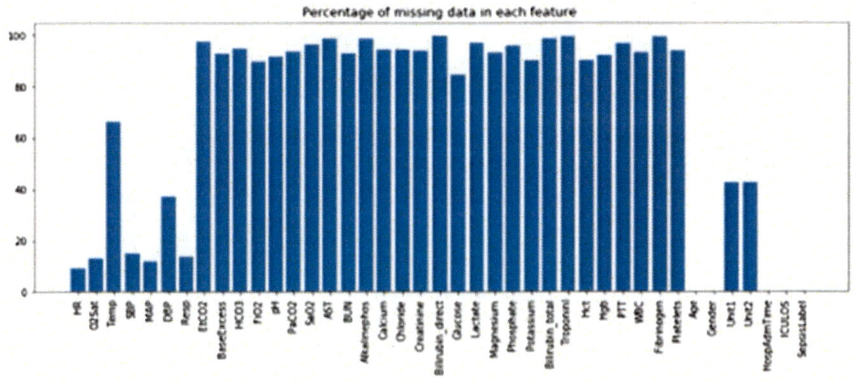

Fig. 3 Sepsis dataset before resampling

3.2 *Preprocessing*

Dataset has been converted from pipe separated file to comma-separated file and with the help of resampling the data have been balanced (Fig. 4).

3.2.1 Feature Importance

The features which are important for early prediction of sepsis are selected with help of Xgboost Algorithm (shown in Fig. 5), benefit of using this is that after the boosted trees are constructed, it is straightforward to get the importance scores for each of the attributes in the dataset. Generally, it provides a score that indicates how useful each feature in the model. The more an attribute is important it will have a higher score of importance. These are ranked based on the comparison of other attributes in the dataset.

Fig. 4 Sepsis database after resampling

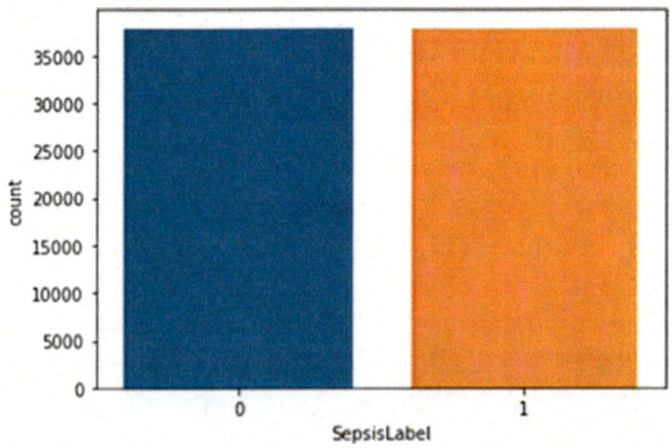

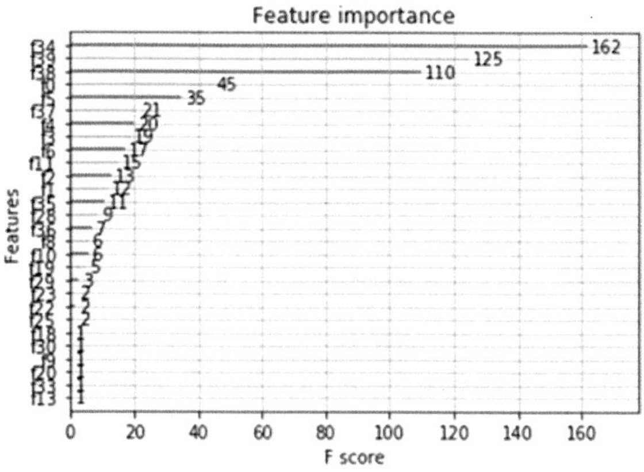

Fig. 5 Feature Importance Score using XGBoost

3.3 *Model Selection*

The selection of the model is a crucial factor, as we use machine learning algorithms to forecast the best outcomes.

The supervised approach is used as a preparation for a collection of input and output couples and to test the input and output association pattern. Supervised learning problems are grouped into two issues:

I. Regression analysis: Actual and Constant values in the target or output variable.
II. Classification: Problems not needed for filtering results.

We have a dataset of 40 Dependent variables (parameters or features) and an indie variable is the aim or performance variable, which determines whether or not patients become sepsis. Various classification algorithms are used in this project which are described as follows. This dataset is validated using data mining methods such as the tool for classifying the target groups mentioned above.

Classifiers: a guided learning technique that helps computers to learn from the knowledge. This knowledge was given and then used to identify new findings. The dataset has been analyzed using the following classifiers: (1) MLP, (2) AdaBoostClassifier, (3) Gradient Boosting Classifier, (4) GaussianNB, (5) LDA, (6) QDA.

MLP Classifier:

MLP Classifier also known as multi-layer Perceptron classifier which itself suggests a Neural Network. MLP Classifier relies on an elemental Neural Network to perform the classification task. It comes under ANN. The phrase MLP is used ineptly, sometimes roughly to refer any feedforward ANN, occasionally strictly referring to networks consisting of multiple layers of perceptrons (with threshold activation). Multi-layer perceptrons now and then are vernacularly referred as "vanilla" neural networks, notably if they contain a single hidden layer avoiding long time-taking lab results. It is very flexible and can be used generally to learn a mapping from inputs to outputs (Figs. 6 and 7).

The model that is being built using MLP Classifier, the data which is obtained after preprocessing is given to the model and the preprocessed data is divided such that eighty percent for training the model and twenty percent used for testing the trained model. With this MLP classifier, we could achieve an accuracy of 94%, with a total of six layers in which one is input layer, four layers are considered as hidden layer and finally the last layer is the output layer, tanh as activation function and, max_iterations up to 5000.

Ada-boost or Adaptive Boosting: It associates multiple classifiers in order to increase the veracity of classifiers and is an iterative ensemble method. We acquired 79.8% accuracy and 0.67 value of log loss through this technique.

ALGORITHM:

1. Take the sepsis dataset and perform re-sampling of the information by re-sampling method so as to balance the dataset.
2. Utilizing Xgboost Algorithm, get the component significance and take out the less significant highlights.
    2.1 Use XGBClassifier() and store it as a variable
    2.2 fit(X,y) where X and Y are input and output labels respectively
3. Take this Preprocessed dataset isolate it into Training and Testing dataset as X_train,Y_train and X_test,Y_test respectively,
    3.1 Train the MLPClassifier model with the Training dataset(X)
        3.1.1 MLPClassifier Function i.e., MLPClassifier() with following
            3.1.1.1 In this function fixing the following values for the parameters
                hidden_layer_sizes =(,4),
                activation='tanh',
                solver = 'lbfgs',
                max_iter = 5000
            3.1.1.2 Putting away the arrival esteem into a variable
        3.1.2 Fit the train information into MLPClassifier using fit(X,Y) where X and Y are input and output labels respectively
    3.2 Validate the prepared model with Testing dataset
    3.3 Print the accuracy

Fig. 6 Algorithm

```
Confusion Matrix :
[[6886  703]
 [ 173 7416]]
Accuracy Score : 0.9422848860192383
Report :
              precision    recall  f1-score   support

           0       0.98      0.91      0.94      7589
           1       0.91      0.98      0.94      7589

    accuracy                           0.94     15178
   macro avg       0.94      0.94      0.94     15178
weighted avg       0.94      0.94      0.94     15178
```

Fig. 7 Confusion Matrix of MLP Classifier

Gradient boosting: It constructs a prediction model in the form of a collection of weak prediction models, more often than not, decision trees. We acquired 91.39% accuracy and 0.31 value of log loss through this technique.

Guassian Naive Bayes: It is an uncomplicated procedure for building classifiers: models that designate class labels to problem instances, expressed as vectors of factor values, where these class labels are taken from some finite set. We acquired 57.76% accuracy and 2.12 value of log loss through this technique.

Linear Discriminant Analysis (LDA): It is a technique of dimensionality reduction. As the name entails, this technique reduces the total number of dimensions (i.e., variables) in a dataset while confining as much knowledge as possible. We acquired 72.4% accuracy and 0.55 value of log loss through this technique.

Quadratic Discriminant Analysis (QDA): It is a variation of LDA in which a singular covariance matrix is predicted for each and every class of observations. It cannot be used as a technique of dimensionality reduction which is a downside of QDA. We acquired 50.92% accuracy and 14.83 value of log loss through this technique.

4 Experimental Results and Discussion

This work explains the simulation results of the proposed model for sepsis early detection. Performance of a classifier is estimated using various evaluation metrics.

4.1 Simulation Setup

The implementation of the classifier is done in PYTHON 3.6 with the configurations of the PC that it has an Intel I7 processor that runs on Windows 10 OS with RAM of 16 GB.

4.2 Evaluation Metrics

The technique that is proposed in this paper is evaluated based on Accuracy and Log Loss as metrics. Evaluation Metrics are explained as follows:

$$\text{Accuracy} = \frac{\text{Number of Correct Predictions}}{\text{Total number of Predictions}}$$

$$\text{logloss} = \frac{1}{N}\sum_{i=1}^{N}\sum_{j=1}^{M} y_{ij}\log(p_{ij})$$

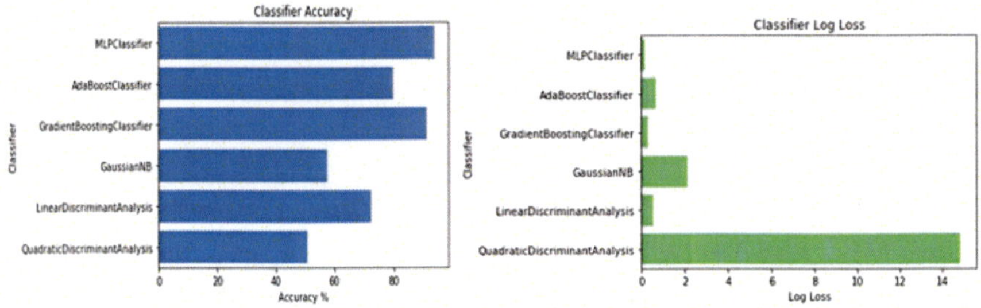

Fig. 8 Graph of Accuracy and Log Loss Classifier

where,

N No of rows in Test set.

M No of fault delivery classes.

Y_{ij} 1 if observation belongs to Class j; else 0.

p_{ij} predicted probability that observation belongs to Class j.

4.3 *Comparative Analysis*

This section presents the comparative analysis of the various classifiers for the sepsis detection (Fig. 8).

The results show that the classification algorithms like MLP Classifier, AdaBoost-Classifier, GradientBoostingClassifier, GaussianNB, LDA, and QDA give the accuracy of prediction like 94%, 71.8%, 91.3%, 57.7%, 72.4%, and 50.9%, respectively. Finally, the results showed that the Multi-Layer Perceptron (MLP) gives the best accuracy in identification of sepsis at early stages with help of clinical data available.

5 Conclusion

Sepsis is a hazardous condition brought about by an infection of the body. So as to prevent fungi, virus, or bacteria, the body generally discharges the chemicals into the circulatory system. Sepsis happens as the body responds to these chemicals out of control, which induces changes that can affect the structures of many organs. This paper has presented a description of Sepsis and its history in the international scenarios and in the national scenario (in the context of India). The symptoms of the disease, signs, complications, and treatment for the disease are presented. This paper also presents the detection of this disease at early stages with higher accuracy using various classifiers mainly on using MLP classifiers in order to develop good prediction models which help in avoiding long time-taking lab results.

References

1. Nachimuthu, S.K., Huag, P.J.: Early detection of sepsis in the emergency department using dynamic Bayesian networks. In: Proceedings of the 2012 AMIA Annual Symposium, (pp. 653–662). Chicago, IL, USA. 3–7 November 2012. [PMC free article] [PubMed] [Google Scholar]
2. Sutherland, A., Thomas, M., Brandon, R.A., et al.: Development and validation of a novel molecular biomarker diagnostic test for the early detection of sepsis. Crit. Care **15**, R149 (2011). https://doi.org/10.1186/cc10274
3. Giuliano, K.K.: Physiological monitoring for critically ill patients: testing a predictive model for the early detection of sepsis. Am. J. Crit. Care. **16**(2), 122–130 (2007). https://doi.org/10.4037/ajcc2007.16.2.122
4. Anderson, S.J., Haney, D.J., Waters, C.A.: Early detection of sepsis. U.S. Patent No. 7,465,555. 16 Dec. (2008)
5. Fairchild, K.D.: Predictive monitoring for early detection of sepsis in neonatal ICU patients. Curr. Opin. Pediatr. **25**(2), 172–179 (2013)
6. Fu, M., Yuan, J., Lu, M., Hong, P., Zeng, M.: An ensemble machine learning model for the early detection of sepsis from clinical data. In: 2019 Computing in Cardiology (CinC), Singapore, Singapore (pp. 1–4) (2019)
7. Kam, H.J., Kim, H.Y.: Learning representations for the early detection of sepsis with deep neural networks. Comput. Biol. Med. **89**, 248–255 (2017)
8. Taylor, R.A. et al.: Prediction of in-hospital mortality in emergency department patients with sepsis: a local big data–driven, machine learning approach. Acad. Emerg. Med. **23**(3), 269–278 (2016)
9. Fagerström, J., Bång, M., Wilhelms, D., et al.: LiSep LSTM: a machine learning algorithm for early detection of septic shock. Sci. Rep. **9**, 15132 (2019). https://doi.org/10.1038/s41598-019-51219-4
10. Shimabukuro, D.W., Barton, C.W., Feldman, M.D., et al.: Effect of a machine learning-based severe sepsis prediction algorithm on patient survival and hospital length of stay: a randomised clinical trial. BMJ Open Respiratory Res. **4**, e000234 (2017). doi: https://doi.org/10.1136/bmjresp-2017-000234
11. Desautels, T., Calvert, J., Hoffman, J., Jay, M., Kerem, Y., Shieh, L., Shimabukuro, D., Chettipally, U., Feldman, M.D., Barton, C., Wales, D.J., Das, R.: Prediction of sepsis in the intensive care unit with minimal electronic health record data. Mach. Learn. Approach JMIR Med. Inform. **4**(3), e28 (2016). https://doi.org/10.2196/medinform.5909
12. Tang, C.H.H., Savkin, A.V., Middleton, P.M.: Non-invasive sepsis patient classification using least squares support vector machine. Biosignals (2009)
13. Lauritsen, S.M. et al.: Early detection of sepsis utilizing deep learning on electronic health record event sequences. Artif. Intell. Med. 101820 (2020)
14. Islam, M.M. et al.: Prediction of sepsis patients using machine learning approach: a meta-analysis. Comput. Methods Prog. Biomed. **170**, 1–9 (2019)
15. Hsu, P., Holtz, C.: A comparison of machine learning tools for early prediction of sepsis from ICU data. In: 2019 Computing in Cardiology (CinC), Singapore, Singapore, (pp. 1–4) (2019)
16. https://www.healthline.com/health/sepsis
17. https://physionet.org/content/challenge-2019/1.0.0/

A Study on Onsite–Offshore Data Security Model for Big Data Applications

T. N. Manjunath, S. K. Pushpa, Ravindra S. Hegadi, and R. A. Archana

1 Introduction

Heterogeneous huge data volume is getting generated from the information technology systems around us in any business domain at faster rates. The data is a combination of structured, semi-structured and unstructured with the features of volume, velocity and variety. Big data frameworks are the need of the hour to handle enormous data integration which helps big data administrators and information security specialists. The development frameworks to address the poor permeability of data and identification of sensitive data. Privacy rules and guidelines involve an exact understanding of data hazards dependent on information residency, expansion, assurance and utilization across frameworks and topographies. Big data security ensures the characteristics of information to drive 360-degree standpoint to use data provenance and assurance of sensitive data. The data under consideration for usage guarantee consistency check with the support of various corporates and industry guidelines information covering at the organization level [1]. The structure gives a typical framework which enables further development and validation activities empowering versatility for the applications under scan. To accomplish reusability

T. N. Manjunath (✉) · S. K. Pushpa
BMS Institute of Technology and Management, Bengaluru, India
e-mail: manju.tn@bmsit.in

S. K. Pushpa
e-mail: Pushpask@bmsit.in

R. S. Hegadi
Central University of Karnataka, Kalaburagi, India
e-mail: rshegadi@cuk.ac.in

R. A. Archana
Sai Vidhya Institute of Technology, Bengaluru, India
e-mail: archana.ra@saividya.ac.in

C. K. Mai et al. (eds.), *Machine Learning Technologies and Applications*,
Algorithms for Intelligent Systems,
https://doi.org/10.1007/978-981-33-4046-6_23

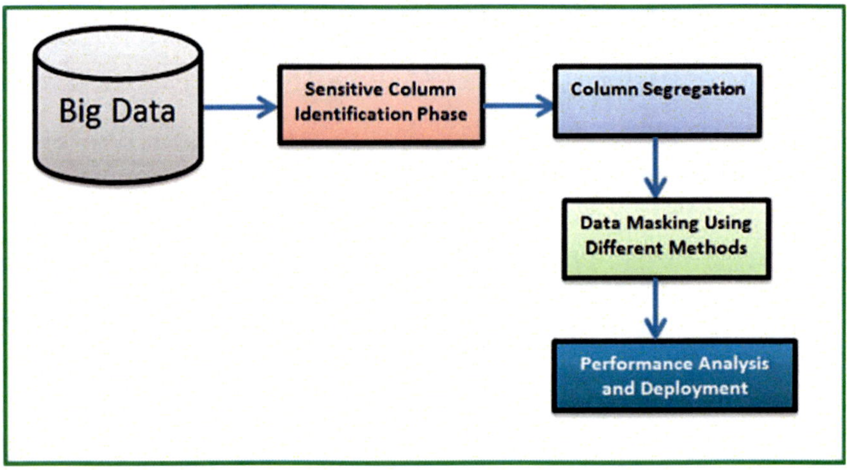

Fig. 1 Big data model for data masking using column segregation

and adaptability by means of the utilization of procedure situated metadata that characterizes the way in covering is to be completed for everything of information as far as sub-setting, encryption, control, etc. Utilized transformation data mechanisms and operational contents of the metadata brings about a totally standard veiling process at whatever point anybody in the association needs to cover a given bit of information for a given reason. Figure 1 shows the big data model for data masking using column segregation.

2 Literature Review

The IBM Security and Ponemon Institute Report of 2019 reveal that there was a 130% increase in data breaches from 2006 to 2019 [2]. The global average breached records are about 25,575 and with the Middle East is at the highest 38,800 records. The report published by Ponemon institute in 2015 on Data Security Intelligence Report reveals that information technology is most concerned where delicate information is present is just growing [3]. Pass on fit-for-reason tremendous data, with a versatile and data quality condition to handle big data challenges and changes the way associations improve and demonstrate their operational methodology. The key challenges that the information quality issues emphasize various properties such as dependability and trustiness in massive data assets. Any request for data quality is a noteworthy issue and obstacle to an affiliations' ability to choose quick decisions, diminish costs, make improvement and advanced improvement. Significant, auspicious and reliable information is fundamental for progress. The Big Data Quality enables your organization to adopt an all-encompassing strategy for overseeing information quality and utilizing the intensity of Hadoop. The data item changes the data quality techniques to be a mutual effort between business customers and IT. Making authentic data that commits to support better business dynamic decisions and finds the information size, course of action at various stages. It passes on the principles that data generators will deploy the big data applications onto the Hadoop cluster on-local system or to the cloud systems. The various big data quality frameworks upgrades and systematize more data, which can be scaled and enable the business in the organization [4]. It prepares and offers the data which is fit and validated for use in bits of information data discovery and profiling of big data quality focus on data disclosure and profiling issues for perceiving fundamental data issues concealed over the endeavour. Astounding and versatile gadgets acquire the data at different points of sale and perceive quality issues of the data, design the business rules appropriately which meet the requirements, similarly as proactively screen the data quality strategy. Information Analyst will use the data which is an instrument that draws in the business to adequately check out improving the idea of data, without the requirement for IT intercession. Eclipse-based information quality advancement condition that improves IT efficiency by streamlining the way toward making complete information quality guidelines [5, 6]. A rich set of data quality transformations offers complete help for all informational variations, so you can convey confided in client, item, money related, or resource information to any information coordination, ace information the executives, or information administration venture [7]. It highlights institutionalization, coordinating, overall location purifying and flexible information quality administration for all undertaking types. It enables you to pass on preassembled data quality standards to improve quality over the endeavour. Experian's 2015 information quality benchmark report highlights that 26% of their data to not be right.

3 Big Data Masking

Here the consideration is on each field/attribute in a table structure, where information security on each field/attribute relies upon the domain of information type. For instance, the attribute position in a table structure for protecting the data through encryption and tokenization for the MasterCard business is considered here. Pseudonymization is used for names, randomization for numbers and sample character masking on various attributes such as national ID is one of the delicate data that supports privacy law consistency for information sensitivity coverage limits the accessibility for databases and files [8, 9]. Let us consider PII (Personnel Identifiable Information) in Excel format, observe the IRI CellShield to find the sales fragile data in various sources which need to be encrypted with any open libraries by strategies for ensuring about masking using appropriate method and conditions. The FieldShield attribute generated audit logs which can ensure and request the response and permit your accreditations and consistency with data privacy laws. FieldShield is an attribute that spread data for validation by considering IRI RowGen for data safety, referential integration for the test data without any orchestration, especially in the

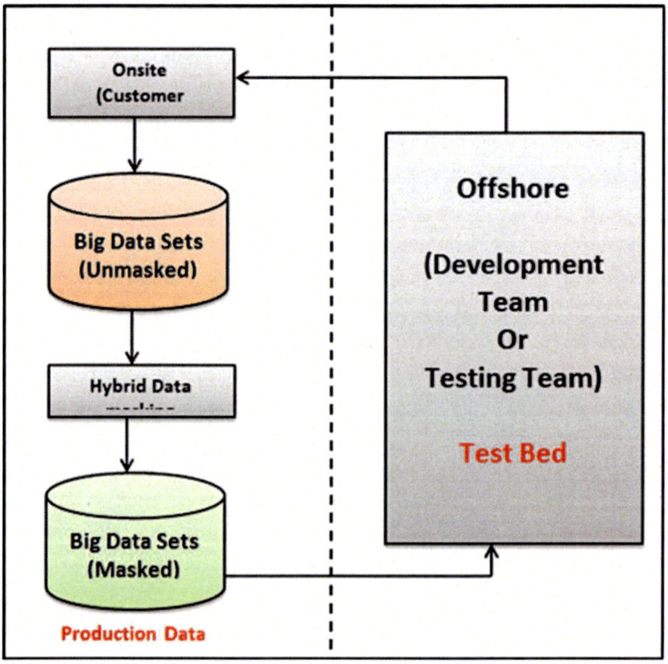

Fig. 2 Migration between production site and test site

event that you cannot get to create data. Keep your own business domain to help, security quality, reversibility and appearance with the help of key management [3, 10]. Randomization can deal with the oversee anonymize or derecognize in the long run personnel data. Figure 2 shows the migration between the production site and test site.

Programming the FieldShield attribute can help IRI Data Protector assortments to enable the data access which is not reversible with respect to irregular information and which is non-deterministic. By adding the special field for the column which is under protection from the data source by detaching the relationship with other attibutes in the source table [11, 12]. This is done using the IRI workbench, where FieldShield attribute of the customer table can address any source record field or database section, alongside data covering limits. Data rearranging resembles possible through custom sporadic limits or match method of reasoning any one can character in the IRI workbench verbalization designer. Occasionally, there is more than one message for every summary, since the encoding technique is not as better as encryption which can be continually reversible, it is every so often sufficient for masking few attributes. Even more commonly, in any case, hashing is utilized with encryption. IRI supplies SHA1 and SHA2 hashing tallies with FieldShield, close to two or three encryption limits. Hash limits are manner used to make checksums [12, 13]. When the checksum function is over, and its substance experienced a close to hash ability to build alternate MAC. To oversee, the data being created in the record was undisturbed. Monster data covering support on Hadoop utilize the properties allowing customers to know the Hadoop cluster size to understand the masking process effectively. The Hadoop gives us direct performance and conspicuous nature when covering occupations inside the Hadoop cluster. Set aside data which is masked continually with each other data sources to the cover Hadoop condition stays with the condition of congeniality to ensure about data sources [14, 15]. The latest appearance of enterprise in like manner presents a best in class decision that helps social data covering endeavours, achieving execution grows 15–80 times faster than the past feature. Furthermore, execution updates are ordinarily and outstandingly applied, growing throughput without the fundamental for blue-print changes, recognizing ordinary settlement and by a wide margin reduced covering empower windows [16]. Data covering is the best way to deal with oversee hold quickly to data security laws, dishonour the effects of a data break, and invigorate the risk and controls structure of your undertaking. IRI FieldShield quickly fulfills the data unmistakable confirmation, affirmation and check necessities of your data stewardship, regulatory consistency and data misfortune expectation programs [17]. There is an epitome for on an spot offshore data security model to keep up a vitally good way from data ruptures which are happening as a result of insiders' activities.

3.1 Dynamic Data Masking

Dynamic data protection regulates steadily to keep advantaged work power, for instance, DBAs, creation of support staff individuals and business customers from getting to delicate and irrefutable information which is not required for them to use for their activities. To estimate the rate of dynamic data masking (DDM) depends on the ability to apply different spreads to different combinations of data attributes in new databases [18]. Data coverage is applied reasonably subject to customer occupations and a bit of user levels, simply individuals which can totally reveal data subsequently. In different divisions, this would infer that an administrator doesn't have the choice to see insisted Social Security numbers, explicit under-study assessments, or tenants' changed adjusted gross compensation figures considering the way that these traits and other definitely unmistakable information would be unequivocally blended, hashed, spread [19]. Data protection can be used for security to unstructured and semi filtered through data. Steady data veiling can in like manner be used identified with encryption to make mixed data legitimately sensible scanning for progress and validation of data. Disguising data rather than encoding it applies basically no presentation discipline [12]. DDM has been appeared and shown to guarantee sensitive information without altering information base execution. Before executing DDM method, the provider had to be completed routinely dependency modules on time with its client's mystery information [20]. While trying to explore the business model issue with various frameworks that must be examined now none of them watched out for the vendors for security and execution issues. Encryption, for example, was deterred as a result of execution degradation in the creation condition.

This could be required for different alterations with persistent changes to database application which is a prohibitive endeavor given that tremendous applications the collusion using were packaged with data models. The affiliation required a positively earth shattering, world class method. As portrayed before the present moment, Dynamic Data Masking uses an obvious visual execution system [21]. DMM allowed specific data to ensure in the association's business customers, beginning late chose and other specialists, staff under contract, and redistributed and IT workers allowing them to use information while consenting to "need to-know" data get the opportunity to blueprints. Despite significantly lessening the risk of a data break, the thing sup-utilized the trades provider with the flexibility to quickly change data veiling limits concerning different authoritative or business necessities. Rule prompting equipped fast security across fundamental creation, planning, and nonproduction conditions. Commenting by using the impact of DMM programming, the association's focal information security official imparted, "In only an enormous segment of a month, the accessible Platform unmistakably made sure about particular information on our charging, CRM, and custom application screens and bundled reports in progress and nonproduction conditions [20, 22]. The thing is straightforwardly a foundation of our hazard the board and consistence methodology." Due to the troubles, affiliations which are powerfully clear profound data veiling method to the breaks and acknowledge data security. Such an answer should interface with IT relationship to: Mask the data which is sensitive revealed with progress condition. DMM urges relationships to achieve these upsetting undertakings, proactively keeping an eye out for information security challenges viably. As the major affirmed eminent data covering thing is accessible, DMM derecognizes data and controls unapproved access to creation conditions. It has diverse focal centers, among the key central inspirations driving data and its chance through imagining that alterations should happen for new databases under validation. This suspicion covering can be applied quickly and inconspicuously to guarantee private data over an association, paying little brain to check. Data stowing ceaselessly is in like manner granular, in that it attracts relationship to unequivocally cover data down to the line, part, or cell level. In addition, data disguising progress can support with existing attestation blueprints, including Active Directory, LDAP and it supplements other data affirmation degrees of progress, for instance, encryption, database development watching (DAM), and security information and even the board with everything considered giving wide data assurance accreditation.

Steps-1 Security for Big Data Using DMM Method

Input: Set of Big Data
Output: Big Data Sets with Privacy

1. Consider a big data set consists of R records BD={r1,r2,…..rn}
2. Each record in R consists of set of columns R={c1,c2,…..cm}
3. Record Validation process
4. Identify appropriate data masking methods to be applied for every column
5. DM={S,KR,M,R,Shuf}
6. store all masked data dynamically
7. Repeat step 1 to 5 for all the files in the big data sets under test

The input will be raw big data sets which will get generated from heterogeneous sources in the business environment to move the data between platforms in the business environments data has to be protected from breaches, classify the data based on nature, for example, textual data, image data and video data then apply different available data masking such as shuffling, substitution, multipliers, randomness and after applying it store the masked data in the temporary table using the concept of key management. This is repeated for different tables or files under consideration.

4 Results and Discussion

The proposed strategies give adaptability for the information will be veiled and guarantee that rules of the business are protected from specific venture application won't be affected. After information isolation, the veiling type will be chosen dependent on the information, for example, substitution, substitution, multiplier, randomizer and rearranging, the equivalent is outlined underneath with model. Figure 3 shows how different masking is used for data security with examples.

The proposed technique is a general methodology that manages the necessities of security issues looked at by different associations when on-location seaward business conveyance model is utilized. Our half and half information concealing model system

Masking Type	Example
Substitution	Ravi becomes Manju
Replacement	Z123456 becomes A999123
Multiplier	22/11/1978 becomes 15/02/1986
Randomizer	City will become xxxxaaa
Shuffling	Ravi → Bindu Mohan → Mahesh Manju → Madhu

Fig. 3 DMM method for data security

guarantee two standards while activity is completed (i) Masking isn't reversible. It is highly unlikely to figure out the first information from the veiled information and (ii) Masked information is usable. For instance, when testing substantial locations, the concealed information must incorporate legitimate postal divisions, not irregular numbers that fit the information type. Correlation investigation of statistical execution of the first information and adjusted information, in request to compute the factual properties, for example, mean, difference, and standard deviation for unique information and changed information. The table shows that after the change likewise the measurable properties are the same as the first. Figure 4 shows the data masking using rules on columns. Figure 5 shows the comparison between original data and masked data.

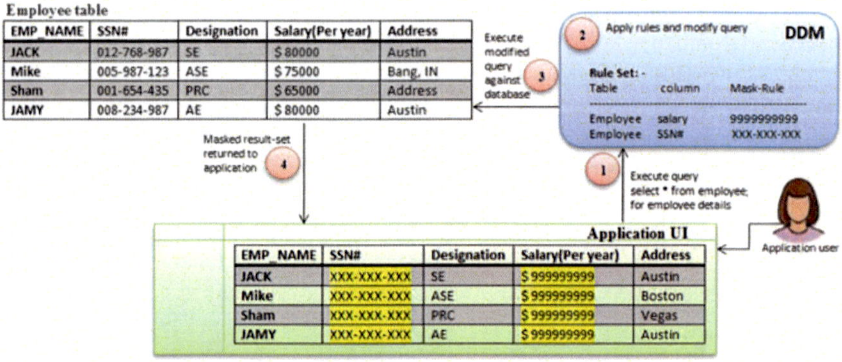

Fig. 4 Example-2 of data masking using rules on columns

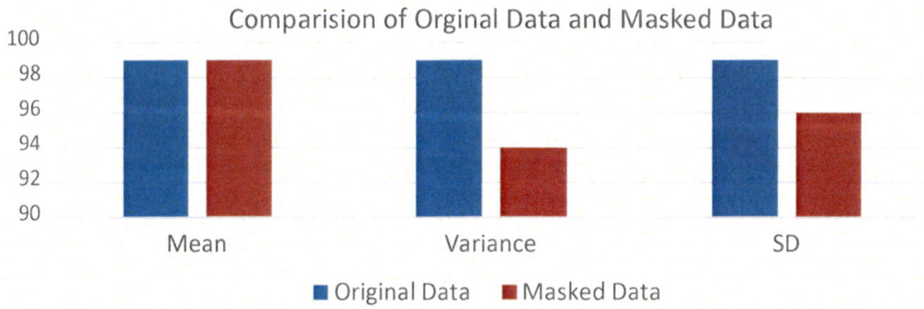

Fig. 5 Comparison of original data and masked data

References

1. Ravikumar, G.K., Justus Rabi, B., Manjunath, T.N., Hegadi, R.S., Archana, R.A.: Design of data masking architecture and analysis of data masking techniques for testing. IJEST **3**(6), 5150–5159 (2011)
2. Aslanyan, Z., Boesgaard, M.S.: Privacy analysis of format-preserving data-masking techniques. In: 2019 12th CMI Conference on Cybersecurity and Privacy (CMI), Copenhagen, Denmark (2019) (pp. 1–6)
3. Obukata, R., Cuka, M., Elmazi, D., Sakamoto, S., Oda, T., Barolli, L.: Performance evaluation of an AmI testbed for improving QoL: evaluation using clustering approach considering distributed concurrent processing. In: Advanced Information Networking and Applications Workshops (WAINA) 2017 31st International Conference on, (2017) (pp. 271–275)
4. Siddartha, B.K., Ravikumar, G.K.: Analysis of Masking Techniques to Find out Security and other Efficiency Issues in Healthcare Domain. In: 2019 Third International conference on I-SMAC (IoT in Social, Mobile, Analytics and Cloud) (I-SMAC), Palladam, India (2019) (pp. 660–666)
5. Understanding and Selecting Data Masking Solutions—Creating Secure and Useful Data-Securosis, L.L.C. Data Masking: What You Need to Know What You Really Need To Know Before You BeginA Net 2000 Ltd. White Paper
6. Dreibelbis, A., Hechler, E., Milman, I., Oberhofer, M., van Run, P., Wolfson, D.: Enterprise master data management: an SOA approach to managing core information. Dorling Kindersley (India) Pvt, India (2008). ISBN 978-0-13-236625-0
7. Gahi, Y., Guennoun, M., Mouftah, H.T.: Big Data Analytics: Security and privacy challenges. In: Computers and Communication (ISCC) 2016 IEEE Symposium on (2016) (pp. 952–957)
8. Ali-Ozkan, O., Ouda, A.: Key-based reversible data masking for business intelligence healthcare analytics platforms. In: 2019 International Symposium on Networks, Computers and Communications (ISNCC), Istanbul, Turkey (2019) (pp. 1–6)
9. Siddartha, B.K., Ravikumar, G.K.: A novel data masking method for securing medical image. In: 2019 International Conference on Smart Systems and Inventive Technology (ICSSIT), Tirunelveli, India (2019) (pp. 30–34)
10. Win, T.Y., Tianfield, H., Mair, Q.: Big data based security analytics for protecting virtualized infrastructures in cloud computing. IEEE Trans. Big Data **99**, 1–1 (2017). https://doi.org/10.1109/tbdata.2017.2715335
11. Khorshed, M.T., Sharma, N.A., Dutt, A.V., Shawkat Ali, A.B.M., Xiang, Y.: Real time cyber-attack analysis on Hadoop ecosystem using machine learning algorithms. In: Computer Science and Engineering (APWC on CSE) 2015 2nd Asia-Pacific World Congress on (2015) (pp. 1–7)
12. Almeida, C., Harshitha, K., Manjunath, T.N.: A study on column segregation for data security. IJRCSIT **2**(2), (2014)
13. Manjunath, T.N, Hegadi, R.S.: Data quality assessment model for data migration business enterprise. Int. J. Eng. Technol. (IJET) **5**(1), (2013)
14. Manjunath, T.N., Hegadi, Ravindra S.: Statistical data Quality model for data migration business enterprise. Int. J. Soft Comput. **8**(5), 340–351 (2013)
15. Ravikumar, G. K., Manjunath, T.N., Hegadi, R.S., Umesh, I.M.: A survey on recent trends, process and development in data masking for testing. IJCSI Int. J. Comput. Sci. **8**(2), (2011)
16. Manjunath, T.N., Hegadi, R.S., Ravikumar, G.K.: Analysis of data quality aspects in data warehouse systems. (IJCSIT) Int. J. Comput. Sci. Inform. Technol. **2**(1), 477–485 (2010)
17. Kimball, R., Caserta, J.: The Data Warehouse ETL Toolkit. Wiley Publishing, Inc. Data Quality: Concepts, Methodologies and Techniques. Data-Centric Systems and Applications - Batini, Scannapieco (2006)
18. Manjunath, T.N., Hegadi, R.S., RaviKumar, G.K.: Design and analysis of DWH and BI in education domain. IJCSI Int. J. Comput. Sci. **8**(2), (2011)
19. Manjunath, T.N., Hegadi, R.S., Mohan, H.S.: Automated data validation for data migration security. Int. J. Comput. Appl. **30**(6), 41–46 (2011)

20. Satish, M., RamakrishnaMurty, M.: Document clustering with Mapreduce using Hadoop framework. Int. J. Recent Innov. Trends Comput. Commun. **3**(1), 409–413 (2015)
21. Ryu, K.S., Park, J.S., Park, J.H.: A data quality management maturity model. ETRI J. **28**(2), (2006)
22. Muralidhar, K., Batra, D., Kirs, P.: Accessibility, security, and accuracy in statistical databases: the case for the multiplicative fixed data perturbation approach. Manage. Sci. **41**(9), 1549–1564 (1995)

Acoustic Characteristics' Heart Sounds S1 and S2 of Single-Level Autoencoder with DNN

P. Jyothi and V. DilipVenkata Kumar

1 Introduction

The heartbeat sound, blood and lungs listened to with a stethoscope invented by Laennec is heart auscultation which is essential to provide information regards to measure defects of heart functioning. Acoustic sound waves converted to signals of electrical through electronic stethoscopes and processed for optimal listening, as due to the deficiency of automated tools for analyzing heart auscultation and interpretation depend on human ear training; hence it makes doctors perform a task on their own ears. Heart conditions are inspected through ECG and ultrasound methods, which are less cost-effective, and more demands on hardware when compared to auscultation [1]. Thus, auscultation is most suitable and is simple with low cost for primary health care heart examination method.

Through repeating cycles of Systole and Diastole the functions of the heart depend, during systole blood is ejected by the heart chambers and during diastole, it is filled with blood. We can also observe this periodicity in heart sounds where S1 is the first sound produced at systole beginning and the second sound which is S2 made at beginning of diastole that is audible which is caused due to the vibrations produced in the vascular system and other sounds S3 and S4 are a category of sounds called murmurs that might be caused because of blood flows irregularity and which can be visually depicted in phonocardiogram (PCG). PCG can then be analyzed through a technique called digital signal processing.

Segmentation of heart sound through phonocardiogram (PCG) is an essential role of examining sounds. Identification of S1 of systolic or S2 of diastolic requires FHSs

P. Jyothi (✉)
CSE Department, VNRVJIET, Hyderabad, India
e-mail: Jyophani.reddy@gmail.com

V. DilipVenkata Kumar
CSE Department, PVKK IT, Anantapuramu, India

C. K. Mai et al. (eds.), *Machine Learning Technologies and Applications*,
Algorithms for Intelligent Systems,
https://doi.org/10.1007/978-981-33-4046-6_24

prerequisite. S1 and S2 occur in ventricular depolarization immediately after R-peak and end-T-wave of the ECG at the end of ventricular depolarization, respectively. Heart sound segmentation is simple in recordings which are noise free but it is complex when the recordings are taken in noise such sources are breathing sounds, motion artefacts, endogenous and also other physiological sounds such as splitting of FHSs, clicks and murmurs. Cardiac function and its activities are evaluated through physical examination of cardiac auscultation. In cardiac cycle, two heart sounds normally occur in sequence for healthy adults.

Diagnostic information is provided through phonocardiogram (PCG) to evaluate cardiac abnormalities including heat failure and defects of the heart. During the cardiac cycle, PCG provides vibratory sounds of electronic recordings which are happening within the heart consists of S1 and S2 are the first and second heart sounds. S1 duration is longer with low pitch sound and S2 duration is shorter with high pitch sound. The interval S1–S2 which is called systole is shorter when compared to interval S2–S1 which is called diastole. Sounds made by heart can also include S3 which is the third heart sound and S4 which is the fourth heart sound which includes other sounds called murmurs, rubs and clicks are produced when the structure of cardiac is affected and heart values are defective. Automatic analysis of PCG signals is measured through parameters for normal and abnormal conditions. Murmurs can further be categorized into EDM (Early-diastolic murmurs), MDM (Mid-diastolic murmurs), LDM (Late-diastolic murmurs) and HDM (Holo-diastolic murmurs).

Cardiac function and its activities are evaluated by physical examination of CARDIAC auscultation. In cardiac cycle, there is an occurrence of two normal heart sounds for healthy adults in which S1 sound occurs starting of the systole phase of ventricular resulting as atrioventricular valves closing tricuspid and mitral valves and at starting of ventricular diastole phase second heart sound occurs which is called S2 resulting as closing pulmonic and aortic valves. An S1 sound will be there for a longer duration with low pitch and S2 sound will there for a shorter duration with a high pitch. Systole is calculated through S1–S2 interval shorter when compared to diastole which is calculated through S2–S1 interval. An acoustic fingerprinting detection framework is proposed in this paper by applying a classifier for S1 and S2 recognition. To classify acoustic incidents, HSAD(Heart Sound Activity Detection), MFCC(Mel-frequency cepstral coefficients) are used and their extraction of features can be applied (Fig. 1).

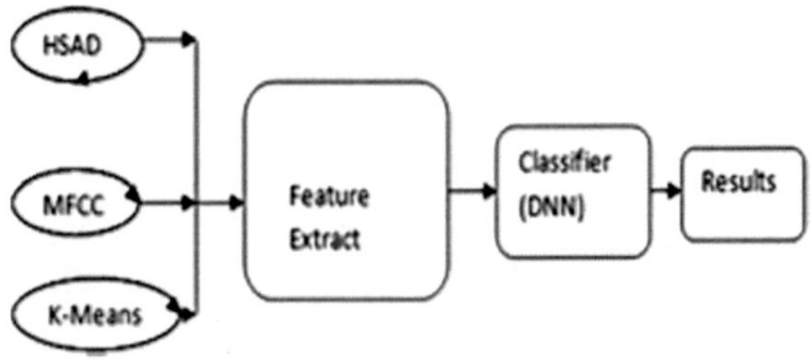

Fig. 1 Flowchart of S1 and S2 recognition system

Sounds of heart can be categorized into two groups using K-means algorithm and with these groups, there is a formation of a center feature vector and by concatenating these feature vectors to make a super vector. To classify S1 and S2 super vector is fed with DNN classifier. Compared with other classifiers such as GMM (Gaussian mixture model), KNN (K-nearest neighbor), LR (Logistic regression) and SVM (support vector machine), the DNN classifier with DAE of the proposed method has higher accuracy to recognize S1 and S2 when compared with the existing methods. Section 2 contains the proposed system overall architecture and its related works for the recognition of S1 and S2. Section 3 contains Heart Sound Classifier of the proposed system. The remaining sections such as Sect. 4 contains discussion and Sect. 5 contains the conclusion.

2 Recognition of S1 and S2 Sounds

2.1 Related Works

To recognize the sounds of heart, S1 and S2, various algorithms of machine learning are implemented. Differentiation of S1 and S2 durations will be used by the proposed system for achieving performance in heart sounds of arrhythmia. The proposed system with spectrograph results were compared with ECG signals [2]. Heart sound diastolic periods and time intervals are used to devise features, which are given as input to the artificial neural network and to other classifiers.

Extraction for the Proposed System
Feature extraction process consists of heart sound activity detection, mel-frequency cepstral coefficients, and K-means processes.

A. Mel-Frequency Cepstral Coefficients (MFCC)

Feature extraction of MFCC consists of six operations, which are FFT (Fast Fourier Transform), Pre-emphasis, Windowing, Nonlinear transformation, Mel-filtering, and DCT(Discrete Cosine Transform). Given signal divided by windowing operation into frames of the sequence. Designing of mel-filtering to form one energy intensity through the integration of mel filter band and human perception. This frequency is efficient in speech recognition [3], acoustic pattern tasks of recognition [4], and speaker recognition [5].

B. K-Means Algorithm

K-means algorithm is used for the determination of representative data points which are called as Population Centers taken from a large number of data points. This algorithm is used mainly for data compression and for pattern classification and K-means algorithm method shown below.

Initialization
Materials used for training are divided as ui, where $i = 1, 2, 3, \ldots N$ into K groups and initial population center denoted μ for each group.

Calculation for Recursive:

Let ui for each find population center which is nearest and assign to it by

$$K^* = \arg_k \min d(u_i, \mu_k),\ i = 1, 2, \ldots N$$

In which $d(.,.)$ represents the distance.

Calculate μ_k population center again, all u_i belongs to the kth group which will make a new group. After calculating the new group compare and check if new groups which formed are the same as that of population set which is original then there is the completion of training otherwise original population groups are replaced by new population groups.

Step ii is repeated for calculations which are recursive.

Acoustic features are clustered in the K-means algorithm for each heart sound into two groups and μk of population center group of each computed and super vector is formed by concatenating these two vectors, which is final which represents a segment of heart sound and these super-vectors are used for classifiers to built and S1/S2 recognition can be performed.

C. Shannon Energy

Shannon energy is used to extract Heart Sound (HS) which will reduce the low amplitude of noise thus it will become easier to distinguish low-amplitude signal. This energy produces signals which are moderate amplitude, when compared with signal of high amplitude the low-amplitude signals are very weak. Shannon energy is good at minimizing envelopes difference of high and low-amplitude signal, makes signals which are low amplitude are found more likely through this reduction.

Average Shannon energy defined as

$$\text{Es} = -1/N \sum_{i=1}^{N} x_{\text{norm}}^2(i)\ \ \log x_{\text{norm}}^2(i)$$

where x_{norm} is normalized signal and N is the length of signal window.

3 Proposed System Heart Sound Classifier

3.1 Denoising Autoencoder

Artificial Neural Network is a mathematical model and executes regression or classification tasks mimics the neural network of biological structures. To solve various

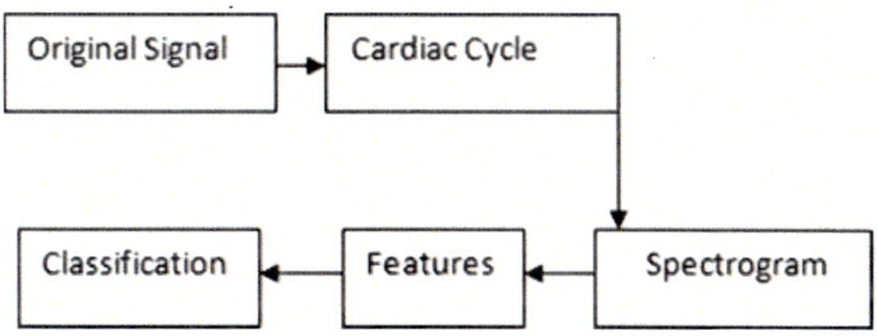

Fig. 2 Flow chart of Cardiac cycle into spectrogram

problems, scholars proposed a diverse neural network model. When multiple layers are used with neural networks then it is known as Deep Neural Network, which shows efficient performance in a variety of tasks which are speech processing [6], automatic speech recognition [7], visual pattern recognition [8]. DNN [9] operating principles whose output layer of current which is used as input for a hidden layer of next. To strengthen regression or classification capability, this algorithm uses a huge number of hidden layers. DNN model structure contains input and output of the first hidden layer relationship that is shown as

$$a_1 = F(W_1 x + b_1)$$

where x—input feature,

a_1—output of first hidden layer,
W_1—weight matrix,
b_1—bias matrix,
$F(.)$—activation function.

As shown in Fig. 2, signals are divided into Cardiac Cycle by applying Shannon energy to Original signal. Applying frequency spectrum to obtain spectrogram in which Cardiac Cycle is used as input. Due to high dimensionality in spectrogram to reduce dimensionality Denoising Autoencoder (DAE) [16] used to reduce noise in signal and single Autoencoder can be used for feature extraction, single Autoencoder implemented in proposed system along with DAE to improve accuracy. Finally, Deep Neural Network (DNN) applied for classification of heart sound.

In this proposed work DAE and 1-AE integrated with DNN used for implementation of heart sound classification.

3.2 *Extracting Features with AE (Autoencoder)*

The Artificial neural network which tries the way to simulate how brain thinks and each node represents a neuron and neuron connection for communication between nodes and whose degree is calculated through connection weight values. Unsupervised learning learns whole data set [10] of useful information through directly models and it does not need to know data label. Unsupervised learning applied to HS feature extraction, which extracts end to end without signal processing understanding and HS signal characteristics.

Three layers of unsupervised learning Autoencoder of an artificial neural network structure are input layer, hidden layer and output layer [11].

4 Discussion

The process of feature extraction collocated with DNN and Autoencoder model to build classification system. S1 and S2 data were labeled medical doctor segmentation at initial stage require experience. To obtain efficient heart sounds analysis to reduce the burden to doctors and for automatic detection both S1 and S2 segments are collected and used for DNN classifier training through KNN (K-nearest neighbor) [12], LR (Logistic Regression) [13], SVM (Support Vector Machine) [14], DNN (Deep Neural Network) classifiers and comparisons are tested. Supervised method is used for training as it can't train by itself. As per single layer Autoencoder no need to train each neuron but they will learn by themselves through unsupervised methods which use Shannon Energy to reduce low-amplitude noise for low-amplitude signal distinction [15] (Table 1).

Deep Neural Network cannot efficiently perform feature extraction and reduce noise in signal so DAE used to reduce noise and to perform feature extraction single layer AE used with the help of DAE and 1-AE integration, Deep neural network performs more accurate results in heart sound classifications compared to previous methods as shown in Fig. 3.

5 Conclusions

This paper proposed a method of classification for Heart Sound (HS) based on integrating Deep Neural Network with Denoising Autoencoder (DAE) and Single layer Autoencoder (1-AE). Spectrogram consists of frequency domain and time domain features which are fed into an AE for removing noise and feature extraction. The high-level features are extracted through unsupervised learning of autoencoder. The results show the proposed method DNN with spectrogram through Autoencoder is more efficient when compared to previous methods acoustic characteristics of

Table 1 Models Classification rate Accuracy

Methods	KNN (%)	GMM (%)	LR (%)	SVM (%)	DNN (%)	DAE & 1-AE (%)
Accuracy	78.11	86.98	87.57	90.53	91.12	99.23

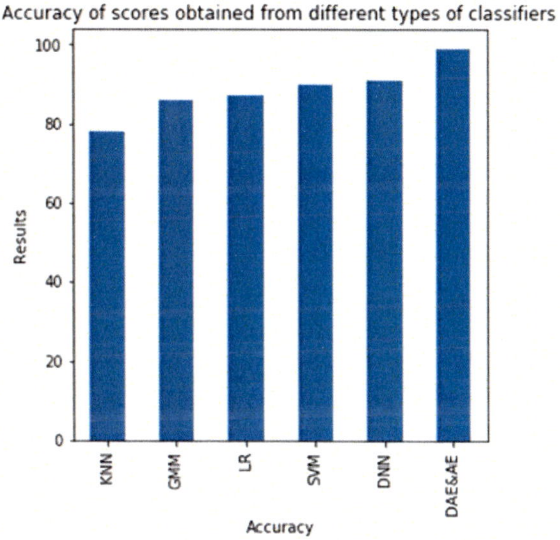

Fig. 3 Performance of various classifiers

cardiac auscultation for classifying S1 and S2 heart sounds. Single Autoencoder with Denoising Autoencoder through DNN provides efficiency of 99.23% when compared with DNN for heart sound classification of S1 and S2. Multi-level Autoencoder with Denoising Autoencoder integration of DNN can be further extended in the future.

References

1. Gussak, I., Antzelevitch, C., Hammill, S.C., Shen, W.K., Bjerregaard, P. (Eds.).: Cardiac Repolarization: Bridging Basic and Clinical Science. Springer Science & Business Media (2003)
2. Kumar, D. et al.: A new algorithm for detection of S1 and S2 heart sounds. In: IEEE International Conference on Acoustics Speech and Signal Processing Proceedings, vol. 2. IEEE (2006)
3. Davis, S., Mermelstein, P.: Comparision for monosyllabic word recognitions for monosyllabic word recognition in continuously spoken sentences. IEEE Trans. Acoust. Speech Signal Process. **28**(4), 357–366 (1980)
4. Logan, B.: Mel frequency cepstral coefficients for music modelling. In Ismir, vol. 270 (2000) (pp. 1–11)
5. Furui, S.: Cepstral analysis technique for automatic speaker verification. IEEE Trans. Acoust. Speech Signal Process. **29**(2), 254–272 (1981)
6. Xu, Y., Du J., Dai, L.R., Lee, C.H.: A regression approach to speech enhancement based on deep neural networks. IEEE/ACM Trans. Audio, Speech, Lang. Process. **23**(1), 7–19 (2014)
7. Hinton, G., Deng, L., Yu, D., Dahl, G.E.M., Jaitly, N., Kingsbury, B.: Deep neural networks for acoustic modelling in speech recognition: The shared views of four research groups. IEEE Signal Process. Mag. **29**(6), 82–97 (2012)
8. Simard, P.Y., Steinkraus, D., Platt, J.C.: Best practices for convolutional neural networks applied to visual document analysis. In: Icdar, vol. 3 (2003)
9. Larochelle, H., Bengio, Y., Louradour, J., Lamblin, P.: Exploring strategies for training deep neural networks. J. Mach. Learn. Res. **10**, 1–40 (2009)

10. Fritzke, B.: Growing cell structures- a self- organizing network for unsupervised and supervised learning. Neural Netw. **7**(9), 1441–1460 (1994)
11. Vincent, P., Larochelle, H., Bengio, Y., Manzagol, P.A.: Extracting and composing robust features with Denoising autoencoders. In: Proceedings of the 25th International Conference on Machine Learning (2008) (pp. 1096–1103)
12. Altman, N.S.: An introduction to kernel and nearest-neighbor nonparametric regression. Am. Stat. **46**(3), 175–185 (1992)
13. Harrell, Jr, Frank, E.: Regression Modeling Strategies: With Applications to Linear Models, Logistic and Ordinal Regression, and Survival Analysis. Springer (2015)
14. Campbell, C., Ying, Y.: Learning with support vector machines. Synthesis Lect. Artif. Intell. Mach. Learn. **5**(1), 1–95 (2011)
15. Liang, H., Lukkarinen, S., Hartimo, I.: Heart sound segmentation algorithm based on heart sound envelogram. In: Computers in Cardiology (1997) (pp. 105–108)
16. Li, F., Liu, M., Zhao, Y., Kong, L., Dong, L., Liu, X., Hui, M.: Feature extraction and classification of heart sound using 1D convolutional neural networks. EURASIP J. Adv. Signal Process. **59** (2019)

A Deep Learning Approach for Cardiac Arrhythmia Detection

N. Venkata Sailaja, S. Varun, A. Vinuthnanetha, G. Shravan, and K. Abhinav

1 Introduction

1.1 About Cardiac Arrhythmia

With the growing scope of Database Management and Network Technology, people want to get information within seconds. It is now possible to solve medical issues with emerging technologies. Data is everything in the healthcare industry. One of the costliest diagnosis is heart failure. Electrocardiograms (ECG) play a vital role in the diagnosis of Cardiovascular Diseases. The heartbeat of normal humans varies between 60 and 90 beats per minute. Cardiologists generally interpret ECG reports to identify different types of irregularities in the heartbeat. An ECG signal has P wave, QRS complex and T wave. For normal humans, these waves occur with specific amplitudes at the duration of 250–1300 ms. The duration of a P wave, QRS complex and T wave is 80 ms, 80–150 ms and 160 ms, respectively. Any variation in the defined timings is considered as an irregular heartbeat or Arrhythmia. This may lead to a wrong diagnosis for patients who may be in the earlier stages of cardiac

N. V. Sailaja (✉) · S. Varun · A. Vinuthnanetha · G. Shravan · K. Abhinav
Department of CSE, VNR VJIET, Secunderabad, India
e-mail: sailaja_nv@vnrvjiet.in

S. Varun
e-mail: varun.savai23@gmail.com

A. Vinuthnanetha
e-mail: vinuthnanetha99@gmail.com

G. Shravan
e-mail: shravan.gopani@gmail.com

K. Abhinav
e-mail: abhinavkarre.ak@gmail.com

C. K. Mai et al. (eds.), *Machine Learning Technologies and Applications*,
Algorithms for Intelligent Systems,
https://doi.org/10.1007/978-981-33-4046-6_25

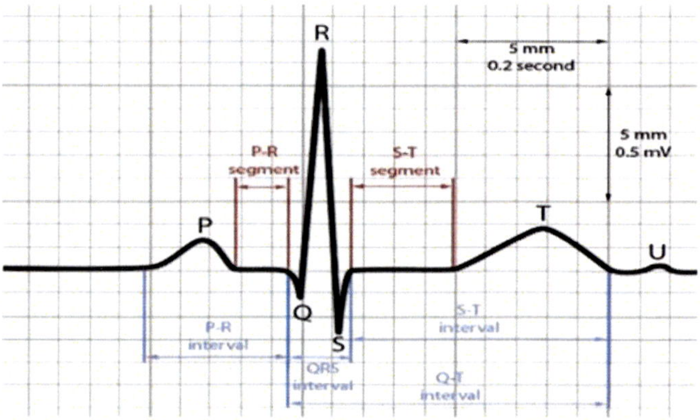

Fig. 1 Structure of ECG signal

disorders. Many feature extraction techniques from ECG signals were suggested in the literature such as linear and nonlinear features of HRV Signal, K-means clustering algorithm, Fuzzy neural networks, Dual-slope method, Dynamic features of ECG signals, Associative Petri net, Complexity analysis.

To have the classification with high accuracy and F1 score, we need to detect the QRS complexes of the ECG signals. PQ interval, QRS complex and T wave represent atrial contraction, ventricular depolarization, and contraction and repolarization of ventricles.

The QRS projection of ECG Signal is shown in Fig. 1.

Here in this work, we have used the database which is already available in Kaggle by de Chazal, where all the values are extracted using the Pam–Tompkins algorithm with every row on this dataset is a QRS Complex. Here we proposed a simple algorithm to identify four different cardiac arrhythmias using the ECG signal. These arrhythmias are namely the Supraventricular Ectopic Beat (SEB), Ventricular Ectopic Beat (VEB), Fusion Beat (FB), and the Normal Beat (NB). To classify ECG signals, we are using this Deep Learning NN method which gives the classification accuracy and F1 score.

1.2 Deep Learning

Deep learning emerged from machine learning family and is a subset of it. Supervised and unsupervised are 2 learning methods. In supervised learning, the classification of arrhythmia is defined by a particular set of classes possible and train the model with those classes to identify those different arrhythmias. The model classifies different types of arrhythmias by calculating the values from the available R-R peak values from the dataset. Different types of arrhythmias can be identified by implementing a CNN model. Trained data is sent to the CNN model to train and test data is used to

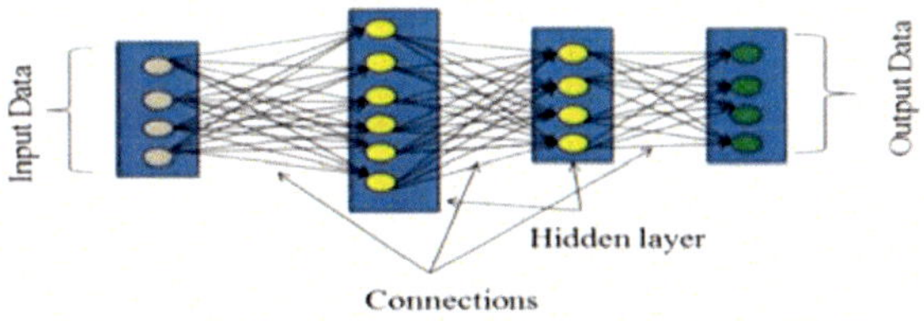

Fig. 2 Structure of any deep learning model

classify different types of arrhythmias. Any CNN model processes the data and classifies arrhythmia with the help of I/O layers, hidden layers and activation functions. The structure of a deep learning model is shown in Fig. 2.

2 Literature Survey

Soni et al. [1] provides a frequent feature choice methodology for cardiovascular disease prediction. The utilization of the fuzzy measure and also the related nonlinear integral provides decent performance. Using characteristics consisting of age, sex, blood stress, and blood sugar will predict the chance of patients obtaining coronary cardiovascular disease. And this improves accuracy and reduces the estimation time.

Mendes et al. [2] provides a decision tree model framework that makes use of a reduced set of 6 binary liability factors. Several health tracking systems use wearable sensors that process continuous knowledge and generate several false warnings. Hence, these systems become inappropriate to be used in clinical operations.

To resolve this drawback, some machine learning methods are described in [3], i.e., data produced by the wearable sensors are collaborated with clinical observations to give premature warning of critical physiological differences in the patients. Incorporating this data with manual observations, the clinical workers make major decisions about the patients.

To diagnose cardiovascular disease and provide appropriate medication Patil et al. [4] used 2 data mining classification techniques like Artificial Neural Network and Naive Bayes. The system is efficient with a simple setup, less power utilization, dominant performance, and time to time reflex.

An algorithm was generated by Azariadiet al. [5] for electrocardiogram analysis and classification of heartbeat and applied it on an IoT embedded platform. The algorithm is a proposal for a wearable ECG detection device, suitable for 24-h continuous observation of patients.

A weighted fuzzy rule-based CDSS was presented by Anooj [6] for the detection of CVD. Useful information from the patient's clinical data is collected automatically. Using the datasets obtained from the UCI repository experiment is carried out and the performance of the system is differentiated with the neural-network-based system using accuracy, sensitivity and specificity.

Ansari [7] implemented work in the year 2011 where a neuro-fuzzy integrated architecture for cardiovascular disease is presented. To display the effective results of the architecture, automated evaluation is performed by making use of the true causes of cardiovascular heart disease. The results show that this type of hybrid architecture is acceptable for recognizing patients with high/low cardiovascular danger.

3 Implementation

3.1 Proposed Predicted Process

Identifying diseases is a very crucial task for doctors in minimum time. Arrhythmia is nothing but a change in the speed or rhythm of the heartbeat. An arrhythmia occurs due to a difference in heart muscle and electric signals which influences the heartbeat. These differences might be by disease, injury, or hereditary damage. To identify arrhythmia ECG test is most commonly used. Since it is very difficult to identify arrhythmia with the naked eye, some algorithms or formulas are implemented to provide accurate results or findings to the doctors. To solve the issue Deep Learning can be used. So, ECG Data is used as a dataset, and using a trained neural network and some data preprocessing methods can resolve the issue. The data collection is obtained from the Arrhythmia Archive of MIT-BIH. 51,002 recordings are collected for the work. The content presented in the database was extracted using the Pam–Tompkins algorithm, representing the QRS Complex with very row on this dataset is a QRS Complex. Pan–Tompkins proposed an algorithm where he used a series of filters to show the frequency content of rapid depolarization of the heart and removes background noise or disturbances if any. Then, the output signal is multiplied twice to increase the QRS contribution. Eventually, adaptive thresholds are used to measure the peaks of the filtered signal. Electrocardiogram morphology, pulse meantime, and R-R peak distances are considered as feature sets. "R-R interval" is measured as the time gap between each beat of the heart and is measured in milliseconds. Even though heart rate focuses on average beats per minute, heart rate variability keeps track of real-time changes between successive heartbeats. Here 0 means Normal, 1 means Supraventricular ectopic beat, 2 means Ventricular ectopic beat, 3 means Fusion Beat. In those 51,002 recordings dataset 89% belong to Class 0, 7% belong to Class 1, 1% belong to Class 2, and 0.08% belong to Class. Later these 51,002 records are split into training and testing.

To achieve the above objectives, we have proposed the prediction process to identify arrhythmia. The architecture of the prediction process is mentioned in Fig. 3.

3.2 Implementation

To implement the above-proposed process we have used Windows 10 Pro, Python 3.7.3, Kaggle as software.

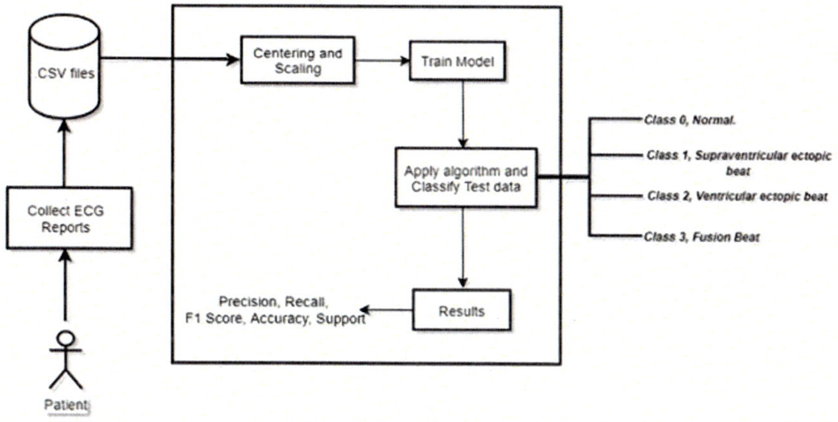

Fig. 3 Architecture of prediction process

From the above figure, the steps involved in the implementation are:

Step 1: Preprocessing

The collected dataset from the MIT-BIH database is already preprocessed and made available by de Chazal [8]. Out of 51,002 recordings available, 89.8% belong to Class 0, 7.4% belong to Class 2, 1.9% belong to Class 2 and 0.08% belong to Class 3. Figure 4 shows the data in each class.

Step 2: Stratified Shuffle Split of data

Split the data to train and test from the dataset. Since there are not many classes with 1 to 3, the data cannot be split randomly. To keep data with these classes in the train and test data, a stratified data split is performed with test size 0.2. Of 51,002 samples, 40,801 are train data samples and 10,201 are test data samples. The percentage of classes present in the test data split will be the same as that of the present in the complete dataset using the stratified data split.

Fig. 4 Percentage of data of each class

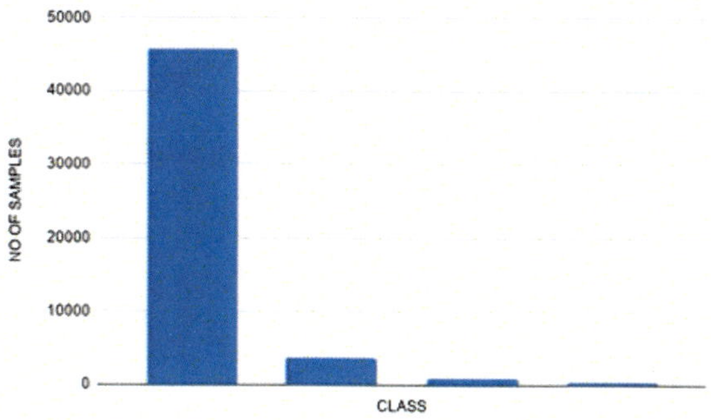

Step 3: Data standardization

Before running the model, standardization has to be implemented on the data. Standardization is to center the available data around 0 and to scale to the standard deviation.

$$x_{\text{standardized}} = \frac{x-\mu}{\sigma},$$

where μ is mean and σ is the standard deviation of the dataset.

Step 4: Build and train the neural model

A neural model is built using TensorFlow and Kera's, which has an input layer, 2 hidden layers, 1 with 200 and the other with 100 neurons. For the output layer 4 neurons are used, one for each class. A ReLu activation function for the hidden layer and a SoftMax function for the output layer is used. This model is trained with 40,801 train samples which were previously divided using *Stratified Shuffle Split.*

Step 5: Test the model

In the testing phase, the remaining 10,201 samples are tested and metrics such as Precision, Accuracy, F1 Score, Confusion Matrix, Support, and Recall are measured.

3.3 Algorithm

1. Take ECG dataset from the MIT-BIH database.
2. Perform stratified shuffle split into the data so that the percentage of classes present in the training and test data will be the same as that of the complete dataset.

    2.1. Use StratifiedShuffleSplit() split data into train and test data.
    2.2. Label each split as trainfeatures, train_labels and test_features, test_labels.

3. Using StandardScaler() to standardize the data.

    3.1. Use fit_transform() on train_features and label as std_features_train.
    3.2. Use transform() on test_features and label as std_features_test.

4. Using Sequential() build and train the neural model on train data (trainfeatures, train_labels).

    4.1. Create a model with 200 neurons for input layer, 100 neurons for hidden layer, and 4 neurons for output layer with ReLu and Softmax activation functions. 4.2 Using to_categorical() convert the labels to categorical.
    4.2. Use SGD() as optimizer. 4.3.1 Fixing parameters in SGD() lr = 0.01 decay = le-6 momentum = 0.9 nesterov = Trule.
    4.3. Using compile() to compile.

4.4.1. Fixing parameters in compile() loss = categorical_crossentropy optimizer = sgd metrics = accuracy.

4.5. Use evaluate() to evaluate to find the scores using train data and batch size.

5. Using predict() test the model on test data 5.1 Print accuracy, f1 score, confusion_matrix, recall, precision and classification_report.

4 Results

The classified ECG signals are provided as input to Deep Neural Network. The 40,801 sample recordings of ECG signals are used for training the DNN and 10,201 samples are used for testing. The network thus classifies the signals into the normal heartbeat and three kinds of cardiac arrhythmia. We conclude that the proposed solution dependably differentiates the four kinds of heartbeats based on the R-R interval of the ECG signals which has been validated over the entire MIT-BIH arrhythmia database and yields an accuracy of 98.7% and 0.9 F1 score. Experimental results are tabulated within the below tables. We investigated the general outline of the model and derived the classification report and metrics for individual classes (Fig. 5) and also metrics such as confusion matrix, precision, accuracy and F1 score (Fig. 6).

Figure 7 shows the distribution of each class predicted. Out of 10,201 samples given as test data to the model 9112, 153, 72,568 samples are correctly predicted as Normal Heartbeat, Supraventricular Ectopic Beat, Ventricular Ectopic Beat, and Hybrid Heartbeat.

Precision–Recall metric is used to estimate classifier result quality. It is a useful measure for success prediction when the classes are very imbalanced. A high area under the curve represents both high recall and high precision, where high precision

Fig. 5 Metric values for individual classes

	Precision	Recall	F1-Score	Support
NB	0.99	1.00	0.99	9165
SEB	0.87	0.71	0.78	195
VEB	0.97	0.98	0.97	758
FB	0.89	0.82	0.86	83

Fig. 6 Metric values for overall model

Metric	Value
Accuracy	0.987
F1 Score	0.90
Recall	0.87
Precision	0.93

ACTUAL

PREDICTED	Normal Heartbeat	Supraventricular Ectopic Beat	Ventricular Ectopic Beat	Fusion Beat
Normal Heartbeat	9112	36	12	5
Supraventricular Ectopic Beat	41	153	1	0
Ventricular Ectopic Beat	15	7	725	11
Fusion Beat	12	0	3	68

NB - Class 0
SEB - Class 1
VEB - Class 2
FB - Class 3

Fig. 7 Confusion matrix

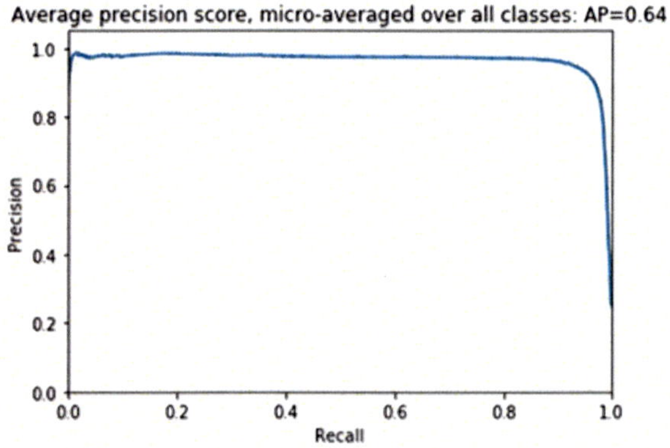

Fig. 8 P-R curves over all classes

relates to a low false-positive rate, and high recall relates to a low false-negative rate. High scores for both show that the classifier is returning accurate results (high precision), as well as returning a majority of all positive results (high recall). The following graphs (Figs. 8 and 9) depict P-R curves for each class and the average precision graph micro averaged over all classes.

5 Conclusion

The irregularity within the heartbeat is known as cardiac arrhythmia. Cardiac arrhythmia leads to issues like heart failure, stroke and additionally cardiovascular arrest that ends up passing away. Therefore, recognizing and treating the heart abnormality at the right time is necessary. If the cardiac arrhythmia is recognized, then the detection of the type of arrhythmia is very feasible. We have developed a model

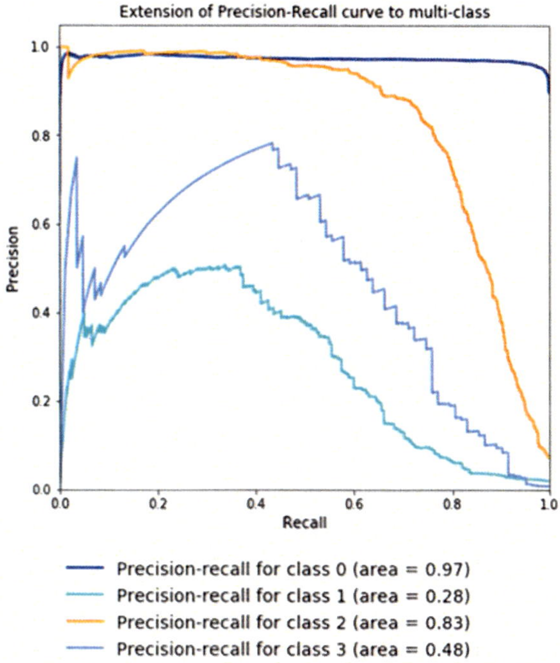

Fig. 9 P-R curves for each class

using deep neural networks that classify the ECG signal recordings into 4 types including normal and Supraventricular Ectopic Beat, Ventricular Ectopic Beat and Hybrid Heartbeat by using MIT-BIH cardiac arrhythmia database.

References

1. Soni, S., Soni, J., Ansari, U., Sharma, D.: Predictive data mining for medical diagnosis: an overview of heart disease prediction. Int. J. Comput. Appl. (IJCA, 0975 8887) **17**(8) (2011)
2. Mendes, D., Paredes, S., Rocha, T., Carvalho, P., Henriques, J., Cabiddu R., Morais, J.: Assessment of cardiovascular risk based on a data-driven knowledge discovery approach. IEEE (2015)
3. Collins, F.S.: Mobile technology and healthcare. https://www.nlm.nih.gov/medlineplus/magazine/issues/winter11
4. Patil, A.B.C., Sonawane, S.S.: To predict heart disease risk and medication using data mining techniques with an IoT based monitoring system for post-operative heart disease patients. In: Sixth Post Graduate Conference for Computer Engineering (cPGCON 2017), Proceedings International Journal on Emerging Trends in Technology (IJETT)
5. Azariadi, D.: Tsoutsouras, V.: Xydis, S.: Soudris, D.: ECG signal analysis and arrhythmia detection on IoT wearable medical devices. In: 5th International Conference on Modern Circuits and Systems Technologies (MOCAST) (2016)
6. Anooj, P.K.: Clinical decision support system: risk level prediction of heart disease using weighted fuzzy rules. Production and hosting by Elsevier

7. Ansari, A.Q.: Automated diagnosis of coronary heart disease using the neuro-fuzzy integrated system. In: Information and Communication Technologies (WICT) (2011)
8. Chazal, D.: Automatic classification of heartbeats using ECG morphology and heartbeat interval features. https://www.ncbi.nlm.nih.gov/pubmed/15248536

Multi-classification for Cardiac Arrhythmia Detection Using Deep Learning Approach

P. Subhash, Pathuri Goutam Sai, Nalla Rohith Reddy, Anurag Pampati, and Sai Keerthan Palavarapu

1 Introduction

Cardiac arrhythmia is the main ongoing ailment that is normally observed in individuals around regardless of the age of the individual. Anomaly in the heartbeat prompts the event of arrhythmia, around 2–3 lakh unexpected passings happen each year which is higher than the individuals confronting demise with lung or bosom diseases. Early defibrillation will assist the individuals with surviving however in the majority of the cases usage, what's more, handling is moderate and that is the primary motivation behind why the passing rate is 15–20% of death tallying every day. These sorts of infections need a quick activity to take and if not, they may prompt end up their life. To keep up a track on the consistency or anomaly of heartbeat and to beat the cardiovascular arrhythmia (changes in heart pulsates) ECG information is taken, handled and the status of the arrhythmia is affirmed.

ECG is an electrocardiogram which is commonly used to quantify the heartbeat rate and produce the signs which are made from waves. On the off chance that the heartbeat is quick, it is considered as tachycardia and moderate heartbeat is delegated bradycardia which will be valuable for an individual to deal with his/her

P. Subhash (✉) · P. G. Sai · N. R. Reddy · A. Pampati · S. K. Palavarapu
Department of CSE, VNR VJIET, Hyderabad, Telangana, India
e-mail: subhash_p@vnrvjiet.in

P. G. Sai
e-mail: pathurigoutamsai@gmail.com

N. R. Reddy
e-mail: nallarohithreddy.nrr@gmail.com

A. Pampati
e-mail: anuragpampati696@gmail.com

S. K. Palavarapu
e-mail: saikeerthan.p123@gmail.com

C. K. Mai et al. (eds.), *Machine Learning Technologies and Applications*,
Algorithms for Intelligent Systems,
https://doi.org/10.1007/978-981-33-4046-6_26

wellbeing conditions. An ECG signal is created with waves which speak to the electrical possibilities of the heartbeat with cathodes set on the chest and appendages. These gather the heartbeat developments and appear in types of waves.

To be more precise the ECG data is initially preprocessed, and the filtered data is considered as actual data, next this data is converted into images and the intermissions of peaks are calculated, then compared with the next peak, if the distance between the peaks is more than it is considered as slow heartbeat which resembles bradycardia else if the distance between the peaks is less than it is considered as fast heartbeat and considered as Tachycardia.

2 Existing System

In the current framework, the information for building the model is taken from the MIT-BIH dataset. Here the prepreparing of signs is done to evacuate gauge, power line, low recurrence clamor, and igh recurrence commotions present in the dataset for progressively precise outcomes. After prepreparing the extraction of the highlights is finished by utilizing a convolutional neural system.

Albeit a large portion of the ordinary example acknowledgment procedures have been recently applied effectively to the ECG arrhythmia identification assignments, ongoing best in class exhibitions got by profound learning strategies, especially Convolutional Neural Networks, in mainstream design acknowledgment moves urged analysts to execute these methods to the field of the clinical picture and sign preparing. The information can be signals or images, however in the current framework the information sources are just signals to the model which they assemble will offer yields to just such signals. The QRS top is recognized, however, there are no divisions of pulses is finished. Here the signs are prepared for building a model. The quantity of groups shaped here are not many.

2.1 Literature Study

Sannino and De Pietro in [1] had at first played out the sign handling by De-noising, Detecting Peaks, portioning it and temporal highlights extraction is performed by expelling power line obstruction and pattern meandering. For the forecast procedure in the PC vision and normal language handling, they have utilized the DNN calculations by changing the quantity of shrouded layers, the enactment work, the quantity of learning steps and arrhythmia is anticipated, and the representation of arrhythmia is finished by WEKA Tool.

Van Zaen et al. in [2] proposed a framework with the keen vest that utilizes two dry bi-anodes to record an ECG signal and the sign is then gushed by means of a portal to the cloud where a neural system distinguishes and groups the heart arrhythmias. Here they have chosen a design that consolidates convolutional and repetitive layers, they intend to total a few databases of ECG signals with beat comments to stretch out the kinds of arrhythmias to improve their exactness.

In [3], Jeong-Hwan Kim et al. have investigated the impact of itemized parameters by introducing a model reasonable for the assessment of ECG arrhythmia by means of different profound learning models. ECG arrhythmia database was utilized to assess arrhythmia grouping with changing of the commencement structure to characterize LBBB (left pack branch square), RBBB (right group branch square), PVC (polyvinyl chloride), and APC (premature atrial edifices) mood and now the classification results contrasted and the best in class PVC and APC pulses as far as rate and contrasted and the past best in class concerning the order of pulses to foresee the arrhythmia.

In [4], C. K. Roopa et al. procured a dataset from MIT-BIH and the investigation on the dataset is finished by Fuzzy-Based Techniques, yet when this is contrasted with neural systems the outcomes are unacceptable consequently, they embraced versatile fluffy ECG classifier for upgrading the exhibition of regular classifiers. Harsh Set Theory and Hidden Markov Models are utilized for diminishing vulnerabilities and uncertainty from the information. Characterization of arrhythmias, for example, Genetic calculations, Fuzzy Logic, Self-Organizing Map, Bayesian, Hidden Markov Models and SVMs to produce ECG signal examples, however, this is far less when contrasted and the precision picked up when neural systems are utilized.

Isin and Ozdalili developed a model [5] in which the ECG signals from the MIT-BIH dataset are preprepared to evacuate DC noise and electrical cable impedance, at that point the QRS fragment or top in the sign is distinguished utilizing the Pan–Tompkins Algorithm. At that point, the signs are changed over to pictures, and the pictures are taken care of to a profound learning system recently prepared on a general picture informational index to complete programmed ECG arrhythmia diagnostics. Moved profound convolutional neural system (in particular AlexNet) is utilized as an element extractor and the extricated highlights are taken care of into a straightforward backpropagation neural system to do the last order.

Izci et al. [6] proposed a significant learning-based novel 2D convolutional neural framework (CNN) approach for the exact request of five remarkable arrhythmia types. ECG signals from the MIT-BIH database were segmented into beats and every one of the beats was changed over into 2D grayscale pictures as a snippet of data for CNN structure. The division is finished by WFDB Toolbox and these photos are urged to LeNET structure for the request methodology. At this moment, process, the signs are not pre-dealt with, and it is considered by changing over them to grayscale pictures.

Sivanantham and Devi proposed [7] that the ECG signals from the MIT-BIH dataset are to be pre-taken care of using a bandpass channel, by then the QRS partition or then again top in the sign is recognized using the Hamilton and Pan–Tompkins Calculation. The part extraction is taking into account beat variability (HRV) that hails by isolating different features in the time region, repeat space besides, nonlinear features from HRV signals, the removed features are continued into a Support Vector Machine to pass on out the last request.

de Albuquerque and Nunes developed an AI approach [8] that uses Optimal Path Forest, a Directed Learning Technique. To choose the best count, they in addition used Support Vector machine estimation which gives high precision than Optimum Path

Forest, But, Ideal Path Forest is progressively capable than support vector machine and never encounters overfitting.

In [9], Xin Gao made a significant learning approach using a convolutional neural framework alongside Back inducing figuring and Long Short-Term Memory (LSTM) strategy. For the classification of significant level ECG Signals, SoftMax classifier is used. At this moment, ECG data tests are acquired to set up a significant learning model.

Niharika Pattanaik and Ipsita Mohapatra executed a significant learning machine [10] where Discrete Wavelet Transform (DWT) is used for QRS partition revelation and SoftMax commencement work is used as a classifier for arrhythmia distinguishing proof.

Warrick and Homsi proposed [11] a structure that is a blend of Convolution Neural Systems and a movement of Long Short-Term Memory units, with pooling, dropout and normalization methodologies to improve their exactness. They contemplated the P, Q, R, S, and T between times. The yield of one CNN layer was given as commitments to 3 LSTM's. They created the CL3 model. They developed this by using Python Keras library and tensor stream as a backend. The entire model is set up by restricting the cross-entropy botch.

In [12], Pooja D. V. Gupta, Surender Jangra used redesigned mutt classifiers. Here every ECG beat was changed into two-dimensional data as information data for the cross-breed classifier. They used diverse improvement frameworks, for instance, genetic figuring, and cuckoo search estimation with perfect objective work. They used MIT-BIH instructive assortments for the appraisal of their classifiers. They contemplated the R-R break and the proposed strategies are to gain perfect R-R between times. Reinforced vector machines and artificial neural frameworks are used close by the GA additionally, CS figurings.

In [13], S. Celin and K. Vasanth used the electrical signs data for foreseeing the time of the ailment. The ECG signal is depicted with different zeniths P, Q, R, S, T, and U. They pondered PR interval QRS complex, QT between time, PR divide, and ST segment. They done the pre-getting ready of the data subject to the degree of chance. Examination of the data consolidates the separating of elements into less troublesome parts. The calm signs are set up through any window, for instance, limit window to gain the QRS between times. They isolated their features by considering the direct besides, nonlinear systems.

In [14], a mix of incorporated extraction and a directed learning computation using CNN is proposed for the portrayal of arrhythmia. The ECG signals from the MIT-BIH dataset are denoised and filtered to remove DC fuss and power line obstacle, by then the QRS segment or top in the sign is recognized using the Bi-orthogonal Wavelet Filter. The features are removed and requested using a significant learning model that includes CNN.

Ji and Zhang developed a significant learning approach [15] using Convolutional Neural System to remove features of ECG images (2D). These ECG pictures (2D) are made from ECG Signals (1D) at the pre-taking care of stage. Using Faster R-CNN, feature map produces 20,000 boxes out of which starting 300 boxes keep up high exactness.

The framework modeled in [16] uses a Convolutional Neural System (CNN), a DL figuring which is capable of portraying signals. Utilizing CNN, features are discovered naturally from the time–space ECG signals. The signs are picked up from the MIT-BIH database. The CNN is readied, had a go at using ECG Dataset gotten from MIT-BIH Database and from the sign 7 sorts of arrhythmia were orchestrated. There is no pre-taking care done by the makers and the CNN is prepared using Stochastic Slope Descent count.

The work [17] has realized a significant learning plan where the main layers of convolutional neurons fill in as feature selectors and, at long last, some totally related (FCN) layers are used for settling on the last decision about ECG classes. The signs from the MIT-BIH database are sent to CNN without pre-getting ready. The CNN goes about as a customized feature extractor and classifier. It isolates the features and endeavors to assemble the ECG signals into 4 gatherings.

The work present in [18] is heart arrhythmia acknowledgment using the blend of a heartbeat changeability (HRV) assessment and "ability of lopsided complex vitality" (PUCK) assessment. From the outset, the R-R between time data were removed from the MIT-BIH database. HRV sees at that point were driven using the joined R-R between time data, including a period space assessment, repeat space examination, likewise, nonlinear assessment. Despite the HRV assessment, PUCK examination was used. A decision tree estimation was applied to all the obtained features for portrayal. Feature extraction execution was improved by including features isolated from the PUCK examination.

Fajr Ibrahem Alarsan and Mamoon Younes developed a Machine Learning model [19] to distinguish arrhythmia using ECG signals. The ECG signals from the MIT-BIH database experienced a Discrete Wavelet Transform (DWT) for the feature extraction process. Features can be assembled into three classes summits features, temporal features, and morphological features. The preplanning is done, and choice tree and random boondocks classifiers are used to portray the ECG signs to distinguish the possible arrhythmias in the patient.

The research carried out in [20] is an assessment from the start of ECG Preprocessing and Beats Segmentation is performed by applying channel with a recurrence reaching to dispense with meandering and commotion, for smoothing of the ECG signal that was finished by applying a moving normal channel which is utilized for clamor decrease channels in signal handling and the higher demand phantom estimations are assessed, spared, and preprepared utilizing convolutional neural systems calculation. Move learning procedures like AlexNet is applied on pre-trained CNN to get the last arrangement.

3 Proposed System

In the model which amassed, the data is taken from the MIT-BIH dataset. The initial step is to preprocess the signs to remove the check, power line, low frequency and high-frequency uproars present in the dataset. By then the division of signs into heart throbs is done by recognizing the QRS peak and the R-R interval time. QRS complex is the most striking waveform inside the ECG. Since it reflects the electrical development inside the heart during the ventricular withdrawal, the hour of its occasion similarly to its shape gives critical information about the flow state of the heart.

An outstanding Pan–Tompkins computation is applied to pass on the QRS acknowledgment. The count fuses a movement of procedures that perform subordinate, making sense of, blend, flexible thresholding and look systems for the acknowledgment of R-peaks of the ECG signal. The beats are changed over into pictures using OpenCV and Matplotlib libraries of Python language. The part extraction is done by the convolutional neural framework which follows the VGG Net designing. The CNN architecture performs automatic feature extraction and classification steps.

The information can be signals or images, so any sort of data sources can be surrendered to the framework for getting the yields. This paper will have a division of heartbeat in the proposed framework, for example, division is a method of masterminding into sub-bunches that ordinarily have separate needs. Heart thumps are changed over into pictures. Here we will have QRS top identification and the recognition of R-R interval. This will have numerous quantities of beats with the end goal that can distinguish the state of the patient decisively. On the off chance that the exactness is less, at that point the design is changed naturally and determined once more. This procedure is done until this show signs of improved precision. Subsequently, the proposed framework has a wide scope of points of interest while contrasted with the current framework.

4 Result

Convolutional Neural Network (CNN) model is built up that consequently performs features extraction and classification. The architecture of the CNN is like that of the VGG Net. To start with, in the wake of downloading the dataset have played out an information division process. The whole ECG signal is isolated into singular beats. In the subsequent, advance, have performed QRS Peak location and R-R Interval identification utilizing the inbuilt capacities in the BioSPPy python library. In the later advance, have performed Images transformation where to convert the ECG portioned signals into pictures. To accomplish better exactness and affectability, we have performed data augmentation in this stage. The information present is enlarged by editing the picture. Each editing technique brings about two of three sizes of an ECG picture that is 96×96. At that point, these expanded pictures are resized to the first size which is 128×128. In spite of the fact that this progression brings about loss of speed of characterization, it improves the accuracy exponentially.

Presently, these pictures are taken care of to the CNN model where the programmed feature extraction and classification happens. This CNN is prepared utilizing the information from the MIT-BIH database. The information given to this model is part of testing and preparing information. Here 80% of the information for preparing and 20% of the information for testing is given. The model prepared

Table 1 Confusion matrix of the CNN model

Predicted	NOR	PVC	PAB	RBB	LBB	APC	VEB
NOR	74,726	151	2	15	18	110	0
PVC	149	6947	1	3	7	11	0
PAB	19	13	6991	2	1	1	0
RBB	83	3	2	7159	1	9	2
LBB	82	32	1	0	7957	1	1
APC	186	17	0	5	1	2337	0
VEB	6	1	0	0	0	0	99

would now be able to arrange the new ECG signals into various classifications of heart issue.

This model is made available on the web through the Python Flask framework. The user can upload their ECG data into the web site and get the results within minutes. The data that is uploaded to the website is stored in a database. The data is then presented to the model to perform classification. After the classification is performed the result is stored in the database and sent to the web server for the user to view his results.

The accuracy of the CNN model is measured when presented the data while testing the model. Here 20% of the data from the MIT-BIH database to find the accuracy of the model is presented. It has achieved 99.23% accuracy using this CNN architecture.

Another technique to test the exactness and the affectability of the model is to compute the confusion matrix. It is a table that is regularly used to figure the presentation of a model. It is commonly utilized for classifiers on a lot of test information. The labels or classes to ascertain the confusion matrix should be known in advance. The confusion matrix of our model is given in Table 1.

The user can upload the ECG signal to the website and get the results within no time. The result will look like a pie chart that is given in Fig. 1 that depicts the percentage or the number of beats that are normal and the percentage of beats that are abnormal and categorize them into subclasses.

The CNN model that was developed is fast and accurate. The speed of the classification depends on the size of the file that uploads into the webserver. Generally, the size of an average ECG signal ranges from 30 to 50 KB. It takes about 3–5 s to process the files of that size. An experiment to find out the time required to classify different sizes of files is performed and observation is illustrated in Fig. 2.

5 Conclusion

Employing automation of the detection and classification of Cardiac Arrhythmia, it is altogether progressively strong for the cardiologists, enabling them to focus more on treatment instead of diagnostics. At present, a capable convolutional neural

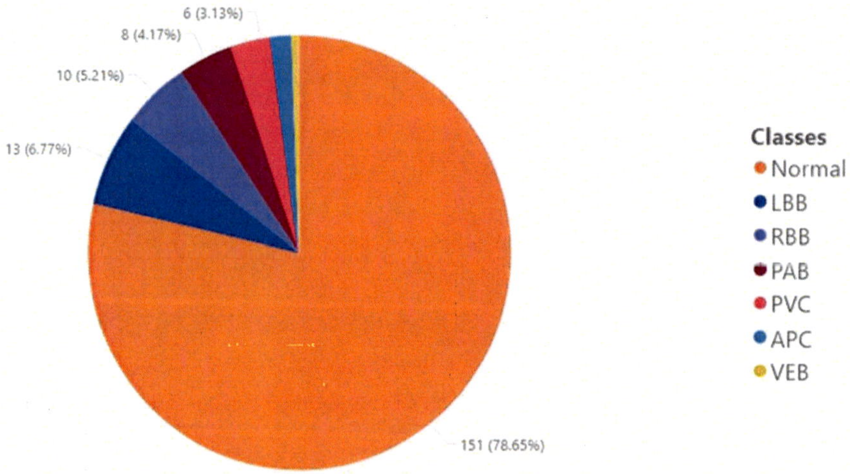

Fig. 1 Result of the classification of different types of arrhythmia

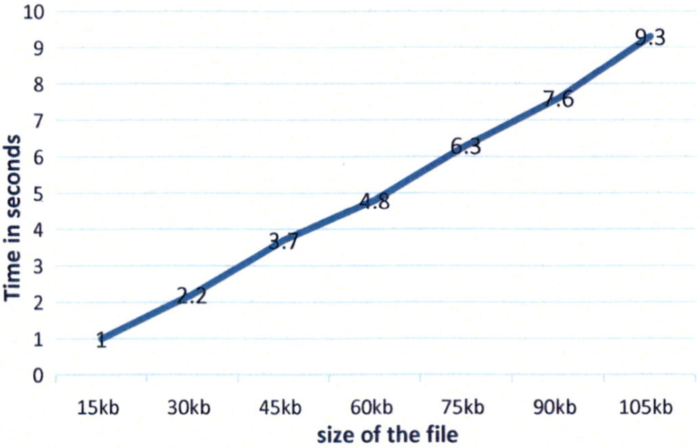

Fig. 2 The average amount of time to classify different file sizes

framework is used for the ECG portrayal. ECG portrayal structure is recognized to do modified ECG arrhythmia diagnostics by describing persevering ECG's. After the ECG records are gotten from the MIT-BIH data store, they are pre-taken care of or isolated by using dish—Tompkins algorithm. This pre-dealing helps in diminishing the uproar data. QRS waves are recognized for the extraction of R-T segments of the ECG. By using the OpenCV which is a Python library the signs are changed over into pictures. Successfully arranged CNN is moved likewise, it is used as a segment extractor for the ECG request. These isolated features are empowered into a back expansion neural framework to aggregate the data ECG R-T divides into anyone referenced condition. Moving a readied significant convolutional neural framework discards the necessity for authority and computational power required for setting up a significant convolutional neural sort out without any planning. With the progressing presentations of significant learning-based therapeutic picture and and sign dealing with procedures, biomedical analysts are one stage closer to PC maintained decisive structure.

References

1. Sannino, G., De Pietro, G.: A deep learning approach for ECG-based heartbeat classification for arrhythmia detection. Future Generat Comput Syst **86** (2018)
2. Van Zaen, J., Chetelat, O., Lemay, M., Calvo, E.M., Delgado-Gonzalo, R.: Classification of cardiac arrhythmia from single lead ECG with a convolutional recurrent neural network. In: Proceedings of the International Joint Conference on Biomedical Engineering Systems and Technologies (BIOSTEC19) (2018)
3. Kim, J.-H., Seo, S.-Y., Song, C.-G., Kim, K.-S.: Assessment of electrocardiogram rhythms by GoogleNet deep neural network architecture. J. Healthcare Eng. (2019)
4. Roopa, C.K., Harish, B.S.: A survey on various machine learning approaches for ECG analysis. Int. J. Comput. Appl. **163** (2017)
5. Isin, A., Ozdalili, S.: Cardiac arrhythmia detection using deep learning. In: International Conference on Theory and Application of Soft Computing, Computing with in Words and Perception, ICSCCW (2017)
6. Izci, E., Ozdemir, M.A., Degirmenci, M., Akan, A.: Cardiac arrhythmia detection from 2D ECG images by using deep learning technique. In: IEEE Medical Technologies Congress, TIPTEKNO (2018)
7. Sivanantham, A., Devi, S.S.: Cardiac arrhythmia detection using linear and nonlinear features of HRV signal. In: IEEE International Conference on Advanced Communication Control and Computing Technologies (ICACCCT) (2014)
8. Hugo, V., de Albuquerque, C., Nunes, T.M., Pereira, D.R.: Robust automated cardiac arrhythmia detection in ECG beat signals. Neural Comput. Appl. (2018)
9. Gao X.: Diagnosing Abnormal Electrocardiogram (ECG) via Deep Learning. Electrocardiography IntechOpen (2019)
10. Pattanaik, N., Mohapatra, I., Mohanty, M.N.: Arrhythmia detection using deep learning. Int. J. Eng. Technol. (2018)
11. Warrick, P., Homsi, M.N.: Cardiac arrhythmia detection from ecg combining convolutional and long short-term memory networks. In: Computing in Cardiology, vol. 44 (2017)
12. Sharma, P., Gupta, D.V., Jangra, S.: Ecg signal based arrhythmia detection system using optimized hybrid classifier. Int. J. Innov. Technol. Explor. Eng. (IJITEE) (2019)
13. Celin, S., Vasanth, K.: Survey on the methods for detecting arrhythmias using heart rate signals. J. Pharmaceut. Res. **9** (2017)
14. Pomprapa, A., Ahmed, W., Stollenwerk, A., Kowalewski, S., Leonhardt, S.: Deep learning of arrhythmia analysis based on convolutional neural network. Int. J. Bioelectromag. **21** (2019)
15. Ji, Y., Zhang, S., Xiao, W.: Electrocardiogram classification based on faster regions with convolutional neural network. Sensors MDPI (2019)
16. Rajkumar, A., Ganesan, M., Lavanya, R.: Arrhythmia classification on ECG using deep learning. In: 5th International Conference on Advanced Computing & Communication Systems (ICACCS) (2019)
17. Pyakillya, B., Kazachenko, N., Mikhailovsky, N.: Deep learning for ECG classification. IOP Conf. Ser. J. Phys. (2017)
18. Mahananto, F., Igasaki, T., Murayama, N.: Cardiac arrhythmia detection using combination of heart rate variability analyses and PUCK analysis. In: 35th Annual International Conference of the IEEE EMBS (2013)

19. Alarsan, F.I., Younes, M.: Analysis and classification of heart diseases using heartbeat features and machine learning algorithms. J. Big Data (2019)
20. Alquran, H., Alqudah, A.M., Abu-Qasmieh, I., Al-Badarneh, A., Almashaqbeh, S.: ECG classification using higher order spectral estimation and deep learning techniques. Neural Netw. World **4** (2019)

Human Age Estimation Using Support Vector Machine

A. Madhavi, G. Bhuvana Sree, V. Shriya, B. Shanmukh, and T. Harshitha

1 Introduction

Age estimation is one of the main approaches to facial image classification. In other words, it can be defined as the determination of a human's age and gender from the picture. Human facial image processing has become an active and interesting research topic for years. Human faces provide a lot of information and have drawn lot of attention and thus have studied very deeply and intensively.

Artificial intelligence is the inducing of human knowledge and capabilities to machines so that the machines can imitate the actions of humans without human intervention. It is one of the most advancing technologies which meet the present-day requirement. Alexa, Google Home, Siri are a few leading examples of AI. Machine learning is one of the applications of artificial intelligence. It elaborates the development of computer programs. Here, we train computers in terms of intelligence such that it acts like humans. It has two parts precisely training and testing the data. In the training part, we train the system in such a way that it acquires all the knowledge that a human possesses. In the testing stage, we try to test the knowledge that the system has after training. The major advantage of machine learning is that it

A. Madhavi (✉) · G. Bhuvana Sree · V. Shriya · B. Shanmukh · T. Harshitha
Department of Computer Science and Engineering, VNR VJIET, Hyderabad, Telangana, India
e-mail: madhavi_a@vnrvjiet.in

G. Bhuvana Sree
e-mail: g.bhuvanasree@gmail.com

V. Shriya
e-mail: velurushriyachowdary@gmail.com

B. Shanmukh
e-mail: shanmukhkrishna459@gmail.com

T. Harshitha
e-mail: harshithatirumani8@gmail.com

C. K. Mai et al. (eds.), *Machine Learning Technologies and Applications*,
Algorithms for Intelligent Systems,
https://doi.org/10.1007/978-981-33-4046-6_27

improves accuracy and efficiency over time and usage. The only major disadvantage is the loss of jobs for unskilled labor. Machine learning has two types of algorithms. They are supervised learning algorithms and unsupervised learning algorithms. Supervised learning has a training mode whereas unsupervised is a self-training algorithm. Supervised model is made to build a prediction. In our approach, we use an algorithm of supervised machine learning known the support vector machine algorithm. This algorithm can be used for classification and regression. In SVM, we can classify data in more than two ways, i.e., it is not restricted to binary classification, unlike many other algorithms. The SVM algorithm helps to improve data accuracy by dividing the classification of data into hyperplanes.

The HaarNet is a global trunk network, like any other trunk network, the HaarNet helps carry data simultaneously through different trunks which converge to form a simple data at the end. HaarNet helps us to effectively evaluate the features of a face. We use this technique to identify the features of the faces that are directly facing the camera. Also, this architecture goes very well with both deep neural networks and also the support vector machine.

The SVM is a machine learning model that consists of numerous kernels. These kernels help us through the training process [10]. As we know our system learns 80% from the training it has acquired, so SVM has a title role to play out in the estimation of age and gender of a person. Since the SVM algorithm has hyperplanes, these hyperplanes help classify the data more accurately which yields in a more precise prediction [1]. Although the SVM is an age old algorithm, for the topic and genre of our system it suits our needs best and also goes well with the rest of the algorithms in the system.

Another algorithm that we made use of in our system is linear discriminant analysis [19]. The linear discriminant analysis is a dimensionality reduction technique. In simple terms, it helps reduce the vast dimensions that the image is spread over into little dimensions without diminishing any major features. The LDA is used in this system because we hope to reduce dimensionality, which in turn makes it easier to process [7]. LDA does this job without eliminating any prominent data in the process. The LDA makes use of Eigenvectors and Eigenvalues to make this possible.

Deep learning is another aspect of machine learning. Deep learning primarily learns from large amounts of data to solve complex problems [19]. Under deep learning, there is an aspect of deep neural networks. Deep neural networks are used majorly to understand and process images. It is very efficient and effective in image recognition and classification. A deep neural network effectively distinguishes one image from the other. Since we process the images of people, we employ DNN for the effectiveness of the algorithm. In our approach, we employ DNN so that we can recognize images in which the face of the person is not in the direction of the camera, i.e., it is sideways. DNN helps to acquire the data of these faces through the process of pooling and mapping. Apart from these, we also make use of linear discriminate analysis, it helps to generate separation between two known categories for better decision-making. It creates a new axis for the results generated. Here, we make use of linear discriminate analysis to achieve proper separation between the obtained predictions according to the specified and trained parameters. Most of these attentions led in the direction of exploring things using face recognition [5]. In our work, the age and the gender classification are done utilizing the Support Vector Machine classifier [6]. Other research topics include predicting face features [9].

There are several applications of age estimation which are.

1. Control of security—An automatic age estimation system which helps to prevent minor from purchasing alcohol or cigarette and also from accessing inappropriate websites.
2. Interaction between the machine and the human: Contents can be displayed to the user based on his/her age. Interaction between the machine and the human: Contents can be displayed to the user based on his/her age.
3. Law Enforcement: Age estimation systems can help in identifying the suspect and also tracking the suspect more efficiently by filtering the database based on the suspect's age. However, despite advances in age estimation, it is still a difficult problem. This is because the age can not only be determined by radical factors, e.g., genetic factors, but also by external factors like lifestyle, expression, and environment.
4. As mentioned earlier in this work, estimation and classification of age and gender from face images have been implemented using SVM. The proposed system has three phases. The first phase is Image Preprocessing. This phase further consists of four stages: gray scaling, histogram equalization, faces detection and cropping and resizing the image. In the second stage, we use the SVM algorithm for classification. Finally, in the third phase which is age estimation phase, we have two stages: estimation and evaluation. The proposed system has been illustrated in Fig. 1. SVM classifier when used with Linear Discriminant Analysis (LDA) gives an accuracy of 84% as an average.

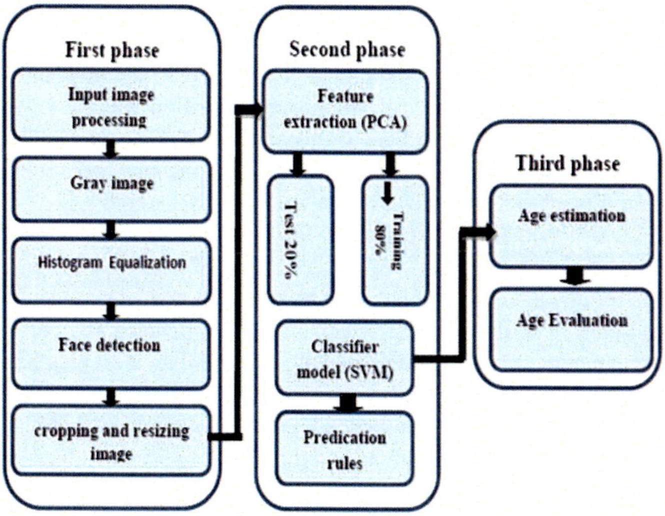

Fig. 1 Block diagram of the proposed system

SVM can also be implemented using Principal Component Analysis (PCA) and Local Binary Pattern (LBP). But the accuracy is less when compared to LDA. It gives an accuracy of 75% and 82% when used with PCA and LBP, respectively.

2 Structure of the Paper

The paper is aligned in the following format beginning with an introduction and concluded by references. It is organized as Introduction, Literature Survey, Proposed System, Image Processing Phase, Gray Image Process, Histogram Equalization, Face Detection Process, Cropping and Resizing image, Data Mining Phase, Feature Extraction, Machine Learning, Experiment Results and Evaluation, Conclusion, Future Scope, References.

3 Related Work

Age estimation is a topic of discussion for nearly a decade now. And there have been numerous approaches on estimating gender and age. With time, the approaches have been more efficient and effective, initially, the accuracy was below 60% but now the approaches promise accuracy above 75%, but trails and approaches are being made every day to improve the accuracy to a maximum level by inclusion of varying attributes and implementation using leading technologies. Few of the approaches and the technologies they have used are listed below:

Chao W. L. 2012. In this research, a new age estimation framework is proposed keeping in mind the essential factors of human ages. Relevant Component Analysis (RCA) is applied in order to realize a proper metric for the purpose of neighbor searching. Locality Preserving Projection (LPP) is trained to reduce the feature dimensionality and get an idea of the connections between features and the aging labels. At last, an age-specific local regression algorithm named KNN-SVR is proposed to capture the complex human old age process. The FG-Net aging database is used for simulating outcome and it is found that the system checks for the lowest Mean Absolute Error (MAE) versus existing methods [4].

Han H. in 2013. A hierarchical method for age estimation is presented in this research. Apart from that, analysis of the influence of old age on the individual face elements by using a component-based representation is also done. Taking the crowdsourced data from the Amazon Mechanical Turk service and performed the test on the human noticing ability on estimating age and thereafter matched with the performance of the proposed age estimation. After examining, the conclusion was drawn that the performance of the proposed system is better or in other ways similar to the age estimations applied by humans on the FG-Net dataset and a subset of the PCSO dataset [10].

Gunay A. in 2015. In this research, a hierarchical estimation of age is presented. The proposed system depends on the decision-level fusion of Active Appearance Model (AAM) and Local Binary Patterns (LBP) are supposed. The main contribution of this research is the decision-level fusion of local and global texture features of facial images. The datasets used by the proposed system are FG-Net and PAL Aging. Preservation of the locality is taken care of by regional LBP histograms and Gabor filters and moreover, these local features are collective with global features of images extracted with AAMs [8].

Liu K. H. in 2015. In this research, a multistep learning shape known as grouping estimation fusion is presented. Different fusions used to enhance the performance are six in number. Experiments conducted used datasets FG-NET and MORPH-II. The experiments also demonstrated the effectiveness of the proposed GEF and also showed that it is possible to enhance the performance of the GEF methods to increase diversity through decisions by having other features or age grouping systems [14].

Jain S. in 2016. It is intended to predict the age of a particular person from the input facial images. The processes that are included in this system are: pre-processing of input image, filtering, facial part detection, edge detection, feature extraction, training the classifier by sending extracted features to the KNN classifier, and finally, testing is done. It is done by passing the test data to classifier in order to obtain the outcome. The database used is the FG-Net database. From the experimental results achieved, the k-NN classifier produces better results for the age-group prediction [11].

After understanding the above existing systems, one primary conclusion we have made is that one algorithm or technique is not going to do the job, for effective working and accuracy, we need to integrate two or more algorithms that complement each other which results in higher efficiency. The maximum efficiency obtained till now in this system was 82% whereas the integration of all the algorithms we have used leads to 84% accuracy.

4 Proposed System

The primary goal of the proposed architecture of the system is to estimate the human age based on the face images without any human intervention. To achieve this, in the proposed system we applied machine learning techniques to estimate the age of a human from extracting features from the facial images using the required algorithms upon the FG-net dataset. The working of the system begins initially with the capture of the live picture of the person that wants his age and gender to be estimated. Now, we have to extract the features from the face and estimate their respective age and gender. After the face of the person or people has been captured, then we try to implement the suitable algorithm for the face according to the position of the face, now we try to equalize the image histogram which basically is contrast correction. After the histogram of the image has been equalized, then we move on to crop the image into the face for a better understanding of the features of the face. Finally, we estimate the age and gender from the training that the system has already undergone. Before the image reaches the user we have to make sure that the image is back into RGB format without any technical discrepancies. This in short is the brief of working of the proposed system. This system aligns with the model of the support vector machine [7].

In this system, we use FG-NET [25] dataset which has a good number of images required for the experiment to perform. It consists of 890 face images of 82 different persons, 12 images per subject. This database was first released in 2004 in an attempt to support research activities related to facial aging. The database FG-NET is divided into seven classes depending upon the age of the person in that image. The first class contains the images of the persons with the age 3–7 years, the second class contains images of persons with the age 8–13 years, third with 14–19, fourth with 20–25, fifth with 26–30, sixth with 31–40, and finally the seventh with 41–50 years. The division is mainly to extract the face features of different age groups. After this step is done, now we implement the three phases of our project.

Image preprocessing Phase

The first phase has four stages in it. They are explained as follows:

Gray image process—In this step, we need to convert the color image into grayscale image. For this, we need to have good knowledge on the color image. We use three steep operations [21]. Get the RGB values of a pixel, perform mathematical operations for turning these values to grayscale values, get the RGB value with the new grayscale value with the help of the following Eq. (1). The RGB image is turned into a grayscale image because the features of a human face more efficiently. Grayscale in simple terms can be known as turning a color image into a black and white image.

$$\text{Grayscale Image} = 0.33\,\text{Red} + 0.56\,\text{Green} + 0.11\,\text{Blue} \tag{1}$$

Histogram Equalization—Histogram Equalization is the process of taking a low contrast image and adjusting the contrast between the image's relative disparities in order to bring out the minor differences in shade and create a high contrast image [24]. The process of histogram equalization is quite necessary for estimation since this determines the features better which helps in better image extraction. Even in simple pictures adjusting the contrast of the pictures helps to sharpen the features of a person (Fig. 2).

Face Detection Process—For detection of a face, there are mainly four algorithms for detection of face in an image or a video. They are Haar Cascade Face Detector, DNN Face Detector which aligns with the support vector machine model. All these algorithms are available in the python library named OpenCV [20]. Among these four algorithms so far mentioned, the DNN algorithm has a good performance when compared to the other three algorithms. DNN is the abbreviated form of Deep Neural Net [19]. So in our work we use the DNN Face Detector for detecting the non-frontal faces in the video. For the frontal face image, we use the Haar Cascade Face Detector. Figure 3 demonstrates the non-frontal face detection using the mentioned algorithms [20]. From (Fig. 3), it is clear that we cannot detect the non-frontal images using the Haar algorithm. So that is why we are using both the algorithms for different purposes. The algorithm places a rectangle for the faces that are identified.

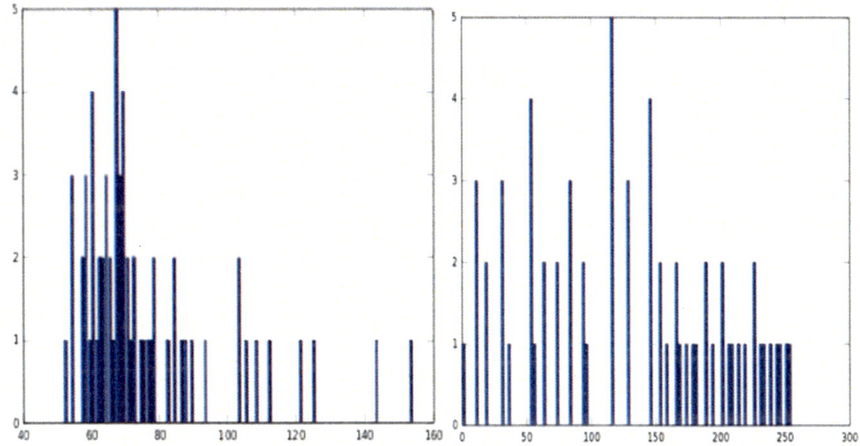

Fig. 2 Represents thc graphs of images before and after applying histogram equalization [24]

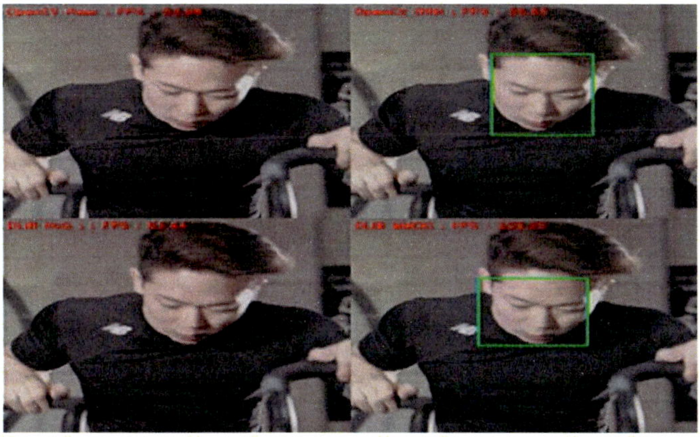

Fig. 3 Demonstration of non-frontal face detection using different algorithms

Cropping and resizing image—After detecting face, the next job that should be done is cropping and resizing of the detected face. This step removes all the extra parts that are included in the rectangular box for making it easy to extract the features.

Data mining phase

This phase is divided into two stages. They are 1. Feature Extraction and 2. Machine learning stage. These stages are explained below in a detailed manner.

Feature Extraction stage—This is the first stage of the data mining phase. In this stage we identify a group of primary features from the faces that are captured in the previous phase. We employ Linear Discriminate Analysis (LDA) [24] for extracting the features. LDA is a technique that takes the input data and transforms it in such a way that the new data will give the statistical properties. The datasets can be converted, tested and the vectors can be diversified by two methods in the converting space [13]. The first transformation is class-dependent: this type of approach involves maximizing the ratio between-class variance and within-class variance [17]. The second transformation is class-independent: this type involves maximizing the ratio of overall variance to within-class variance [17]. Linear discriminate Analysis (LDA) is mainly used to decrease the dimensional representation of the face images. Eigenface is the Eigenvector which is selected from LDA Eigenfaces. In the process of age classification, each face image from training dataset is transformed into a vector. After that, the covariance matrix is computed by multiplying variance image by variance image transform. Eigenfaces (Eigenvectors associated with Eigenvalues) are then constructed, which represent various features of face image as shown in Algorithm 1 [12].

In this first stage, calculation of the average vector mi using Eq. (2), (i = 1, 2, 3) classes:

$$mi = \begin{bmatrix} \mu\omega i(sepallenght) \\ \mu\omega i(sepallenght) \\ \mu\omega i(sepallenght) \\ \mu\omega i(sepallenght) \end{bmatrix}, \text{ with } i = 1, 2, 3 \text{ vector}$$

$$mi = \frac{1}{n}\sum\nolimits_{i=1}^{n} xi \tag{2}$$

After that we need to compute the scatter matrix that includes within-class scatter matrix which is Eq. (3) and class covariance matrices which gives Eq. (4) and between-class scatter matrix from Eq. (5).

$$SW = \sum_{i-1}^{c}(N-1)\frac{1}{N-1}\sum_{x \in Di}^{n}(x-mi)(x-mi)^T \tag{3}$$

$$\sum i = \frac{1}{N-1}\sum_{x \in Di}^{n}(x-mi)(x-mi)^T \tag{4}$$

$$SB = \sum_{i}^{c} Ni(x-mi)(x-mi)^T \tag{5}$$

where m = total average,

C = no. of class,

mi = Sample average,

Ni = Sizes of classes.

Now we move on to the next step, where we solve the generalized Eigenvalue problem for matrix $SW^{-1}SB$, then calculate the Eigenvector-Eigenvalue.

$$Av = \lambda v \tag{6}$$

where $A = \mathrm{SW}^{-1}\mathrm{SB}$, v = Eigenvector, λ = Eigenvalue.

After that, we calculate Y which is given by Eq. (7)

$$Y = X * W \tag{7}$$

where W is a 4×2 matrix that is used to transform the samples into a new sub-space.

The LDA is carried out by calling the LDA–Fisher Faces Feature Algorithm.

Algorithm 1: LDA –Fisher Faces Feature [14]

Input: Crop and resize face image from preprocessing phase
Output: Fisher face feature vectors

Begin

Step 1: Read the transformed face images.

Step 2: Store the transformed face images in a matrix.

Step 3: compute the d-dimensional mean vectors with the help of Eq (3).

Step 4: Compute scatter matrices.

Compute within – class scatter matrix (SW) with the help of Eq (4).

Compute class – covariance with the help of Eq (5).

Compute between – class scatter matrix (SB) with the help of Eq (6).

Step 5: Solve the generalized eigenvalue problem for matrix using Eq (7).

Step 6: Select linear discriminates for the new property sub-space.

Sort the eigen vectors via decreasing eigen values.

Select k eigen vectors which have the biggest eigen values.

Step 7: Transform the samples to the new sub-space with the help of Eq (8).

Step 8: Now, return feature vectors that have been obtained.

End

Machine Learning Stage—In this stage, we implement the machine learning algorithm SVM. Support vector machine so-called as SVM is a supervised learning algorithm which can be used for both classification and regression problems as support vector classification (SVC) and support vector regression (SVR), respectively. SVM works on the idea of finding a hyperplane that best separates the features into different domains. The hyperplane is also referred to as the decision surface. This is the most important stage in our work. The SVM classifier analysis the numerical properties of various image features and then organizes the data into categories. The major reason for SVM being this famous and also including it in our work is its Kernel trick. There are mainly three types of kernels in SVM [22]. They are: 1. Linear kernel, 2. Polynomial kernel and 3. Radial Basis Function kernel (RBF) or Gaussian kernel. SVM, today, is a very popular classifier in various pattern recognition problems, including face recognition and detection problems [12]. In our work, we employ the RBF kernel. This kernel is demonstrated below in the form of an algorithm [1].

Algorithm 2: Radial Basis Function (RBF) Kernel

Input: Feature vector for all of training images, feature vector for Testing images

M: n-classes, attributes

Output: Class test

Begin

Step 1: Read feature vector for all of the training images, with M attributes and feature vector for all Testing images.

Step 2: Obtain the weight w and kernel K by using the classical adaptive scaling SVM.

Step 3: Now, initialize entire data into a dataset.

Step 4: To rank the features use Random Forest Algorithm.

Step 5: Remove the less important features from the dataset and update it so that the size of the dataset can be reduced.

Step 6: Apply SVM with RBF kernel in order to reduce features.

Minimize $(\frac{1}{2}\|w\|^2 + C\sum_{i=1}^{m}\xi)$

Subject to: $\mathbf{Y}_i(\mathbf{W}^T \mathbf{x}_i + b \geq 1 - \xi_i)$, $\xi_i \geq 0, i = 1,2,\ldots\ldots,m$

Where $\mathbf{Y}_i$ is the class label of support vector $\mathbf{x}_i$,

w is a weight vector,

b is bias and variables

ξ is positive slack which is necessary to allow miss classification

Step 7: Consider that parameter C seeks to decision error when searching for the maximum marginal hyper plane.

Maximize $[\sum_{i=1}^{l} \alpha i - \frac{1}{2}\sum_{i=1}^{l} \alpha i\alpha j y i y j\, K(xi, xj)]$

Subject to: $0 \leq \alpha i \leq C$, $1 \leq i \leq 1$, $\sum_{i=1}^{l} \alpha i \quad yi = 0$, where indicates the overall inputs. Where x is a training sample and k is the radial basis func tion is given by $K(xi, xj) = exp(-\gamma \|xj - xi\|^2)$

Step 8: Choose the best parameter for create as a classification model.

End

5 Experiment Results and Evaluation

Experimental results are obtained by using the so-far developed project, i.e., SVM classifier with linear discriminate analysis. So as mentioned, the design process is made up of three phases—image pre-processing, data mining, and age estimation. The transformation of the image in every phase is depicted below (Fig. 4).

The grayscale image in simple terms can be known as an image that has been deprived of colors other than black and white. After we get the gray scale image then we need to transform it into a histogram equalized image. In the histogram equalization, we try to focus and clear out any noises from the image (Fig. 5).

So now we need to crop the image and identify the face of the person and then the data mining and finally the age should be predicted. The results of these steps are mentioned below. The final picture depicts the results (Fig. 6).

The results in the above picture estimate the age of the person to be between 8 and 12 years and their gender to be female. Now, the actual results are that the age of the person in the image is 13 years and their gender is female. Hence, we can see that the accuracy is above 84%. But, on an average, this approach is 84% accurate for human faces. So, in this way, we get the results for the age estimation with the

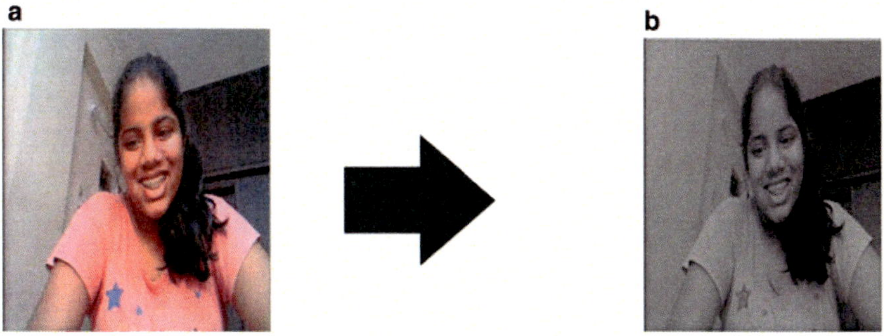

Fig. 4 **a** RGB image. **b** Grayscale

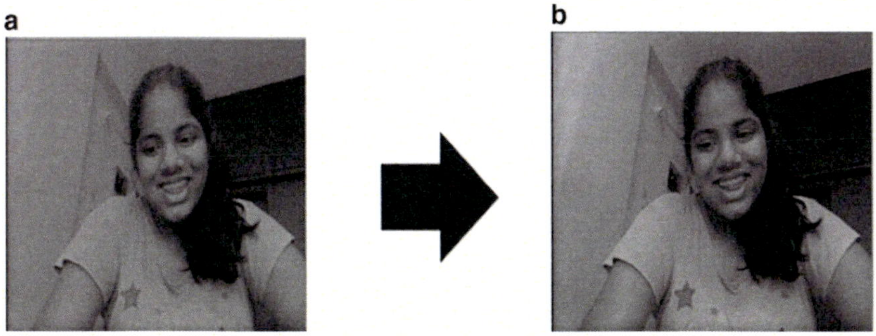

Fig. 5 **a** Grayscale. **b** Histogram equalized

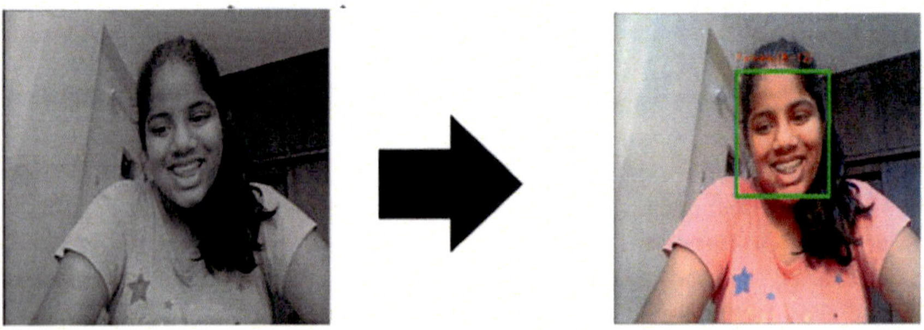

Fig. 6 The final picture depicts the results

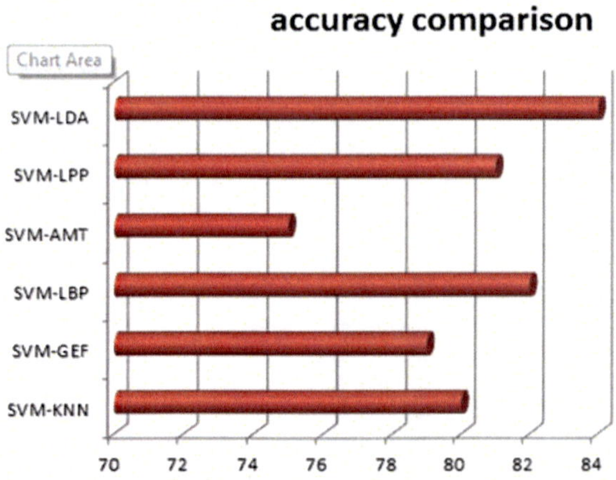

Fig. 7 Comparing accuracies of various methodologies with SVM-LDA

proposed system. Another inclusion in this approach is that it can recognize multiple faces at a time and it can also recognize not just live faces but also human images from a computer or mobile (Figs. 7 and 8).

In this approach, the age of multiple people can be estimated simultaneously but this has a negative effect on the time complexity and accuracy. If the number of people in a frame at a given moment exceeds eight then, the accuracy rate of estimation seems to tumble. The above graph depicts this.

6 Conclusion

Age estimation is a very important activity in facial image classification. Age estimation is defined as the age of a person that can be extracted from a 2-dimensional face image. In this work, the classifier named Support Vector Machine (SVM) is used with the FG_NET dataset. This dataset is divided into seven classes which represent different intervals of age. The classes are 3–7 years, 8–13 years, 14–19 years,

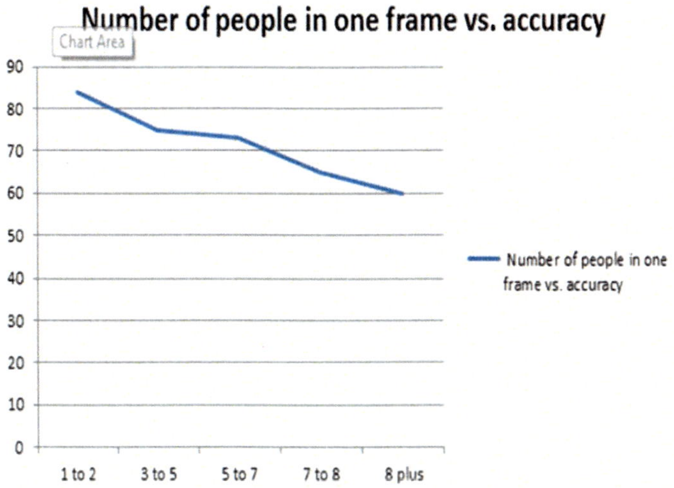

Fig. 8 Comparing accuracies w.r.t number of people in one frame

20–25 years, 26–30 years, 31–40 years, and 41–50 years. The SVM classifier with Linear Discriminate Analysis (LDA) is found to give the results with more accuracy when compared to other classifiers like SVM with Principle Component Analysis (PCA) and SVM with Local Binary Pattern (LBP). SVM with LDA has an accuracy rate of 84% whereas the other two have an accuracy of 75% and 81%, respectively. As this area has many applications in many fields like security, law enforcement, and many more, it has a good scope for future advancements.

7 Future Scope

From the proposed system, we can find the age and gender of a person with facial images. But any research is not limited. We can explore and find many new features. So, for this system also there may be certain issues that are not addressed by the proposed system. The main future activity which can be focused on is working on the development of the Indian Aging Database and making use of it [3]. Apart from this there are many other future research directions, namely, in the regions of Fusion [2], Multi-instance, Ethnic [16], Lifestyle [15], Environment [15], Databases, Profile Face Aging, Multi-sensor. In today's developing world, life style and environment play a huge role in the aging of a person these aspects can be focused in the future, maybe due to pollution or weather conditions also there is an urge developing to know the age of a person. From government jobs to security checks, this technology can help in numerous ways. This can also act as a quick identity proof for senior citizens and can also be used to primarily estimate the fitness range of a person.

References

1. Agrawal, S., Raja, R., Agrawal, S.: Support vector machine for age classification. Int. J. Emerg. Technol. Adv. Eng. (2012)
2. Akinyemi, J.D., Onifade, O.F.W.: An ethnic-specific age group ranking approach to facial age estimation using raw pixel features. In: Proceedings of IEEE Symposium on Technologies for Homeland Security, pp. 1–6. IEEE, Waltham (2016)
3. Angulu, R., Tapamo, J.R., Adewumi, A.O.: EURASIP J. Image Video Process. (2018)
4. Chao, W., Liu, J.: Facial age estimation based on label-sensitive learning and age specific local regression, acoustics, speech and signal processing (ICASSP). In: IEEE International Conference, 25–30 March (2012)
5. Chellappa, R., Wilson, C.L., Sirohey, S.: Human and machine recognition of faces: a survey. Proc. IEEE **83**, 705–740 (1995)
6. Choi, C.: Age change for predicting future faces. Proc. IEEE Int. Conf. Fuzzy Syst. **3**, 1603–1608 (1999)
7. The FG-NET Aging Database 2009 [Online]
8. Gunay, A.: Facial age estimation based on decision level fusion of AAM, LBP and gabor features. (IJACSA) Int. J. Adv. Comput. Sci. Appl. **6**(8) (2015)
9. Han, H., Otto, C., Jain, A.K.: Age estimation from face images: human vs. machine performance, Spain. In: International Conference on Biometrics (ICB). IEEE Explore Digital Library. https://doi.org/10.1109/ICB-6613022. 978-1-4799-0310-8 (2013)
10. Han, H., Otto, C.: Age estimation from face images: human vs. machine performance. In: IAPR International Conference on Biometrics (ICB), Madrid, Spain, June 4–7 (2013)
11. Jain, S., Patil, A.: Human age prediction from facial images. Int. J. Innov. Res. Comput. Commun. Eng. **4**(7) (2016)
12. Jiawei, H., Micheline, K.: Data Mining: Concepts and Techniques, 3rd edn. Morgan Kaufmann Publishers, Simon Fraser University. ISBN 978-0-12-381479-1 (2011)
13. Jie, Y., Hua, Y., Kunz, W.: An Efficient LDA Algorithm for Face Recognition. School of Computer Science Interactive Systems Laboratories, Carnegie Mellon University (2002)
14. Liu, K., Yan, S.: Age estimation via grouping and decision fusion. IEEE Trans. Inf. Foren. Secur. **10**(11) (2015)
15. Ni, B., Song, Z., Yan, S.: Web image and video mining towards universal and robust age estimator. IEEE Trans. Multimedia. **13**, 1217–1229 (2011)
16. Ni, B., Song, Z., Yan, S.: Web image mining towards universal age estimator. In: Proceedings of ACM International Conference on Multimedia, pp. 85–94. ACM Press Beijing (2009)
17. Raschka, S., Mirjalili, V.: Python Machine Learning, 2nd edn paperback. Michigan State University, Michigan (2017)
18. Tharwat, A., Gaber, T., Ibrahim, A., Hassanien, A.E.: AI Commun. (20xx), 1–22. https://doi.org/10.3233/AIC-170729. IOS Press
19. https://en.wikipedia.org/wiki/Histogram_equalization#Implementation
20. https://machinelearningmastery.com/introduction-to-deep-learning-for-face-recognition/
21. https://towardsdatascience.com/linear-discriminant-analysis-in-python
22. https://towardsdatascience.com/support-vector-machines-svm-c9ef22815589
23. https://www.isip.piconepress.com/publications/reports/1998/isip/lda/lda_theory.pdf
24. https://www.learnopencv.com/face-detection-opencv-dlib-and-deep-learning-c-python/
25. https://www.researchgate.net/figure/HaarNet-architecture-for-the-trunk-and-three-branches-25-Max-pooling-layers

Real-Time Credit Card Fraud Detection Using Spark Framework

A. Madhavi and T. Sivaramireddy

1 Introduction

Credit card fraud transaction is an immensely popular and common issue these days worldwide. All credit-card-based companies are spending a huge amount on overcoming such fraud transactions every year but still as increasing technological advances new fraud ways are also increasing exponentially. Capturing cardholder historical behavior patterns helps derive new rules to identify a transaction as fraud or non-fraud [5]. Those behavioral patterns can be extracted in the model while implementing the algorithm. In this study, we primarily targeted on improving model performance and real-time processing of transactions instead of batch processing. The possible ways to improve model performance by feeding more pre-processed training data to an algorithm and enhance the algorithm logic to be more robust [6]. As transaction predictions should be handled in millions, we need to use distributed frameworks to drive such large-scale transactions and identify a transaction is genuine or fraud. We use Spark as a distributed framework that is similar to the Hadoop environment but can be 100 times faster compared to MapReduce processing. Though Hadoop and Spark were developed mainly to process large amounts of data we have a small difference in their selection. Hadoop is used mostly with ETL processing of batch data whereas Spark is used for ETL processing of real-time data with the tremendous performance [7]. We have used Scala programming in a Spark as it is the indigenous language of Spark and proved swift in performance over Python or java.

A. Madhavi (✉) · T. Sivaramireddy
Department of Computer Science, VNR VJIET, Hyderabad, India
e-mail: madhavi_a@vnrvjiet.in

T. Sivaramireddy
e-mail: sivaramireddy.tiyyagura@gmail.com

C. K. Mai et al. (eds.), *Machine Learning Technologies and Applications*,
Algorithms for Intelligent Systems,
https://doi.org/10.1007/978-981-33-4046-6_28

In this work, the data utilized during the experiment are indiscriminately formatted under a normal distribution. As part of pre-processing data, we scale parameters to give a good understanding of the machine learning algorithm and extract two additional features computed from five existing features which help in turning training data to random forest algorithms to give the best accuracy [8]. We used Spark streaming jobs to process collected facts in real-time and Kafka messaging systems to process data with no time these two are the best combination to handle information provided in real-time with distributed systems. Once results obtained from Spark streaming with a random forest model will display fraud transactions in fraud alert dashboard.

2 Related Work

In the existing systems, research is done on a theme study involving credit card fraud screening whereby data is balanced with sampling techniques and applied supervised machine learning algorithms [9]. The substantial problem with fraud detection is finding the best dataset to process, i.e., real-time data is practically not available due to the point of confidentiality. These concerns were not an obstacle to analysts as they were implementing their work in coordination with corporate partners and some were advised to apply a synthetic dataset of transactions [10]. Besides forthcoming genetic algorithms can be used for data generation and accomplishing the uniform data from cumbersome dataset tasks [11] As credit card safeguard technologies getting upsurge proportionally the new techniques were used by fraudsters. To overcome this situation, the data mining tool's performance is less than ideal so implementing fraud detection needs the finest enhancements like adopting big data machine learning approaches [12]. The supervised and unsupervised machine learning models are giving improved performance, in light of this we are using both the algorithms in our system, i.e., K-means clustering algorithm for data balancing and random forest classification algorithm for decision-making whether a transaction is genuine or fraud [13]. In the existing system, k-means algorithm is implemented for clustering accounts based on customer spending habits and then the XGBoost classification algorithm is applied for decision-making [14]. Most research happens on tuning algorithms and trying to improve model performance and reached an efficiency of up to 85%. We can improve the efficiency of any machine learning model with three possible solutions, they are data, algorithm, and cost-sensitive.

2.1 Data Level

The given input data is generated from the existing Sparkov data generator tool [15], by providing customer names it will generate customer data. Referring to customer data the mapped transaction data generated provides transaction category and merchant names to the tool. The generated data for training machine learning may not have both fraud and non-fraud transactions in an equal ratio [16]. We must balance both class data and this can be achieved in multiple ways, i.e., sampling techniques. The popularly known sampling techniques are undersampling and oversampling. The undersampling method takes random samples from a majority class set that approximately match to the number of samples in the minority class set to get a balanced dataset. But we must discard large majority class samples which leads to loss of information and may reduce model potential related to the majority class sample. Conversely, the oversampling technique will synthesize the minority class set and produce samples approximately equal to the majority class which leads to more noise in newly created minority class data. Part of the work we choose undersampling not just with a random selection of samples but with the k-means algorithm to prepare balanced both class data. This data sampling can be implemented as hybrid sampling which is a combination of both undersampling and oversampling techniques [17].

2.2 Algorithm Level

As fraud detection problem solution classifies a transaction as a fraud or non-fraud, a classification algorithm is the best choice to use. In a classification problem, some of the known existing algorithms are implemented with naïve Bayes, k-NN and decision tree models [18]. We are concentrating on a random forest algorithm with the implementation of k-fold cross-validation to validate data with multiple trees [19].

2.3 Cost-Sensitive

Cost-sensitive is part of machine learning that often comes to the picture mainly for imbalanced data classification problem because the wrong prediction of positive or minority class case is a blunder than identifying wrongly for negative or majority class. This can be achieved with imbalanced dataset resampling, algorithm-level modifications, and using ensemble learning methods. Here cost means penalty awarded for the wrong prediction of positive or minority class sample [15]. This aims to minimize the cost of a model on the training dataset whether it is having imbalanced data, it can be achieved by checking the cost of tests, data instability, misclassification errors, etc. The most common and basic metric used to find the cost for an imbalanced dataset is the confusion matrix.

3 Proposed System

The current system focuses on all three solutions to achieve optimal performance, i.e., data-level, algorithm-level enhancements with fraud detection ensemble model, and cost-sensitive learning. As cost-sensitive learning can be achieved with data resampling and algorithm modifications. We mainly focus on data pre-processing and ensemble learning model.

3.1 Random Forest Model

The random forest/ensemble model-derived at the base of the decision tree gives the best results compared to other classification algorithms [16]. Decision tree is an initiative algorithm that decides on the sequence of questions on data features, but it may overfit the mode due to low variance and high bias in nature. This drawback can be overcome with a random forest algorithm having important features as Gini impurity, bootstrapping, and random selection of features for each node in the decision tree. Gini impurity is a measure used by the decision tree to decide on splitting each node, which represents a probability that a sample from a node will classify incorrectly according to the distribution of samples in a node. Bootstrapping which randomly selects samples with replacement. Random set of features while considering split for each node in a decision tree. Combining all these features, random forest is made of multiple decision trees and taking average voting to make predictions make random forest an ensemble model and also a bagging example (see Fig. 1).

We select a random forest model as the best outfit of all classification algorithms for credit card fraud recognition [20].

3.2 Architecture

In the intended system, we want to focus not just on the data and algorithm-level and also on real-time data processing. As real-time credit card transactions will happen in millions, we should process all transactions with no time and decide in millisecond time. To achieve this, we are implementing distributed architecture and deploying random forest models to achieve both performance and model efficiency at the same time.

The proposed architecture is divided into three parts data streaming, data processing, and data representation (see Fig. 2).

Data streaming. Data streaming is achieved using the Kafka tool and Spark streaming job. where Kafka is the best messaging platform to process data from web/custom apps to database/data warehouse, etc., in our context process data is nothing but transactions and each transaction is treated as a message by Kafka. As

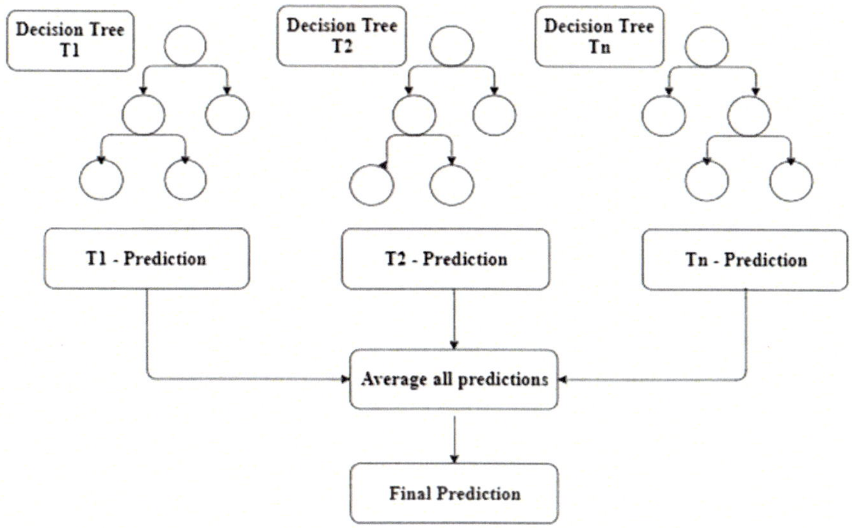

Fig. 1 Random forest model demonstration

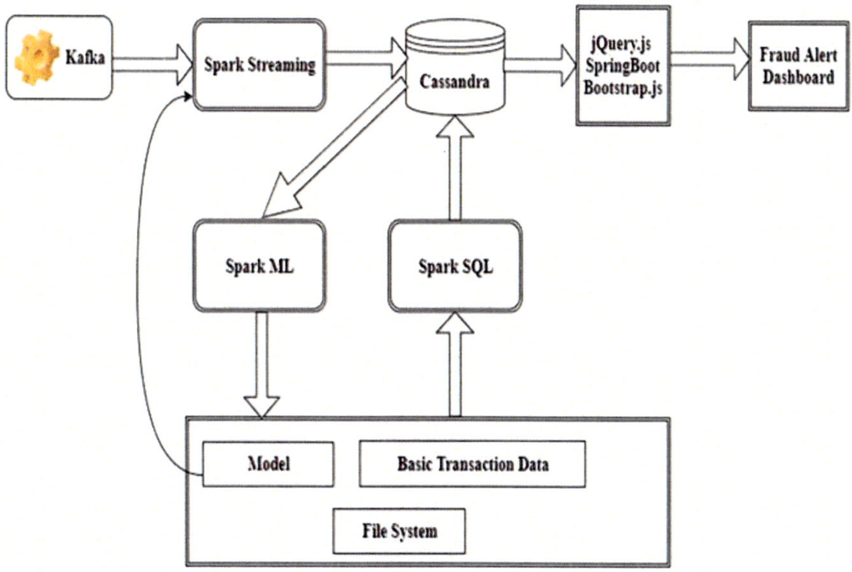

Fig. 2 Proposed system architecture

Table 1 Additional feature extraction information

Actual features	Extracted feature	Description
DOB	Age	Customer age
Customer (latitude and longitude) and Merchant (latitude and longitude)	Distance	Distance between the customer and merchant locations

part of Spark streaming, the messages received from Kafka will apply a random forest model, taking decisions will store transaction details to Cassandra an in-memory database. Cassandra is very scalable and efficient for distributed databases geared toward handling huge data volumes through commodity servers. It suits best for data processing in a distributed environment.

Data processing. Data processing is a crucial part of any machine learning algorithm. Like all algorithms, we also processing raw data in three sub-stages.

Data transformation. Data transformation includes converting string type columns values to numeric value using string indexer in Scala and scaling numeric columns converting to vector and slicing them to get normalized values which should have mean equal to zero and standard deviation to one. These transformations were applied to features before process training data to machine learning models.

Feature extraction. Extracting features is an approach of retrieving new features from existing [13]. This will improve model efficiency and helps to make the best decision on input data. In our system, we are extracting customer age based on his date of birth and distance between customer and merchant based on their latitude and longitude parameters (see Table 1). Both are important features to decide whether a transaction is a fraud or genuine, considering transaction amount value and customer spending habits like frequency of performing domestic and international transactions in the past, etc.

Balanced dataset. Any credit card company data will have 96% genuine transactions and less than or equal to 4% fraud transactions. Similarly, the dataset considered for the proposed system has 97:3 genuine and fraud transactions. Processing complete data as it is will give high variance for the machine learning model results in detecting most transactions as genuine. So, we must balance the dataset before feeding it to the machine learning model for accurate prediction. To achieve a balanced dataset, we will use the k-means clustering algorithm to make the majority class transactions count approximately equal to the minority class [21].

Data representation. We have a user interface called a fraud alert monitoring dashboard used to display any fraud transaction detected in our process. This fraud alert dashboard was developed using jQuery, Spring Boot, Bootstrap, and SockJS technologies and linked with Spark streaming job. Whenever a Spark streaming job predicts any new transaction as fraud, the fraud alert dashboard will get a notification and it will be displayed in the fraud alert monitoring dashboard.

4 Implementation

The proposed system mainly focuses on achieving the best performance while process millions of credit card transactions online. As discussed in the data processing section, the input transaction data has been transformed and applied to data balancing using the k-means algorithm. In the wake of data is balanced, it will be processed through a random forest algorithm for model creation. The streaming job will utilize the created model and will be used for processing new input data and making choices on transaction type, i.e., fraud or genuine. Corresponding fraud transaction notifications will be displayed in the UI dashboard (see Fig. 3). To achieve expected results, we used a distributed system and enhanced data processing with a random forest ensemble model. We have implemented the current system in a Linux environment with the mentioned software. Cassandra database, Spark streaming, Spark MLlib, Spark SQL, Kafka messaging system, Spring Boot, and jQuery.js for UI dashboard. Most of the projects are designed with Spark-based framework to achieve fast processing of transactions and deciding whether a transaction is a fraud or non-fraud.

Detailed process to build the proposed system.

1. Create all table schemas in Cassandra and upload data to them from the file system using Spark SQL.
2. Start pre-processing training data and prepare a balanced dataset having equal no of fraud and non-fraud transactions.
3. Train ensemble random forest model and take metrics like F1 score to decide how efficient this model to process fraud transactions better.
4. Now both the pre-processing model and ensemble model saved to the filesystem for future is used to process new incoming transactions from Kafka.
5. Start Spark streaming job which inputs both data pre-process and ensemble models and waiting for new transactions to process.
6. Kafka messaging system will start publishing new incoming transactions to the created topic.

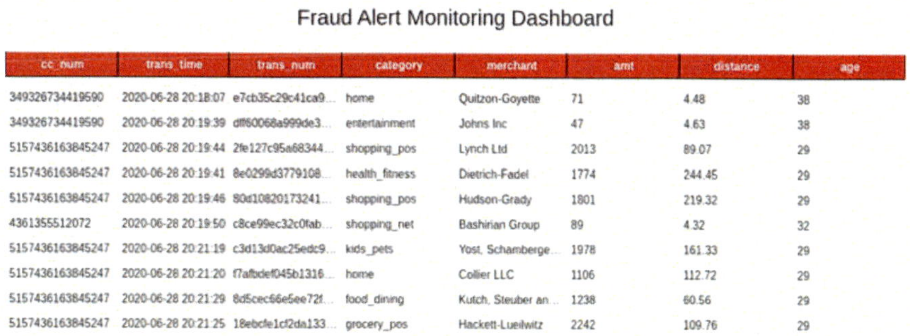
Fraud Alert Monitoring Dashboard

cc_num	trans_time	trans_num	category	merchant	amt	distance	age
349326734419590	2020-06-28 20:18:07	e7cb35c29c41ca9...	home	Quitzon-Goyette	71	4.48	38
349326734419590	2020-06-28 20:19:39	dff60068a999de3...	entertainment	Johns Inc	47	4.63	38
5157436163845247	2020-06-28 20:19:44	2fe127c95a68344...	shopping_pos	Lynch Ltd	2013	89.07	29
5157436163845247	2020-06-28 20:19:41	8e0299d3779108...	health_fitness	Dietrich-Fadel	1774	244.45	29
5157436163845247	2020-06-28 20:19:46	80d10820173241...	shopping_pos	Hudson-Grady	1801	219.32	29
4361355512072	2020-06-28 20:19:50	c8ce99ec32c0fab...	shopping_net	Bashirian Group	89	4.32	32
5157436163845247	2020-06-28 20:21:19	c3d13d0ac25edc9...	kids_pets	Yost, Schamberge...	1978	161.33	29
5157436163845247	2020-06-28 20:21:20	f7afbdef045b1316...	home	Collier LLC	1106	112.72	29
5157436163845247	2020-06-28 20:21:29	8d5cec66e5ee72f...	food_dining	Kutch, Steuber an...	1238	60.56	29
5157436163845247	2020-06-28 20:21:25	18ebcfe1cf2da133...	grocery_pos	Hackett-Lueilwitz	2242	109.76	29

Fig. 3 Fraud notification dashboard

7. Spark streaming jobs will subscribe to the initiated topic and start consuming transactions.
8. All new transaction processing results will be stored in Cassandra fraud and non-fraud transaction tables for future machine learning model training purposes and transaction monitoring.
9. If any transaction is identified as fraud, then an alert will be receiving at fraud alerts monitoring dashboard to display the same.
10. Once in a week or month, the machine learning model will be re-trained, if model efficiency is good than the previous deployed model then the new model will be deployed stopping the old model.

5 Results

We are estimating the results of the implemented model based on evaluation metrics that were more popular known to calculate the accuracy of the machine learning classification problem [10].

5.1 Performance Metrics

The performance metrics represent the cost sensitivity of the developed model, mainly checking the wrong prediction of positive values or minority class. Current results show the very little cost for the evaluated model as we get a true positive rate equal to 1.

Confusion matrix. The error matrix helps to measure the effectiveness of machine learning classification/categorization problem. A classification problem generally differentiates between two classes and the confusion matrix will represent four different combinations of having actual and predicted values (see Table 2). Here N refers to the total number of transactions used for model evaluation. The efficiency metrics shall be evaluated against the error matrix as below.

Heat map. A heat map is a diagrammatical representation of the confusion matrix, which shows results with color presentations, i.e., color darkness will vary from minimum value to maximum value in the confusion matrix (see Fig. 4). The minimum value of the confusion matrix will have a complete light color and the maximum value

Table 2 Confusion matrix for model evaluation

Confusion matrix		
N = 2474	Actual = 1	Actual = 0
Predicted = 1	110	12
Predicted = 0	0	2352

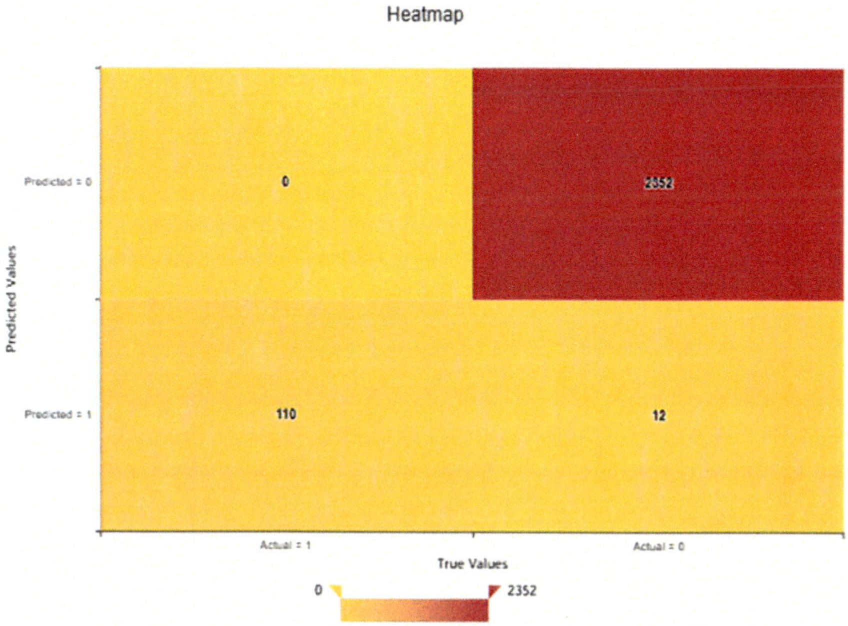

Fig. 4 Heat map representation for the confusion matrix

will be a complete darkness and the other values of color will settle between color shade from low to high based on their actual value.

All the metrics' values will be between 0 and 1 but the expected value will vary from one metric to another.

True positive rate. Total fraud transactions in the test dataset were predicted correctly as fraud. The anticipated value should be at a possible high, i.e., 1 is perfect and 100% fraud transactions were identified.

False-positive rate. Total genuine transactions in the test dataset were predicted as fraud. Presumed value should be at a possible low.

Precision and Recall. Precision is a metric to test how often the model predicts positive. Predictable value should be at a possible high, i.e., obtaining near to 1 is the finest result. The recall is nothing but a true positive rate, identifying how many fraud transactions were correctly classified.

F1 Score. F1 score is an overall extent of model accuracy which is computed using both precision and recall metrics. F1 score resembles having low false positive and false negatives so that we are finding real threats and not deviated with false alarms.

Expecting value should be at a possible high.

ROC. ROC metric is to show the performance of a classification model against problems considering all classification thresholds (see Table 3). This will be evaluated as a true positive rate against a false-positive rate. The intended value should be at a feasible high.

Table 3 Evaluation metrics summary

Measure	Formula	Result
True-positive rate/recall	TP/(TP + FN)	1
False-positive rate	FP/(FP + TN)	0.00471
Accuracy	(TN + TP)/(TP + FP + FN + TN)	0.99547
Precision	TP/(TP + FP)	0.90163
F1 score	2 * (Precision * Recall)/(Precision + Recall)	0.94827
ROC	The TP rate against the FP rate	0.99764

The evaluation of this algorithm is compared to other classification algorithms which provide efficiency up to 85% maximum, whereas the proposed model with the best pre-processing data we got 90% and above in all the times. As part of our study, we got to know if there is not much data it gives better results with classical algorithms than deep networks [22].

6 Conclusion

In the proposed manuscript, we have presented a real-time distributed fraud scrutinizing method for credit card transactions. Data pre-processing with scaling and extracting new features such as age and distance from existing data helps to prepare the best training data set for random forest ensemble learning method which is best for fraud detection out of all classification algorithms. The developed model is utilized in Spark streaming jobs that receive data from Kafka messaging system and process transactions. Transaction results will be stored in Cassandra database and if any fraud transaction comes will be displayed in the fraud alert dashboard. The random forest model gave the best accuracy with optimal scaling of data by introducing new features in it. As model selection also varies sometimes based on specific data processing, we have to decide before deploying the model with cost evaluation on the trained model.

7 Future Scope

The proposed model is the best suite for fraud detection problems though we have tested the model with few simulated datasets results that may vary slightly compared to real-time data testing and will give us more insight to decide on model performance. As an enhancement to this model, we can implement the XGBoost classification machine learning model and for processing real-time data, we can try Apache Flink to replicate real-time processing.

References

1. Popat, R.R., Chaudhary, J.: A survey on credit card fraud detection using machine learning. In: 2018 2nd International Conference on Trends in Electronics and Informatics (ICOEI), Tamilnadu, Tirunelveli, pp. 1120–1125 (2018)
2. Kumar, P., Iqbal, F.: Credit card fraud identification using machine learning approaches. In: 2019 1st International Conference on Innovation in Info and Communication Technology (ICIICT), TN, Chennai, India, pp. 1–4 (2019)
3. Kuruwitaarachchi, N., Bhagyani, C., Mihiranga, S., Premadasa, S., Thennakoon, A.: Real-time Credit-Card Fraud Detect Using Machine Learning. https://doi.org/10.1109/CONFLUENCE.2019.8776942 (2019)
4. Rajeshwari, U., Babu, B.S.: Real-time credit card fraud detection using streaming analytics. In: 2016 2nd International Conference on Applied and Theoretical Computing and Communication Technology (iCATCCT), Bangalore, pp. 439–444 (2016)
5. Zheng, L., et al.: A new credit card fraud detecting method based on behavior certificate. In: 2018 IEEE 15th International Conference on Network, Sensing and Control (ICNSC), Zhuhai, pp. 1–6 (2018)
6. Mittal, S., Tyagi, S.: Performance evaluation of machine learning algorithms for credit card fraud detection. In: 2019 9th International Conference on Cloud Computing, Data Science and Engineering, Noida, India, pp. 320–324 (2019)
7. ARMEL, Zaidouni, D.: Fraud detection using apache spark. In: 2019 5th International Conference on Optimization and Applications (ICOA), Morocco, Kenitra, pp. 1–6 (2019).
8. Xie, Y., Liu, G., Cao, R., Li, Z., Yan, C., Jiang, C.: A feature extraction method for credit-card fraud detection. In: 2nd International Conference on Intelligent Autonomous System (ICoIAS), Singapore, pp. 70–75 (2019)
9. Dhankhad, S., Mohammed, E., Far, B.: Supervised machine learning algorithms for credit card fraudulent transaction detection: a comparative study. IEEE Int. Conf. Info Reuse Integr. (IRI) 122–125 (2018)
10. Puh, M., Brkić, L.: Detecting credit card fraud using selected machine learning algorithms. In: 2019 42nd International Convention on Information and Communication Technology, Electronics and Microelectronics (MIPRO) Opatija, Croatia, pp. 1250–1255 (2019)
11. Randhawa, K., Loo, C.K., Seera, M., Lim, C.P., Nandi, A.K.: Credit card fraud detection using AdaBoost and majority voting. IEEE Access **6**, 14277–14284 (2018)
12. Gyamfi, N.K., Abdulai, J.: Bank fraud detection using support vector machine. In: 2018 year IEEE 9th Annual Information Technology, Electronics and Mobile Communication Conference (IEMCON) Vancouver, BC, pp. 37–41 (2018)
13. Jiang, C., Song, J., Liu, G., Zheng, L., Luan, W.: Credit card fraud detection: a novel approach using aggregation strategy and feedback mechanism. IEEE Internet of Things J. **5**, 3637–3647 (2018)
14. Kasa, N., Dahbura, A., Ravoori, C., Adams, S.: Improving credit card fraud detection by profiling and clustering accounts. In: 2019 year Systems and Information Eng Design Symposium (SIEDS), VA, US, 2019-year, pp. 1–6 (2019)
15. Data generator i.e. https://github.com/namebrandon/Sparkov_Data_Generation
16. Kho, J.R.D., Vea, L.A.: Credit card fraud detection based on transaction behaviour. In: TENCON, 2017 IEEE Region 10 Conference Penang, pp. 1880–1884 (2017)
17. Dighe, D., Patil, S., Kokate, S.:Detection of credit card fraud transactions using machine learning algorithms and neural networks: a comparative study. In: 2018 Fourth International Conference on Computing Communication Control and Automation (ICCUBEA), Pune, IN (2018)
18. Malini, N., Pushpa, M.: Analysis on credit card fraud identification techniques based on KNN and outlier detection. In: 2017 Third International Conference on Advances in Electrical, Electronics, Info, Communication, and Bio-Informatics (AEEICB), Chennai, pp. 255–258 (2017)

19. Kumar, M.S., Soundarya, V., Kavitha, S., Keerthika, E.S.: Credit card fraud detection using random forest algorithm. In: 2019 3rd International Conference on Computing and Communication Technologies (ICCCT), Chennai, India, pp. 149–153 (2019)
20. Varmedja, D., Karanovic, M., Sladojevic, S., Arsenovic, M., Anderla, A.: Credit card fraud detection—machine learning methods. In: 2019 18th International Sympos East Sarajevo, Herzegovina and Bosnia 1–5 (2019)
21. Wang, H., Zhu, P., Zou, X., Qin, S.: An Ensemble Learning Framework for Credit Card Fraud Detection Based on Training Set Partitioning and Clustering, pp. 94–98 (2018)
22. Roy, A., Sun, J., Mahoney, R., Alonzi, L., Adams, S., Beling, P.: Deep learning detecting fraud in credit card transactions. In: 2018 Systems and Information Engineering Design Symposium (SIEDS) Charlottesville, USA, pp. 129–134 (2018)

Deep Learning Model for Recognizing Text in Complex Images

Gnana Prakash Thuraka, Vemparala Sravani, B. Sujatha, and L. Sumalatha

1 Introduction

In the world, there are 2.2 billion people suffering from blindness. In India, approximately 40 million people are visually impaired. These statistics were given by WHO (World Health Organization) and also mentioned that majority of visually challenged people are above the age of 50 years. There might be many inventions/discoveries brought to light especially for the blind but in recent years due to the advancement in technology, tools like reading document, detecting objects, giving information about it etc., are effectively used. Text recognition plays an important role especially for those people who want to read or to know what is written on an object or a building name etc. There are models that are able to recognize the text of documents or simple images. In addition to that, few more models are developed that are explained in Sect. 2. Our model helps in an advanced way with less effort and simplified work. Text recognition is not only helpful for blind people but also has many other applications like vehicle number plate detection, traffic sign guidance, data entry etc.

An image is called a complex image when any one of the following factors exists:

G. P. Thuraka · V. Sravani (✉)
Department of CSE, VNRVJIET, Hyderabad, India
e-mail: vemparalasravani@gmail.com

G. P. Thuraka
e-mail: gnanaprakash_t@vnrvjiet.in

B. Sujatha
Department of CSE, GIET, Rajahmundry, India
e-mail: birudusujatha@gmail.com

L. Sumalatha
Department of CSE, JNTUCEK, Kakinada, India
e-mail: Sumalatha.lingamgunta@gmail.com

C. K. Mai et al. (eds.), *Machine Learning Technologies and Applications*,
Algorithms for Intelligent Systems,
https://doi.org/10.1007/978-981-33-4046-6_29

- The text is present on the graphical background.
- The text and background are similar.
- The image is captured including the environment/surroundings.

Complex images are the images where the background images contain dominating design and effects. These types of images contain graphical background. Examples shown in Fig. 1 are collected from online sources.

Below are the images captured along with the surroundings. In these images, the text may be the name of a hotel or residence or store, etc.…In such type of images, along with the text, many objects are present which distracts the focus of the viewer. Examples shown in Fig. 2 are obtained from Street view text dataset.

In few images, the text may be similar to the background color or surface. These images can be focused on scene images or natural scene images. Examples in Fig. 3 are taken from ICDAR 2013 dataset.

Fig. 1 Text on colorful backdrop

Fig. 2 Text present in natural scene image

Fig. 3 Text like background images

Though there are built-in functions and libraries, these types of images cannot be converted easily into text. When an image is processed using Tesseract, it results in garbage output as the function cannot understand the image and it fails to identify the text due to the noise and complexity present in the image. However, the noise and complexity can be removed from an image using image preprocessing techniques so that the text can be easily extracted. Whereas in the natural scene images, due to the surroundings and other disturbance factors exist in the image even though the noise removal operations are applied. It is a bit difficult to identify the text because in natural scene images the text may be on a simple background but contains the other unnecessary things around it which is a tough task. In such cases text detection is required. Using deep learning model, the text in an image can be located and that particular area is cropped and converted to text. According to our model, when the text in an image is not recognized by OCR or tesseract it is a complex image. Such type of text can be extracted using the proposed model. In our model, we have used OpenCV library and benchmark datasets. OpenCV is used to perform different operations on an image.

2 Existing Work

Here are few models discussed on the text detection and extraction based on the factors like text, background, etc., a few methods are purely concentrated on disturbances of the image like noise, low contrast, etc., and few images are converted to text based on natural scene and graphical background.

Before the implementation of deep learning techniques, the basic methods used for text recognition is connected component analysis, Stroke Width transform. Ranjit [1] worked on the image processing techniques by removing the color of the image and converting it into binarized form. Variance is calculated for the selected regions, as the variance does not remain the same for the area which contains text and background. Canny edge detection is implemented for finding the boundary of the text. Regions containing text and non-text are differentiated using connected components.

Kumuda [2] calculated components for localizing the text. Text and non-text regions are classified using Adaboost classifiers. This method is executed in such a way that the background and text pixels are separated. Character by character the whole text is recognized from the image. Morphological operations are applied and the text is extracted.

For enhancement to such classic models, many machine learning algorithms were used for text recognition. Generally, these are used in classifying generated components and differentiating text and non-text regions, etc. Siyu [3] implemented two random forest classifiers, one for classifying the characters and another for filtering characters that are invalid. For processing the image, basic methods are used similar to the work of [2]. In addition to that, characters are validated using machine learning algorithms.

The methods used in [1, 2] are implemented in deep learning models. Zhang [4] has worked on the connected component analysis, where the candidates are extracted based on the stroke width transform. After calculation, strokes are clustered into groups using a single link algorithm. They used MSER (maximally stable extremal region) for calculating the similarity matrix. When everything is ready along with identified strokes CNN is applied for classifying text candidates. Houssem [5] executed on SVM (support vector machine), CNN for filtering the candidates. Text areas are detected using MSER and from the region, the detected words are grouped.

There are other well-known and popular deep learning methods. Lei [6] worked on bounding the text using Faster RCNN. The text area is directly predicted using region proposal network, with the help of image dimensions, the coordinates of the text region are calculated. RCNN (region-based convolutional neural network) [7] takes a long computing time. Faster RCNN [6] is introduced to resolve this problem but it is generally suitable for text bounding rather than detection. YOLO (you only look once) [7] acquires more accurate results than SSD [8] but cannot detect more regions. Zhi [9] worked on text detection using a connectionist proposal network. They have worked on VGG [7] and RCNN for detecting the text. Instead of detecting the text as individual words, the whole text is analyzed as a sequence. Minghui [10] came up with Text Boxes method. It is an end-to-end recognition system which can detect horizontal text. Text is spotted in the image and recognized using a fully convolutional neural network, but the words containing less than 3 letters cannot be detected by this model. This model has chosen VGG as the base model and then added new layers.

All these models are used to recognize text in complex images, whereas the method differs from author to author. Some of them have executed connected component analysis, sliding window technique, few have used machine learning algorithms, a combination of both methods, deep learning, etc. The performance of these models is compared with our model, namely text region recognition network (TRRN) in Table 3.

3 Methodology

Though there is a change in the methods of solving text recognition, basic image processing techniques are required before executing a model. A detailed framework of our method is mentioned in Fig. 4.

The method explained in this paper comprises both image processing and Convolutional Neural networks (CNN). Image preprocessing techniques are implemented and are used for reducing the image complexity by applying background removal operations. The proposed system contains 3 steps.

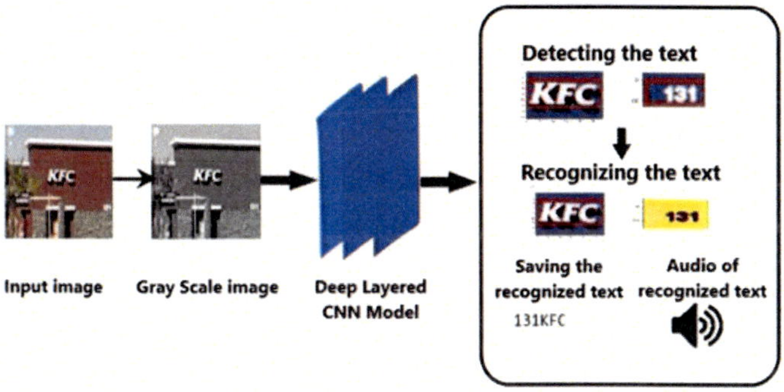

Fig. 4 Proposed model framework

Fig. 5 Background removal of input image

First step: In this stage, basic image processing techniques are implemented which are used for background removal of the complex images. Few techniques are binarizations of image, grayscale of image, etc. Using OpenCV text can be extracted from a few images directly after processing the above techniques. The image browsed by the user is converted into grayscale. Then using the thresholding technique [11], the unnecessary background is removed through which we can easily extract the text. This is the basic step where the complexity of the image is reduced and processed for extraction. Background elimination images are shown in Fig. 5 where the images are taken from online sources and processed.

Second step: This phase comprises training the model with Benchmark datasets. Deep layered CNN is used for image training. Through CNN the images are trained effectively. Images are converted to grayscale and then forwarded for training. Our model named text region recognition network is used for recognizing text.

Here is the algorithm used for calculating text area.

1. Image is taken for the process.
2. Apply image processing techniques
3. Consider dimension of the image.
4. Calculate all possible regions.
5. For r in len(regions)
    i. Calculate the coordinates.
    ii. Rectangle boxes are formed around the detected region.
    iii. Check the probability (p) of the region containing text by setting a threshold value.
    iv. If p > threshold value then
        Add to regions.
    v. Crop the image according to the region dimensions.
    vi. Display the cropped image.
    vii. else
        ignore that region.

Third step: Region of interest is calculated in this step. After training the model, the input shape and the probability are calculated from the model. Image processing operations such as removing the noise of the image, backdrop complexity is applied using Otsu binarization. After this implementation, the image is resized and probability is predicted from the processed image. Then the text regions are calculated by getting the shape of the text regions. In few cases, text may be recognized but not exactly drawn around the text region. To avoid this, few extensions are provided to the regions that act as buffer in the above cases. While predicting the text regions, few regions are overlapped. To overcome this, the overlapped regions are calculated and merged into one single box if there are multiple regions around the same text region then it forms as one single region. If the chances of the region containing text are low then those regions are ignored by adjusting the probability value. As our model focuses on the horizontal text detection and slightly aligned text, rectangle-shaped regions are formed around the text area which is displayed in Table 1.

The main role of a model lies in finding the text region. Once the text regions are formed, they are cropped. This cropped image is processed further for extraction using Tesseract. Tesseract is a conversion tool for converting simple or basic images into text. It supports only the plain background images and cannot give the result for complex images. After extraction, the text is stored in a file and converted into audio. Here the text may not be converted even though cropped because of the disturbance factors so once again the processing techniques are performed on the cropped image, text visibility increases and the extraction is successful from the image. Our model results are displayed in Table 2 and the output is compared with another model in Table 4.

Table 1 Text regions that are cropped for processing

Input image	Cropped image	Input image	Cropped image

	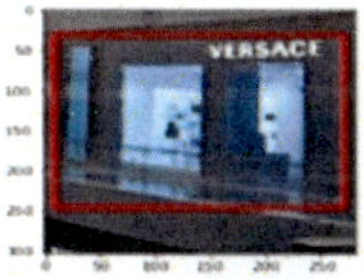 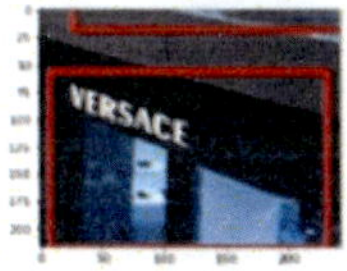		
	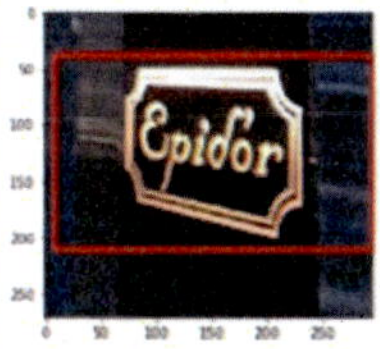		

4 Experiments and Results

The input model is trained with benchmark datasets such as Street view Text dataset (SVT) containing 350 images in which 200 are for training and the remaining are for testing. ICDAR 2003 [12] contains 1000 images containing cropped words. ICDAR 2015 contains images that are used for text localization, end-to-end text recognition. ICDAR 2013 [13] images are cropped words of the text present in the images which are useful to obtain accurate recognition. The total text includes high-resolution images and are captured in a natural scene. These are the datasets used for text recognition. To calculate the performance of the model there are few model evaluation metrics namely Precision (P), Recall (R), F-Score (F). These can be determined using a confusion matrix.

The overall model evaluation is defined using performance evaluation metrics. And also, our model is compared with other models which have obtained good results in text recognition. Goel [14] implemented using EAST model [7]. The authors have not implemented any post-processing techniques but have used output layers for calculating regions in image. This is the popular model that has obtained good

Table 2 Final results of the proposed model on benchmark datasets

Input image	Output
	simple image conversion ...... FLATS 61to69 text is saved Playing Audio
	 complex image text is .... mountains text is saved Playing Audio..
	 Except for access and buses text is saved Playing Audio..
	 GUESS text is saved Playing Audio..

results when compared to other models. 88% precision is obtained for their model. Our model has achieved 91.3% precision.

5 Limitations

In a few cases, there may be similarities between numerical and alphabets. When the model detects the text, few alphabets look like numbers or vice versa. Font styles of different numerals are considered as alphabets. This leads to wrong interpretation and recognizes the alphabet as numerical. Our model detects only the English language.

Table 3 Calculation of benchmark datasets

SVT				ICDAR 2003			
Model	**P**	**R**	**F**	**Model**	**P**	**R**	**F**
Gupta [15]	0.65	0.6	0.62	Wu [16]	0.74	0.63	0.68
Andrei [17]	0.47	0.63	0.54	Zhang [18]	0.74	0.64	0.68
Tian [9]	0.68	0.65	0.66	Li [19]	0.79	0.64	0.71
Tang [20]	–	0.76	–	Feng [21]	0.8	0.69	0.74
Zheng [22]	0.68	0.53	0.6	TAS [23]	0.82	0.66	0.73
(TRRN) Proposed Model	**0.84**	**0.76**	**0.8**	Jaderberg[24]	–	–	0.79
				TRRN	**0.874**	**0.731**	**0.796**
ICDAR 2013				**ICDAR 2015**			
Model	**P**	**R**	**F**	**Model**	**P**	**R**	**F**
Mitra [25]	0.77	0.65	0.7	Tian [9]	0.74	0.52	0.61
CASIA NLPR [13]	0.78	0.68	0.73	Shi [26]	0.731	0.768	0.75
Pan [8]	0.8	0.6	0.68	Long [27]	0.849	0.804	0.826
USTB Text Star [13]	0.885	0.664	0.759	Dan [28]	0.85	0.82	0.83
Wei [22]	0.88	0.74	0.8	Liao [29]	0.878	0.785	0.829
Text Spotter [13]	0.87	0.64	0.74	Dai [30]	0.886	0.8	0.841
TRRN	**0.89**	**0.78**	**0.83**	**TRRN**	**0.918**	**0.88**	**0.898**

As we have focused on the image background, if the text font is small our model can detect but cannot recognize the text.

6 Conclusion

This paper discusses parameters of complexity and the disturbance factors causing the text unable to extract. Our model is implemented for recognizing text, and then converting the text into audio. CNN model is used for text recognition. If the text is directly extracted using Tesseract from the image then that image is called a simple image. Though the OCR online tools are available, they are effective to some extent in processing the image complexity. Our model can recognize text from simple as well as complex images. It is an enhancement for existing systems to overcome their limitations up to some extent. Our main aim for the implementation is guiding the visually challenged people. They cannot see or read anything until and unless they use special mechanisms. For example, if a visually impaired person desires to read a building name. Here our model can be used in assisting the person by detecting the text from the captured image, recognizing the text, and storing the text. This stored text is converted into audio and the person can hear that text which is present in the captured image. This can be done with product labels, posters, etc. The proposed

Table 4 Comparisons of text recognition results

Input image	Other Model Results	Proposed Model Results
	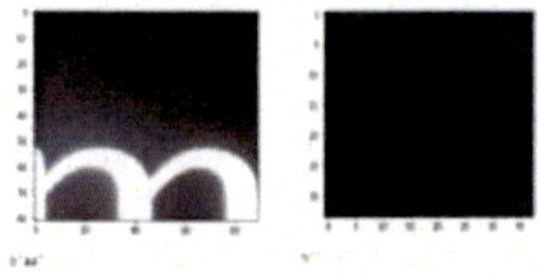	
	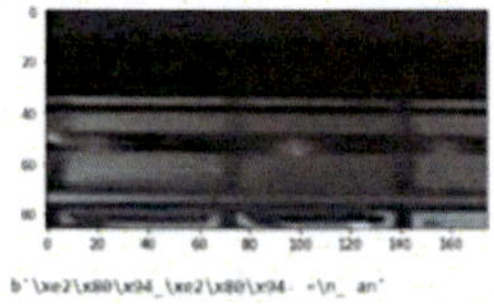	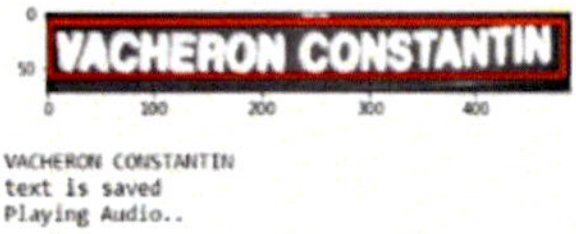
	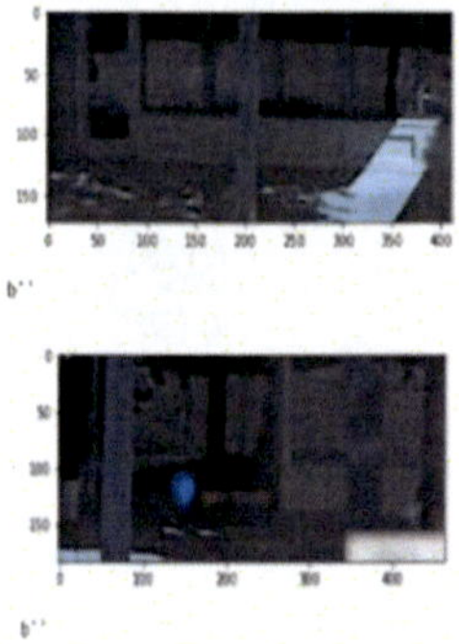	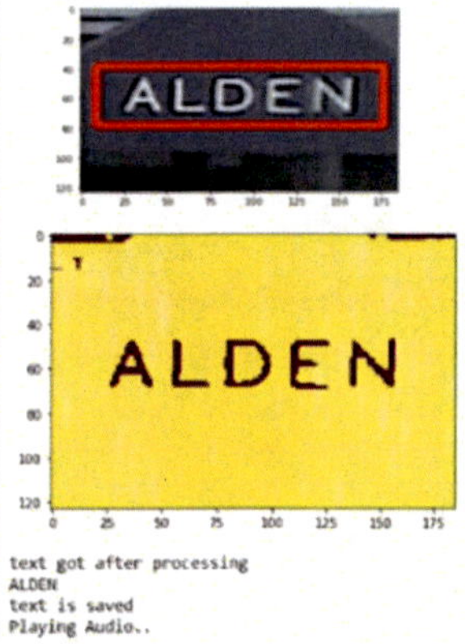

model can be extended further in analyzing traffic instructions, adding languages, etc.

References

1. Ghoshal, R., Roy, A.: A novel method for binarization of scene text images and its applications in text identification. Springer (2018)
2. Basavaraja, K.T.: Edge based segmentation approach to extract text from scene images. IEEE (2017)
3. Richard, S.Z.: A text detection system for natural scenes with convolution learning and cascaded classification. IEEE (2016)
4. Zhang, X., Gao, X.: Text detection in natural scene images based on color prior guided MSER. Elsevier (2018)
5. Turki, H., Halima, M.: Text detection based on MSER and CNN features. IEEE (2017)
6. Sun, L., Zhong, Z., Huo, Q.: Improved localization accuracy by Locnet for faster R-CNN based text detection in natural scene images. Elsevier (2019)

7. Sun, Y., Dawut, A.: A review: text detection in natural scene image. IEEE (2018)
8. He, P., Huang, W., He, T.: Single shot text detector with regional attention. IEEE (2017)
9. Tian, Z., Huang, W.: Detecting text in natural image with connectionist text proposal network. Springer (2016)
10. Lao, M., Shi, B., Bai, X.: Textboxes: a fast text detector with a single deep neural network. (2016)
11. Gnana Prakash, T., Anusha, K.: Text extraction from image using python. (2017)
12. Lucas, S.M., Panaretos, A., Sosa, L.: ICDAR robust reading competitions. IEEE (2003)
13. Karatzas, D., Shafait, F., Uchida, S.: ICDAR 2013 robust reading competition. IEEE (2013)
14. Goel, V., kumar, V.: Text extraction from natural scene images using open CV and CNN. (2019)
15. Veldaldi, A., Gupta, A., Zisserman, A.: Synthetic data for text localization in natural images. IEEE (2016)
16. Wu, H., Zou, B.: Scene text detection using adaptive color reduction, adjacent character model and hybrid verification strategy. Springer (2015)
17. Plozounov, A., Ablavatski, A.: Word fence: text detection in natural images with border awareness. IEEE (2016)
18. Zhang, G., Huang, K., Zhang, B.: A natural scene text extraction method based on the maximum stable extremal region and stroke width transform. (2017)
19. Li, Y., Jia, W., Shen, C.: Characterness: an indicator of text in the wild. IEEE Trans. (2014)
20. Tang, Y.: Scene text detection and segmentation based on cascaded convolutional neural networks. IEEE Trans. (2017)
21. Feng, Y., Song, Y., Zhang, Y.: Scene text detection based on multi-scale SWT and edge filtering. IEEE (2016)
22. Zhang, Z., Shen, W., Yao, C.: Symmetry based text line detection in natural scene images. IEEE (2015)
23. Soni, R., kumar, B.: Text detection and localization in natural scene images based on text awareness score. Springer (2019)
24. Jaderberg, M., Simoyan, K., Vedaldi, A., Zisserman, A.: Reading text in wild with convolutional neural networks. (2015)
25. Behzadi, M., Safabakhsh, R.: Text detection in natural scenes using fully convolutional dense nets. IEEE (2018)
26. Shi, B., Bai, X.: Detecting text in natural images by linking segments. IEEE (2017)
27. Long, S., Ruan, J.: Text snake: a flexible representation for detecting text of Arbitrary shapes. Springer (2018)
28. Deng, D., Liu, H., li, X.: Pixel link: detecting scene text via instance segmentation. (2018)
29. Liao, M., Shi, B.: Text boxes ++: a single shot-oriented Scene text detector. IEEE Trans. (2018)
30. Dai, Y., Huang, Z.: Fused text segmentation networks for multi oriented scene text detection. IEEE (2018)

Machine Learning Approach to Track Malnutrition in Children with Rural Background

S. Nagini, Sravani Nalluri, B. Rachana Reddy, and Andukuri Lekha

1 Introduction

Malnutrition is an important issue that affects kids worldwide. Given that, the social determinants of health are outlined as "the conditions within which folks are born, grow, live, work and age, as well as the health system". In India, almost 44% of the children under the age of 5 are underweight and prone to malnutrition. Malnutrition in children occurs due to some complex factors like poverty, maternal health illiteracy, home environment, dietary practices, hygiene practices, gender imbalances, poor sanitary and environmental conditions, improper maternal, baby and kid feeding, and care practices, limited access to quality education, health and social care services. Children of today are citizens of tomorrow, and hence improving nutritional status of children becomes extremely important. Early childhood constitutes the most crucial period of life, when the foundations are laid for cognitive, social and emotional, language, physical/motor development and cumulative lifelong learning. Malnutrition in Bharat conjointly persists attributable to the antique patterns of social and economic exclusion. Over four-hundredth of youngsters from regular Tribes and regular Castes are scrubby [1]. On the point of four-hundredth of youngsters from the opposite Backward categories are scrubby. A basic definition of deficiency disease is

S. Nagini (✉) · S. Nalluri · B. R. Reddy · A. Lekha
Department of Computer Science, VNR Vignana Jyothi Institute of Engineering and Technology, Hyderabad, India
e-mail: nagini_s@vnrvjiet.in

S. Nalluri
e-mail: sravani_n@vnrvjiet.in

B. R. Reddy
e-mail: bh.rachana11@gmail.com

A. Lekha
e-mail: lekhaandukuri2215@gmail.com

C. K. Mai et al. (eds.), *Machine Learning Technologies and Applications*,
Algorithms for Intelligent Systems,
https://doi.org/10.1007/978-981-33-4046-6_30

"lack of the minimum quantity of proteins, carbohydrates, lipids, vitamins, minerals, and different nutrients essential for health and correct growth".

An Android-based food recognition application that is used as a health awareness tool for non-healthy individuals. The application lets the user take the image of the food and show its nutrition contents [2]. A system using the concept of Data Mining for detecting malnutrition is proposed. It consists of a number of roles to study the data and predict malnutrition [3]. A Smart Log system using IOT is an automated nutrition tracking system that uses a food weighing sensor to quantify the nutrients consumed by the user [4]. Through the IoT, an automatic food monitoring system with predictions to deliver a balanced meal is proposed [5].

An application called img2 calories uses advanced image recognition technology to tell you the cost of food. It'll be able to establish any foods that anyone captures in photos and associate an amount of calories to every item [6]. A similar application offers a technique for dietary assessment towards automatically finding the sort of food from completely different photos captured throughout consumption occasions. This food recognition system may be simply integrated into dietary assessment applications [7].

A system model is developed via numerous sensors as well as cameras to come up with 3D pictures for food volume estimation [8–10]. Another project pitched and evaluated a unique fully automatic-linguistic segmentation way for pixel-level classification of food on a plate having a deep convolutional neural network (DCNN) [11]. A preliminary approach for the amount of estimation with depth data is used. Sparse coding is used in SIFT and local binary pattern feature descriptors, and these options combined with Gabor and color options are used to represent food ingredients [12].

In a system where the user's will login to the system via the web is used. Adults and caretakers are going to be the main users of the system. Caretakers are going to be the direct physicians that are concerned with child growth [13]. Another system planned a solution which could acknowledge the contents of a meal from a picture, and so predict its nutrition contents, like calories, that additionally collects the images offline and apply the classifier to predict what types of items are present in a meal and looks at the nutrition facts [14].

2 Existing System

Malnutrition is a serious issue in India, especially in rural areas. There were different applications and projects taken up to solve malnutrition in India. Out of which most of them require hardware equipment like watches and sensors to detect the person's health conditions. Devices like watches measure the heart rate, pulse rate, Blood Pressure, there are also sensors that gives the body mass index of a person. According to current analysis within the world, there's a mobile multi-agent-based system to watch e-health. Some of the applications have a nutritionist recommendation who looks at all the details of the patient and proposes a diet suitable for the patient [2, 9, 15, 16].

There are also applications, where the person has to upload a picture of the food item every time they have a meal, the application uses image processing to identify the ingredients present in the meal. In extension of that, they collect the 3D picture of the food to calculate the volume a child consumes [15, 17]. In Systems like MHSC, the food identification is done by integrating features such as texture and colour [8].

Another approach suggested is by linking open-source application Program Interfaces (APIs) through the barcode to get the nutrient values [7, 18, 19]. As mentioned above, all the applications have some hardware machines or nutritionists to track the nutrients a child takes where these are not possible in the rural area [20, 21]. They may not be having enough knowledge to operate the machines so these methods cannot be implemented in such areas.

3 Proposed System

Malnutrition Tracker is an application that predicts the possibility of malnutrition before it occurs. The prediction is based on the food intake of the child and the living conditions such as area, temperature, climatic conditions, etc. and income of their parents. This is done by comparing the nutrition values of the foods taken by the child to the minimum requirements of that nutrient based on age and gender. If the application detects that the child is not obtaining enough nutrients, it notifies that he could have malnutrition. Once it is detected, it recommends various foods that can be given to compensate for the nutrients that are lacking in the child. It also provides info on the nutritional values of those foods so they can be chosen based on availability.

The application provides guidance to the parents in maintaining the health of the child by regularly tracking their height and weight. This can be used as an initial step in reducing malnutrition as regular visits to the hospitals is not possible in a few areas. For this, we have visited a few government schools and collected the data of students and their diet plans. There is an additional attribute considered that is the income of parents who are not present in the existing systems. The reason it is considered because the government will be providing only mid-day meals but for the rest of the day they depend on their homes. If a parent is earning enough to provide proper food to their family, there would not be any malnutrition observed else it does affect their children's health. In existing systems, there is the involvement of hardware machines to detect malnutrition but it needs some expertise to use it which may not be expected at all places especially in rural areas. So we proposed this system which requires basic knowledge to enter the data of food intake which is an easy task.

4 Methodology

The data was collected by visiting 6 government schools in the area of Suraram, Hyderabad by observing the meal habits of the students studying there. It was nearly 20,000 samples from different schools there. 30% of which are used for testing and the other 70% for training. Initially, 10 case studies have been considered. Each case study deals with a person consuming food items like dairy products, vegetables, meat, and food with carbohydrates. These food items have been considered into each day's food intake for 15 days. For every food item, the minimum requirement of the nutrients of each item has been compared with the food taken by the person. Table 1 represents the format of sample data collected from government schools.

We present a system which recognizes the contents of the above-mentioned meals and then predict if the person has malnutrition. For the dataset, each food quantity **100 g was considered. The nutrients namely protein, carbohydrates, fat and fibre** have been observed in every 100 g of the food intake [22, 23] for 15 days and income plays a major role too so these are the attributes considered to run the algorithm and the class label used is YES/NO [24].

For example, if we consider 100 g of chicken contains 27 g of protein. It is tabulated for 15 days. So the total proteins would be 15 * 27 = 405 g. The algorithm implemented here using a decision tree (Fig. 3) calculates the nutrient content taken by the children. It is calculated separately for every nutrient according to its content in every food item [25, 26]. The main attributes considered while collecting the data are vegetables, rice, fruits meat, dairy products and pulses. Below displayed is the **flowchart (Fig. 1) to represent the work flow of the application.**

Decision Tree (Fig. 3) is the technique used for the prediction of accuracy. And Fig. 2 shows the results of the work carried out. Different algorithms are applied like SVM (as shown in Fig. 4) and Naïve Bayes on the same data sets so as to compare

Table 1 Sample data of children's intake in Government Schools

Name	Protein	Carbohydrates	Fat	Fiber	Malnutrition	Annual income
Child 1	1797.45	2064	697.5	367.5	No	150,000
Child 2	2208.48	2777.6	744	392	No	120,000
Child 3	1437.96	2083.2	558	294	No	100,000
Child 4	1318.13	1909.6	511.5	269.5	Yes	75,000
Child 5	1198.3	1736	465	245	Yes	80,000
Child 6	1078.47	1562.4	418.5	220.5	Yes	95,000
Child 7	958.64	1388.8	372	196	Yes	100,000
Child 8	838.81	1215.2	325.7	171.5	Yes	80,000
Child 9	1677.62	2430.4	651	343	No	115,000
Child 10	1557.79	2256.8	604.5	298.5	No	90,000
Min Req	1050	2550	600	270	–	–

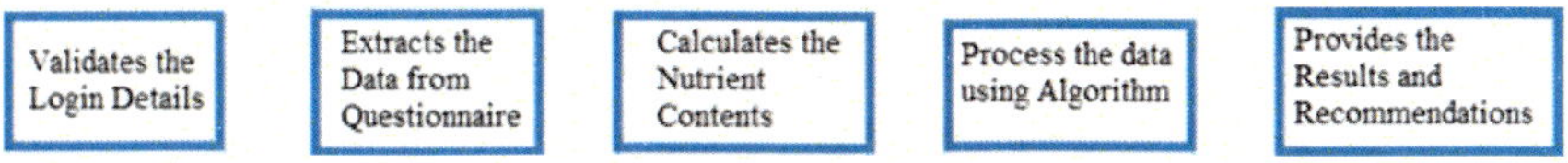

Fig. 1 Flowchart to represent workflow of the application

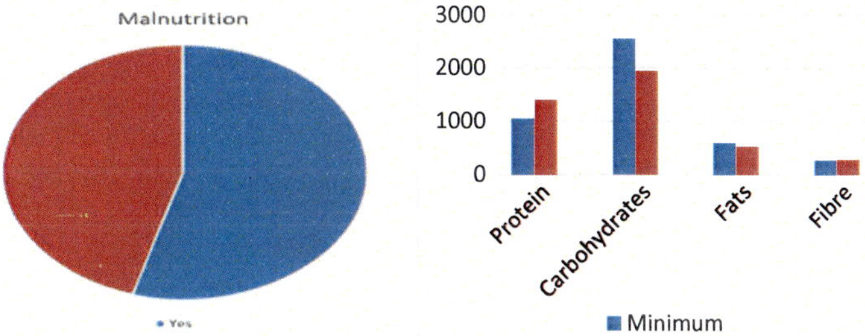

Fig. 2 Pictorial representation of the results observed for the dataset considered

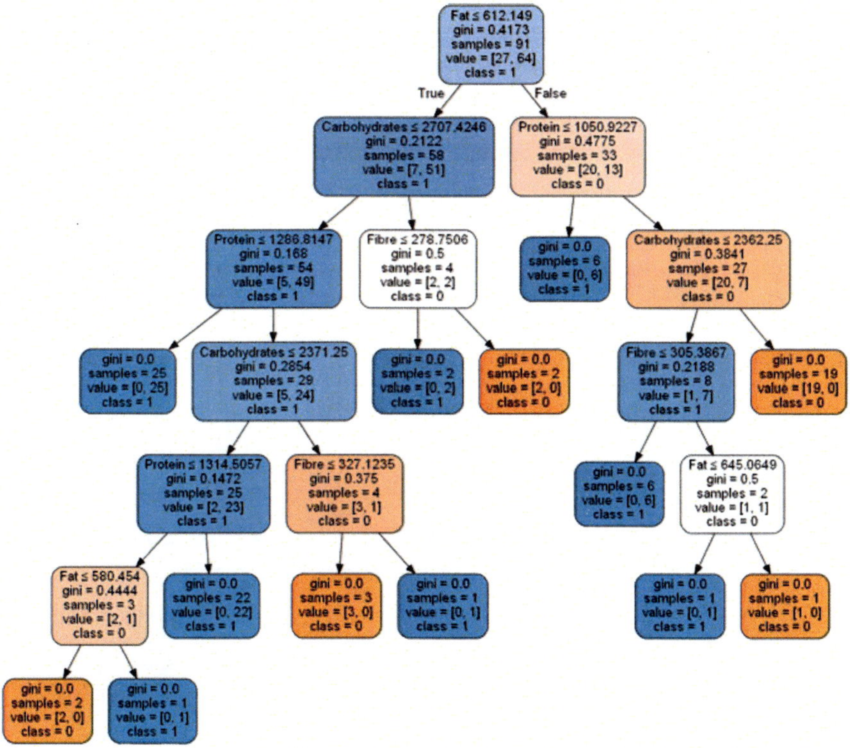

Fig. 3 Decision tree

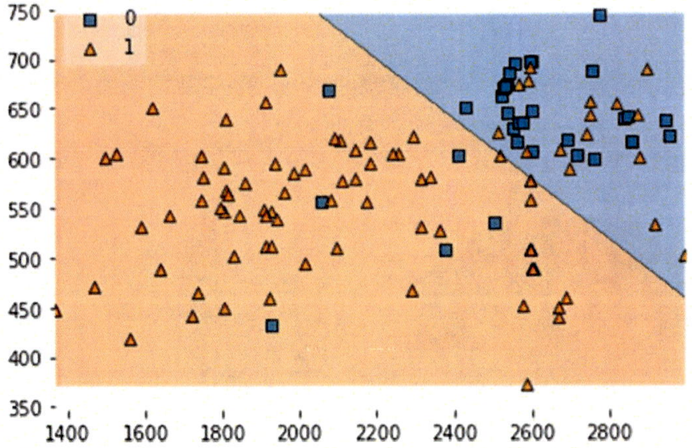

Fig. 4 SVM decision region boundary

the efficiency of accuracy that are derived. Table 2 and Fig. 5 shows the percentage of accuracies obtained when two different datasets are used.

Decision tree algorithm (Fig. 3) which uses tree-like graph structure implementation provides accurate results than the Naive Bayes algorithm which calculates the probability of happening and Support Vector Machine algorithm which uses hyper-planes to classify the result. Hence, Decision Tree algorithm is opted for more accurate results.

Table 2 Comparison of accuracies of different algorithms

Dataset	Decision trees (%)	Naive Bayes (%)	SVM (%)
Dataset 1	97	89	87.17
Dataset 2	94	80	83

Fig. 5 Data visualization comparing the datasets

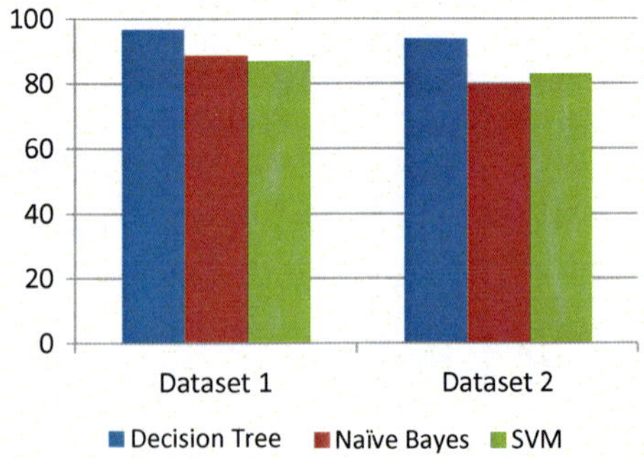

5 Conclusion

The proposed system focuses on the problem of malnutrition among school going children from a rural area in India. It provides effective services to address malnutrition and is targeted to the areas where the malnutrition is highest. As the system doesn't require any hardware equipment and requires a smart mobile phone which is available to everyone in the rural area nowadays it is easy to implement this procedure in rural areas. It educates the people thereby providing useful information about nutrition and also recommends ways approaches to prevent malnutrition. Definitely this process if implemented successfully ensures the next generations to be strong and healthy.

6 Future Scope

The existing systems need technical machines to detect malnutrition but the proposed system helps with a simple algorithm to detect it. The information will be stored and piled up according to the deficient nutrients and can be shared to the NGO's to provide rural children the necessary help to eliminate malnutrition. Government sector can collect the application outcome to make necessary changes in the mid-day meals being offered. Tie up with doctors to monitor the health condition of a child from time to time will also help in eliminating malnutrition.

References

1. Rao, V.S., Krishna, T.M.: A design of mobile health applications. Am. J. Eng. Res. (AJER)
2. Sun, L., Wang, Y., Greene, B., Xiao, Q., Jiao, C., Ji, M., Wu, Y.: Facilitators and barriers to using physical activity smartphone apps among Chinese patients with chronic diseases. BMC Medi.
3. Wang, Q., Egelandsdal, B., Almli, V.L., Oosdtindjer, M.: Diet and physical activity apps: perceived effectiveness by app users. JMIR mHealth uHealth **4**(2), 3–5 (2016)
4. Ambhore, K., Dawande, N.A.: Measuring calories and nutrition from food image. Int. J. Adv. Res. Comput. Commun. Eng. **5**(6) (2016)
5. Hasman, L.: An introduction to consumer health applications for the iphone. J. Consum. Health Internet
6. Online article: Presidents council on fitness, sports & nutrition. Eat healthy-why is it important?
7. World Health Organization Report (2011) mHealth: new horizons for health through mobile technologies: second global survey oneHealth
8. Kawano, Y., Yanai, K.: Food image recognition with deep convolutional features. ACM 978-1-4503-3/14/09
9. Reichert, L., Billings, C., Brown, J., Brown, P.: Evidence based guidelines on health promotion for older people
10. Duan, P., Wang, W., Zhang, W., Gong, F., Zhang, P., Rao, Y.: Food image recognition using pervasive cloud computing. In: IEEE International Conference on Green Computing and

Communications and IEEE Internet of Things and IEEE Cyber, Physical and Social Computing (2013)
11. Merler, M., Wu, H., Uceda-Sos, R., Nguyen, Q.B., Smith, J.R.: Snap, eat, repeat: a food recognition engine for dietary logging. MADiMa 2016, ACM. ISBN 978-1-4503-4520-0/6/10
12. Chang, T.: Food fight: a social diet network mobile application
13. Kitamura, K., Yamasaki, T., Aizawa, K.: Foodlog: capture, analysis and retrieval of personal food images via web. In: Proceedings of the ACM Multimedia 2009 Workshop on Multimedia for Cooking and Eating Activities, pp 23–30. ACM (2009)
14. Knez, S., Sain, L.: Food object recognition using a mobile device: state of the art
15. Maduabum, F.O.: Nutritional awareness of bank workers in Lagos State Nigeria
16. Lis, K., Reichert, M., Cosack, M., Billings, J.R., Brown, P.R.: Evidence-based guidelines on health promotion for older people
17. Probst, Y., Nguyen, D.T., Tran, M.K., Li, W.: Dietary assessment on a mobile phone using image processing and pattern recognition techniques: algorithm design and system prototyping
18. Zhu, F., Bosch, M., Woo, I., Kim, S.: The use of mobile devices in aiding dietary assessment and evaluation. IEEE J. Select. Topics Signal Process. **4**(4), 1–6 (2010)
19. Plessis, K.: Diet and nutrition: a literature review of factors influencing blue-collar apprentices. Nutrients (2015). ISSN 2072-6643
20. Story, M., Neumark-Sztainer, D., French S.: Individual and environment influences on adolescent eating behaviors. J. Am. Diet. Assoc.
21. Cecchini, M., Sassi, F., Lauer, J.A., Lee, Y., Guajardo, V., Chisholm, D.: Tackling of unhealthy diets, physical inactivity, and obesity: health effects and cost-effectiveness. Online medical article
22. Deshpande, S., Basil, M.D., Basil, D.: Factors influencing healthy eating habits among college students: an application of belief model
23. Robinson, E., Aveyard, P., Daley, A., Jolly, K., Lewis, A., Lycett, D., Higgs, S.: Eating attentively: a systematic review and meta-analysis of the effect of food intake memory and awareness on eating
24. Yang, S., Chen, M., Pomerleau, D., Sukthankar, R.: Food recognition using statistics of pairwise local features
25. Johnson, N.B., Hayes, L.D., Brown, K.: CDC National health report leading causes of morbidity and mortality and associated behavioral risk and protective factors
26. Hippocrate, A.A.E.: Food weight estimation based on image processing for dietary assessment. Master's thesis. Nara Institute of Science and Technology

Survey on Multimodal Emotion Recognition (MER) Systems

Bhanusree Yalamanchili, Keerthana Dungala, Keerthi Mandapati, Mahitha Pillodi, and Sumasree Reddy Vanga

1 Introduction

In today's world, the rapid growth of artificial intelligence has escalated the need for better and natural interaction between humans and machines. Emotion recognition systems can be deployed into diverse applications. The information obtained from these systems is being used in fields like health, education, tourism, commerce, etc. Unimodal systems cannot provide more information about the user and various exterior factors. This has led to the development of multimodal systems rather than unimodal systems. The ways to recognize the emotions of users are asking from a user, Voice recognition, Tracking implicit parameters, Facial expression recognition, Gesture recognition, Vital signals, and Hybrid methods. Physiological features such as respiratory volume, skin temperature, heart rate, respiration pattern, EDA, PPG, and EMG can also be used to determine emotion. For multimodal systems to give accurate results using different types of data, fusion techniques are used. Various fusion techniques which can be used are feature-level, hybrid multimodal, decision-level, rule-based, classification-based, model-level, and estimation-based fusions. The databases accumulated by researchers for these systems can contain only text, only audio, only image, only video, audio and video, physiological data or audio, text, and video data. The compilation of work done in the area of emotion identification is represented in this paper.

B. Yalamanchili (✉) · K. Dungala · K. Mandapati · M. Pillodi · S. R. Vanga
VNRVJIET, Hydrabad, India
e-mail: bhanusree6912@gmail.com

C. K. Mai et al. (eds.), *Machine Learning Technologies and Applications*,
Algorithms for Intelligent Systems,
https://doi.org/10.1007/978-981-33-4046-6_31

2 Survey Study

2.1 Modularity and Scope

Yoon et al. [1] identified emotion using both text and speech data as emotional dialogues are composed of both sound and spoken content, the data is assigned individually to the four emotion categories happy, sad, angry, and neutral. Tripathi and Beigi [2] did emotion recognition based on data from text, speech, motions which are captured from expressions on the face, rotation, and movements of hands, they considered only for 4 types of emotions sadness, anger, neutral, and excitement (happiness).

Sahu [3] emotion is recognized based on audio and text, he considers 6 types of emotions angry, happy, sad, fear, surprise, neutral. Majumdera et al. [4] obtained the audio, video, and text features and hierarchical fusion method is used, the emotions identified are anger, happiness, sadness, neutral, excitement, frustration, fear, surprise, and other. Lian et al. [5] designed a multimodal emotion recognition system which considers audio, video, and text information, the various emotion categories identified are happy, angry, fear, disgust, neutral, surprise and sad.

Shoumy et al. [6] reported a wide-ranging survey by merging the physiological modalities with other modalities. Arana et al. [7] have focused on the Moodies app, designed for identifying the speaker's emotions using voice. Ren et al. [8] recognized emotion using audio and video. In this paper, they considered 7 types of expressions from facial emotions they are happiness, surprise, sadness, fear, anger, neutral, and disgust. Imani et al. [9] did a survey on emotion recognition with emphasis on E-learning methods which includes speech, video, and text, they considered 6 types of emotions which are fear, happiness, anger, disgust, sadness, surprise in their survey on emotion recognition. Jiang et al. [10] emotional cognitive system is proposed which can examine and expect depression based on EEG data, voice data, expression data, and smartphone data. Happiness, disgust, sadness, fear, anger, surprise, and neutral class are labels of emotions recognized. The paper written by Lalitha et al. [11] focused on the study of emotion detection, it constitutes the emotions happy, angry, neutral, disgust, sad, fear, boredom, and anxiety. In Mustafa et al. [12], analyzed research in speech emotion recognition to find the current effort of research, and areas in which research is missing, the main objective was to study what is being done in this field of research. Cummins et al. [13] this is the first review into the automatic analysis of speech for use as an objective predictor of depression and suicidality, emotional symptoms such as sadness or hopelessness are identified.

Table 1 Datasets

Dataset	Audio	Video	Text
IEMOCAP	✓	✓	✓
CMU-MOSI	✓	✓	✓
AFEW	✓	✓	✓
Fernandez	✓	×	×
RECOLA	✓	✓	×
eNTERFACE	✓	✓	×
Berlin	✓	×	×
Ravdess	✓	×	×
Fer2013	×	✓	×

2.2 Datasets

IEMOCAP is a dataset which is used by Yoon et al. [1] used this dataset for analyzing emotion using audio, video, and text, the same dataset is used by Tripathi and Beigi [2] and Jiang et al. [10] used the same dataset.

CMU-MOSI is also a dataset used by [10], and they achieved different efficiency with it, Majumdera et al. [4] used it for emotion recognition using audio, video, text (Table 1).

AFEW dataset is used by Lian et al. [5] for audio, video, and text emotion recognition, whereas the same dataset is used by Ren et al. [8] for only audio emotion recognition and in [10] again they used this dataset for audio and visual emotion recognition.

Fernandez is another dataset that is used by Arana et al. [7] for emotion recognition using voice.

Berlin dataset is used by Lalitha et al. [11] for emotion recognition using audio.

2.3 Features

Yoon et al. [1] identified MFCC, prosodic, transcripts features that hold sequential audio information, statistical audio data, and textual data, respectively. These kinds of information are the same as those used in the TRE and ARE cases. MFCC feature which is used by them contains a total of 39 features. Tripathi and Beigi [2] used Energy-based features mel-frequency cepstral coefficients(MFCC) of entire 34 features and Fourier frequencies. These include 13 MFCC, 8-time spectral features like zero-crossing rate, short-term energy, 13 chromatogram based. Sahu [3] used audio features such as pitch, harmonics, speech energy, pause central moments are used. Text features like term frequency, inverse document frequency. Majumdera et al. [4] extracted utterance level features for three modalities. A deep convolutional neural network was functional on each utterance to extract textual features. OpenSMILE is used which unconsciously extracts pitch and voice intensity. 3D-CNN is used to extract structures from video frames and to model temporal features across frames.

Lian et al. [5] used OpenSMILE toolkit to extract audio features, ASR bottleneck features are extracted from two models: the English ASR acoustic model and the Chinese ASR acoustic model, SoundNet Bottleneck features are extracted from SoundNet network and VGGish network is used to extract VGGish Bottleneck features. Multiple video features are extracted such as handcraft video features, CNN bottleneck features, C3D features, background features, and identity features. Term Frequency-Inverse Document Frequency and Word Vectors are used to extract features from raw texts. Shoumy et al. [6] discussed feature extraction using apache spark and nose extraction methods called divide-and-conquer linear discriminant analysis (Div-ConLDA) and the divide-and-conquer principal component analysis (Div-ConPCA). Jose et al. [7] focussed on the key features or workings of speech for detecting emotions single out pitch or essential frequency, time, and voice quality.

Ren et al. [8] The method employed by them for mining audio and video types are 2D-CNN and 3D-CNN respectively. Imani et al. [9] From speech signal audio feature extraction takes place, MFCC features are applied on the ASR model which derives linguistic features, video features are extracted from video signal which is applied on LSTM model for emotion prediction. Jiang et al. [10] discussed different feature extraction techniques for different modalities such as EEG features, visual features, audio features, and text features. Lalitha et al. [11] considered the perception features like mel-frequency cepstral coefficients (MFCC's), bark frequency cepstral coefficients (BFCC), perceptual linear predictive cepstrum (PLPC), revised perceptual linear prediction coefficient's (RPLP), mel-frequency perceptual linear prediction cepstrum (MFPLPC), and inverted mel-frequency cepstral coefficients (IMFCC). Earlier, discoverers in the arena of speech dispensation have achieved numerous speech features like excitation source features, vocal tract parameters, prosodic features, and fusion features for the task. Mustafa et al. [12] extracted features from the speech are of two types, low-level-descriptors (LLDs) and useful features. LEDs consist of prosodic features, and spectral features and their spinoffs like pitch (F0), formants, energy, mel-frequency cepstral coefficients (MFCCs) and linear prediction cepstral coefficients (LPCC). Functionals consist of arithmetical results such as mean, average, and mode, and the zero-crossing rate. Nicholas et al. [13] brush up that prosodic features signify the long-time (phoneme level) differences in alleged rhythm, stress, and modulation of speech. Three prevalent samples consist of the speaking rate, the pitch (auditory perception of tone) and volume, and energy dynamics (Table 2).

Table 2 Features

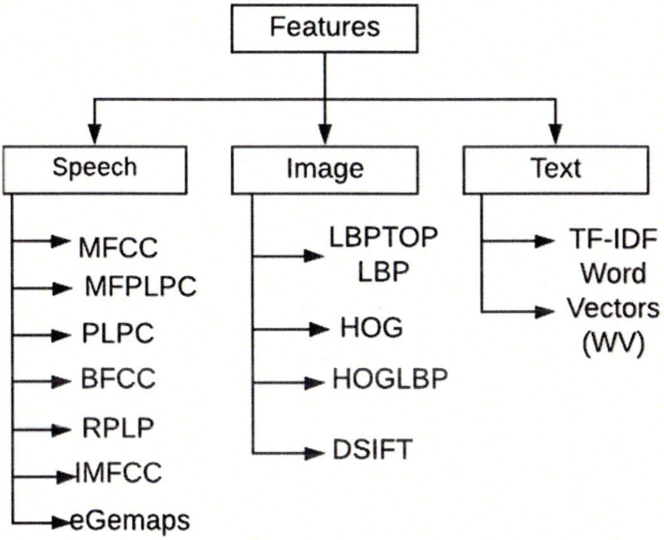

2.4 Models

Yoon et al. [1] proposed a model that encodes the records from audio and text series using dual RNNs and then pools the info after these sources to predict the class in which this emotion belongs to. This model analyses the speech files as of the signal level to the semantic level, and it thus uses these statistics in the files more broadly than mockups that concentrate on audio features. Wide tests are directed to examine the efficiency and possessions of the projected model and achieved an accuracy of about 68.8–71.8% using IEMOCAP. Tripathi and Beigi [2] uses practices that used HMMs (Hidden Markov Models), SVMs (Support Vector Machines) and a few other low learning methods and Sahu [3] used two approaches, in the first method they trained traditional machine learning classifiers, namely, SVM's, naïve Bayes, random forests and logistic regression. In the second approach, they built a multi-layer perceptron and an LSTM classifier.

Hierarchical fusion is used by Majumdera et al. [4] to improve the multimodal fusion mechanism. The proposed fusion strategy functions in a hierarchical manner by fusing the modalities two at a time first text + video, text + audio, audio + video), and then fusing all three modalities (audio, video, and text). BS-Fusion is used by Tao et al. [5]. In this approach, features from different modalities are trained discretely based on different classifiers. AFEW dataset for audio, video, and text emotion recognition and achieved 60.34% accuracy. [6] Feature-level fusion, conclusion level fusion, crossbreed multimodal fusion, model-level fusion, rule-based fusion, cataloging-based fusion, and estimation-based fusion techniques are discussed. Arana et al. [7] detected not only a simple emotional attitude in the voice

Table 3 Models

Models	Datasets	Data	Accuracy (%)
RNN	IEMOCAP	Audio, video, text	71.8
CNN	CMU-MOSI	Audio, video, text	49.17
CNN + LSTM	AFEW	Audio, video, text	60.34
RNN	RECOLA	Audio, video	76
CNN + LSTM	eNTERFACE	Audio, video	90.85
CNN + RNN	Berlin	Audio	88.9
RNN	Ravdess	Audio	89.02
CNN	Fer2013	Video	82.19

but also the better nuances. They used Fernandez for emotion recognition using voice and achieved 50% efficiency (Table 3).

Ren et al. [8] utilized a 2D CNN model for speech signal feature extraction, and for video signal extraction they used 3D CNN and LSTM. Imani et al. [9] studied models which use ASR and LSTM models in their survey Jiang et al. [10] comprehensively explained the data-driven multimodal emotion information fusion and examine the hitches and forthcoming research direction in that field. They used the CMU-MOSI dataset and they achieved an efficiency of about 49.17%. Berlin dataset is used by Lalitha et al. [11] for emotion recognition using audio and achieved 88.9% accuracy. In [13] It is understood that an alike intellectual process causes utmost suicide tries. Philosophies try to clarify this process comprise the Interpersonal Theory, the Cry of Pain Model, and the Stress-Diathesis Model). Mutual refrains amid these prototypes include stressful life events, mental health issues, feelings of hopelessness and impulsivity, and violence.

3 Results

IEMOCAP dataset used in [1] and [2] gives similar accuracy in both papers, 71.8% and 71.04%, respectively. The ARE model used in [1] for audio gives the base performance as minimal features of audio are used. The TRE model gives high performance compared to the previous ARE model. Next, the MDRE model gives a similar performance growth. Though the MDREA model is introduced which even deals with the attention parameter, it does not give more efficient results than MDRE. The reason for this is insufficient data available. Tripathi and Beigi [2] considered only four emotions which are anger, excitement, sadness, and neutral. In [3], the audio-only model used reveals that detecting "neural" or distinguishing between "angry", "happy" and sad is the most difficult for the model. In the text-only model, "sad" is the toughest. By joining text and audio features gives us a boost of ~14% for all metrics. The main objective of this paper was to prove that lighter machine learning models can achieve accuracy close to deep learning models. Majumder et al. [4] presented a fusion strategy that outperforms the early fusion method, which is widely used. Contextual hierarchical fusion performs 1–2% better than hierarchical fusion. The quality of text unimodal features was less compared to audio and video features and only four distinct emotions were considered. Lian et al. [5] used BS-Fusion in which a subgroup of emotions is selected based on the classification performance on the dataset. The accuracy achieved is 60.34%. This approach had high performance in anger, happy, and neutral emotions but disgust and surprise emotions were confused with others. Shoumy et al. [6] presented the summary of multimodal emotion recognition models and the future scope of computer-human interaction. A large number of papers are reviewed including researches done using big data analysis frameworks. In [7], soundtracks in four different languages- English, Spanish, Italian, and French were used, and it was observed that the emotion remained almost unchanged when the language was changed. The average performance of Positive emotions was 58.48% of the total and Negative emotions, obtaining the figure of 89.47%. In [8], only audio model achieved 48.74% accuracy when 2D-CNN(TL + CL) is used, the only visual model achieved 58.62% accuracy when 3D-CNN(LSTM + TL + CL) is used, when a Fusion of audio and video is done, they fused the two different modal features by defining the weight based on statistic method to generate more robust feature for the classification task. In [9] no experiment is done, only a survey on different approaches is done. In [10], the paper summarizes and reviews the main technologies in the domain of multimodal data fusion for emotion recognition. In [11], average performance achieved in the categorial dimension when their system is applied is 88.9%, preprocessing of sample data is done for better efficiency, DNN with three hidden layers resulted in better performance in classifying seven emotions of Berlin speech emotion. In [12] clarifying the issues and enhancing performance was the objective. In [13] prediction of depression and suicidality are done, the importance of using LSTM is discussed.

4 Conclusion

In recent years, emotion recognition has become an active research area. In this paper, we have discussed different approaches done by various people in this field of emotion recognition using audio, video, and text. We have discussed what are the different datasets used by different people, what features can be extracted, and which models can be applied for efficient accuracy.

So after reading all the papers, we came to the conclusion that we could achieve an accuracy of about 0.89 when RNN model is applied on Ravdess dataset for speech emotion recognition and about 0.82 accuracies can be achieved when CNN model is applied on Fer2013 dataset which are maximum accuracies achieved among all. It is also observed that when fusion is done the efficiencies achieved are far less than efficiencies achieved when experiments are performed on individual data.

References

1. Yoon, S., Byun, S., Jung, K.: Multimodal speech emotion recognition using audio and text. Accepted as a conference paper at IEEE SLT 2018
2. Tripathi, S., Beigi, H.: Multi-modal emotion recognition on IEMOCAP with neural networks. Submitted on (arXiv:1804.05788) 16 Apr 2018
3. Sahu, G.: Multimodal speech emotion recognition and ambiguity resolution. arXiv:1904.06022, 12 Apr 2019
4. Majumdera, N., Hazarikab, D., Gelbukha, A., Cambriac, E., Poriac, S.: Multimodal sentiment analysis using hierarchical fusion with context modeling. Elsevier, 28 July 2018
5. **Lian, Z., Li, Y., Tao, V., Huang, J.: Investigation of multimodal features, classifiers and fusion** methods for emotion recognition (2018)
6. Shoumy, N.J., Ang, L.M., Seng, K.P., Rahaman, D.M.M., Zia, T.: Multimodal big data affective analytics: a comprehensive survey using text, audio, visual and physiological signals
7. **Arana, J.M., Gordillo, F., Darias, J., Mestas, L.: Analysis of the efficacy and reliability of the** moodies app for detecting emotions through speech: does it actually work?
8. Ren, M., Nie, W., Liu, A., Su, Y.: Multi-modal correlated network for emotion recognition in speech. Vis. Informa. **3**(3) (2019)
9. Imani, M., Montazer, G.A.: A survey of emotion recognition methods with emphasis on e-learning environments. J. Netw. Comput. Appl. (2019)
10. Jiang, Y., Li, W., Hossain, M.S., Chen, M., Alelaiwi, A., Al-Hammadi, M.: A snapshot research and implementation of multimodal information fusion for data-driven emotion recognition. Elsevier, 9 June 2019
11. Lalitha, S., Tripathi, S., Gupta, D.: Enhanced speech emotion detection using deep neural networks. Int. J. Speech Technol. (2019)
12. Mustafa, M.B., Yusoof, M.A.M., Don, Z.M., Malekzadeh, M.: Speech emotion recognition research: an analysis of research focus. Int. J. Speech Technol. (2018)
13. Cummins, N., Scherer, S., Krajewski, J., Schnieder, S., Epps, J., Quatieri, T.F.: A review of depression and suicide risk assessment using speech analysis. Elsevier, 30 March 2015

Cancer Classification Using Mutual Information and Regularized RBF-SVM

Nimrita Koul and Sunilkumar S. Manvi

1 Introduction

1.1 DNA Microarray Technology

The normal function of the genetic material called DNA inside a cell is vital to normal health of the individual. Any kind of abnormal mutation in the working or structure of a gene can lead to various diseases and abnormalities like cancers [1]. If we can identify these mutations in the genes, we can diagnose diseases at very early stages and even predict them. It has been observed that mutations in the gene BRCA1 related to human breast cancers are 800 in number and this gene is responsible for around 60% of breast cancers. The mutations can be of different types and each has a different biological marker. DNA microarray technology [1] is a method of identifying mutations in DNA by looking at expression values of genes comprising the DNA. A DNA chip is a glass slide like a computer chip. On the slide surface, are placed thousands of single strands of normal DNA composing the genome of an organism, and also on the slide are placed the strands of the genes suspected of having a mutation of the same genome. In this paper, we have presented an approach to distinguish between the four types of SRBCT [2] tumors using the steps of feature selection and classification.

N. Koul (✉) · S. S. Manvi
REVA University, Bangalore 560064, India
e-mail: nimritakoul@reva.edu.in

C. K. Mai et al. (eds.), *Machine Learning Technologies and Applications*,
Algorithms for Intelligent Systems,
https://doi.org/10.1007/978-981-33-4046-6_32

1.2 Feature Selection

The classification and clustering tasks of machine learning [3] benefit in terms of computational resources requirement and model accuracy if the input dataset contains a minimal number of relevant features only. Feature selection, therefore, is the process of selecting the subset of most optimally relevant genes from the original high-dimensional dataset. The feature selection approaches are classified as univariate and multivariate depending on whether each feature is considered individually for evaluation of relevance to the target variable or in groups. Information-theoretic methods like signal-to-noise ratio, threshold of miss-classification, correlation coefficient, various distance metrics, information gain, mutual information, t-statistic, Wilcoxon statistic are all examples of univariate methods [4]. In univariate methods, an input data set with ‘p’ number of genes needs to evaluate ‘p’ sets by combinatorial search over the space of all possible subsets of features. Multivariate methods, consider groups of features and search through the solution space using techniques like recursive feature selection, genetic algorithms, etc. Another way of categorization of feature selection algorithms is a filter, wrapper and embedded methods.

2 Literature Survey

In this section, we have given a survey of related papers that deal with the similar problem of classification of gene expression datasets. Since the proposed method is a filter method, we have focused on papers that also use filter methods and work on gene expression data related to cancers. The authors of [1] were the first to show that gene expression data can be used for successful classification of cancers. They applied information theory concepts to distinguish two subtypes of Leukemia into ALL and AML. The authors in [2] used artificial neural networks for the classification of SRBCT into four classes using the same dataset as the focus of this paper. In [3], researchers have presented the results of application of an information theory method called partial least squares regression as a feature selector for biological data for the problem of multiclass classification. In [5], the authors applied artificial neural networks and a method called minimum redundancy maximum relevance for feature selection and classification from biological datasets. The authors in [6], applied a hybrid method called bee colony optimization as feature selector from cancer datasets of gene expression. Bee colony optimization is a bioinspired evolutionary algorithm for feature selection. The authors in [7], employed a correlation-based method for feature selection. The distributed correlation approach was used to identify the most relevant features from the gene expression dataset. The paper [8] covers a survey of prominent feature selection methods as applied to cancer gene expression datasets for the purpose of classification. In [9], researchers have applied logistic regression as feature selector algorithm to the expression data set. The authors in [10] have applied in a pipeline genetic algorithm and mutual information to gene expression data set for feature selection. The authors in [11], have used kernel support vector machine as feature selector for cancer diagnosis from gene. In [12], the authors used an ensemble and hybrid technique consisting of cellular automata with ant colony optimization on gene expression datasets for feature selection. In [13], the authors created groups of the gene expression dataset to divide it into folds. From each fold, best genes were selected and then combines to find the best overall genes. In [14], the authors employed a combination as a hybrid of recursive feature elimination and support vector machine for feature selection and classification from gene expression data. In [15], authors have used recursive feature elimination with SVM as evaluator for gene selection from gene expressiond data with very good results and very few genes. In [4], the authors have presented an excellent comparison of the univariate and multivariate methods of feature selection.

3 Proposed Method

In this paper, we have proposed a method for the classification of cancer gene expression datasets using a filter information-theoretic method called mutual information and radial basis kernel support vector machine with regularization. The method was tested on SRBCT [2] dataset and the results are presented in the next section. The algorithm for the proposed method is given below:

Algorithm 1: SRBCT_Classification
Input: Gene Expression Data Matrix, M
Output: Set of Labels, L, for each row in test data
Step 1: Impute the missing values in M using mean values
Step 2: Standardize the data matrix to mean 0 and standard deviation 1
Step 3: Compute mutual information between each pair of genes in M
Step 4: Sort the gene pairs in decreasing order of their mutual information values
Step 5: For each gene subset 'S' with sizes in (5, 10, 20, 30, 40, 50):
  a. Use the subset S to train regularized RBFSVM classifier
  b. Record the corresponding classification accuracy in list –A
Step 6: Select the subset 'F' corresponding to maximum value in A
Step 7: Return F as subset of genes with maximum classification accuracy

Figure 1 shows the schematic representation of the proposed method.

4 Results and Discussion

The proposed method was tested on SRBCT [2] cancer gene expression data set. Using mutual information as the metric we identified 5, 10, 20, 30, 40 and 50 most relevant genes from the original dataset containing 2308 genes. The implementation

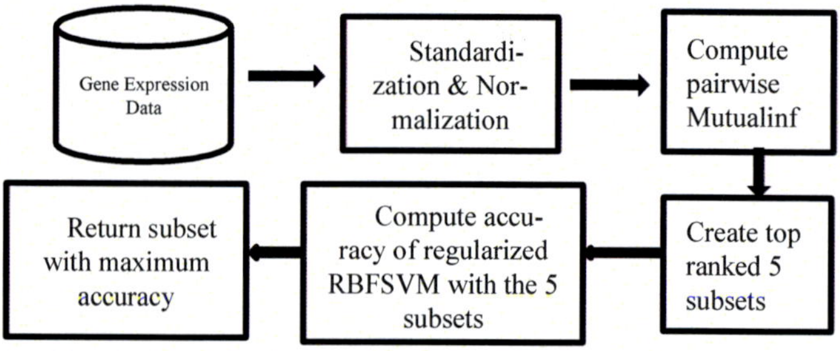

Fig. 1 Schematic of the proposed method

was done in Python on a Windows 10 machine. The results about classification accuracy with full feature set are compared with the classification accuracy with selected subsets of genes and the results show that 20 selected genes give almost 100% accuracy.

Figures 2 and 3 present the comparison of classification score and training times proposed RBF-SVM with gamma = 0.001 and Naïve Bayes classifier with the full feature set of all 2308 genes and with reduced feature subset of 20 genes only.

Figures 4 and 5 show the confusion matrix of classification with a full feature set and with 20 selected genes. From the figures, it is clear that using 20 selected features we get an acceptable classification accuracy for SRBCT data.

Figure 6 shows the values of precision, recall, F1-score and support for each of the 4 classes in the SRBCT dataset with 20 selected genes.

Figures 7 and 8 show the comparison of learning curves for the full feature set and 20 selected genes with two classifiers Naïve Bayes and the proposed RBF-SVM with a regularization parameter of 0.01.

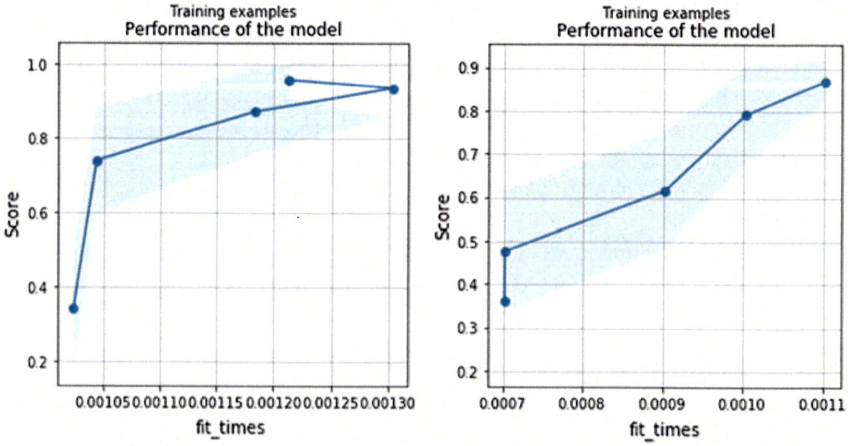

Fig. 2 Performance comparison between the proposed method and Naïve Bayes for all features

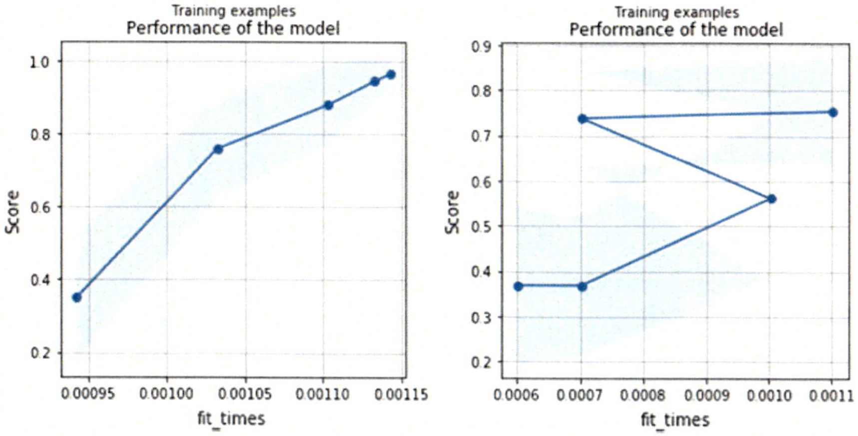

Fig. 3 Performance comparison when 20 selected features are used for training

Fig. 4 Confusion matrix with full feature set

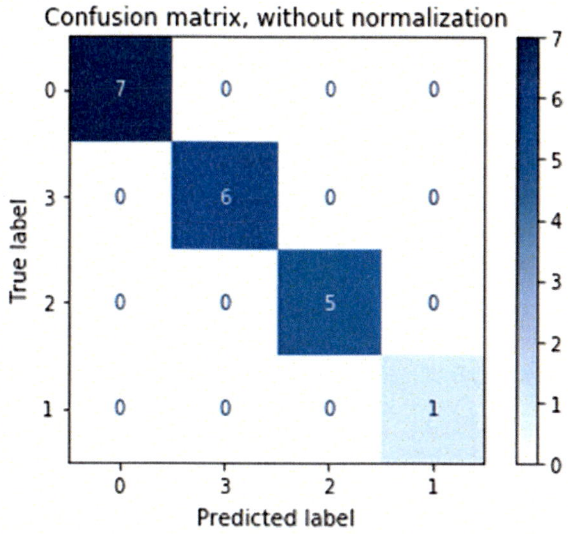

5 Conclusion

Filter approaches to feature selection from gene expression datasets have been demonstrated to perform better as compared to the wrapper and embedded approaches. This is primarily because of the high computational demands of the later classes of methods. The proposed method is a filter approach that selects a feature based on the unique information it brings to the data with respect to the target variable value. We have used various configurations of support vector machine with radial basis function kernel and various values of gamma and C to identify optimal values

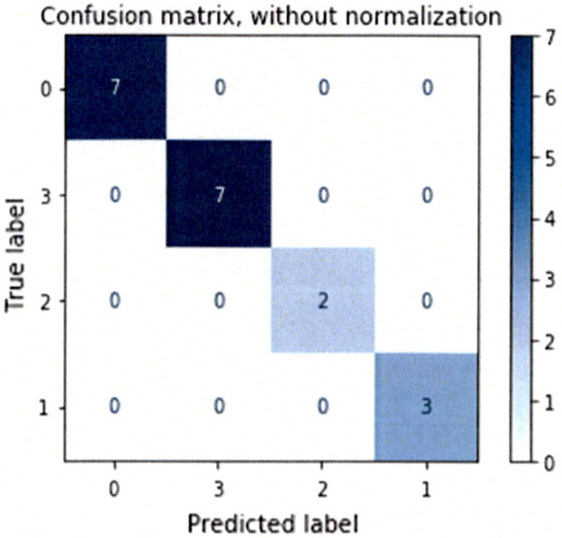

Fig. 5 Confusion matrix with only 20 selected features

	precision	recall	f1-score	support
0	1.00	1.00	1.00	8
1	1.00	1.00	1.00	5
2	1.00	1.00	1.00	4
3	1.00	1.00	1.00	2
accuracy			1.00	19
macro avg	1.00	1.00	1.00	19
weighted avg	1.00	1.00	1.00	19

Fig. 6 Values of precision, recall, and F-score for 20 selected features

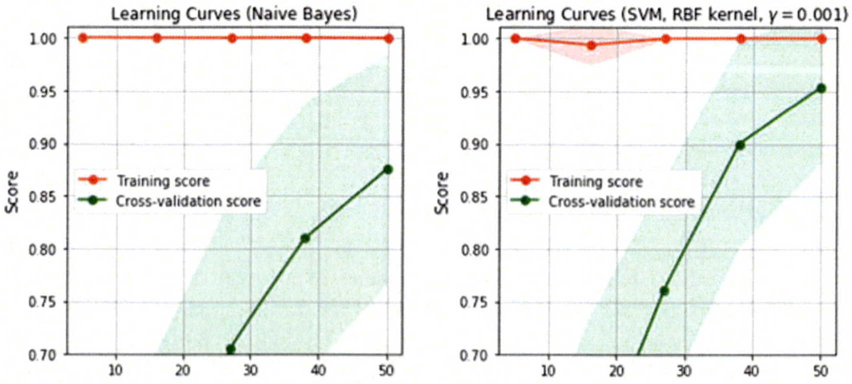

Fig. 7 Comparison of Learning curves of SVM RBF and Naïve Bayes for the full feature set at gamma = 0.001

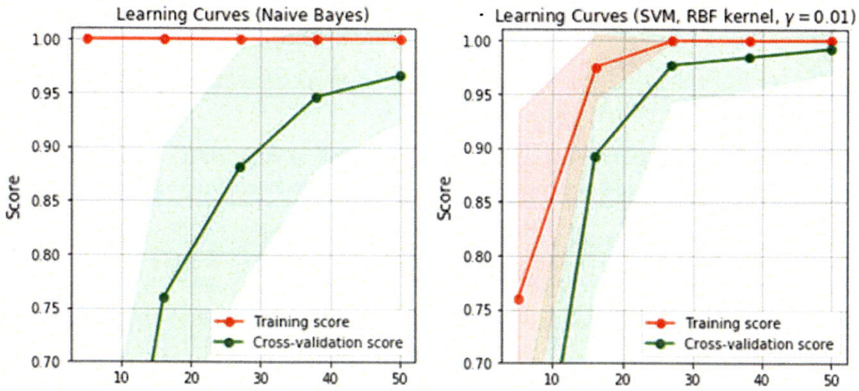

Fig. 8 Comparison of learning curves at 20 selected genes at gamma = 0.01

of these regularization parameters in order to obtain the best classification performance of the SRBCT dataset. The results demonstrate that the proposed method works very well and the classification accuracy of 100% is obtained with a small subject of features of size 30. In future, the proposed method is to be applied on other gene expression datasets and other multi-modal datasets as well.

Acknowledgements The authors would like to thank the Department of Science and Technology (DST), Government of India, for financially supporting this work under the scheme DST-ICPS 2019.

References

1. Golub, G.T.R., Slonim, D.K., Tamayo, P., Gaasenbeek, M., Huard, C., Mesirov, J.P., Coller, H., Loh, M., Downing, J.R., Caligiuri, M.A., Bloomfield, C.D., Lander, E.S.: Molecular classification of cancer: class discovery and class prediction by gene expression monitoring. Science 531–537 (1999)
2. Khan, Javed., Wei, Jun S., Ringnér, Markus., Saal, Lao H., Ladanyi, Marc., Westermann, Frank., Berthold, Frank., Schwab, Manfred., Antonescu, Cristina R., Peterson, Carsten, Meltzer, Paul S.: Classification and diagnostic prediction of cancers using gene expression profiling and artificial neural networks. Nat. Med. **7**(6), 673–679 (2001)
3. Guyon, I., Weston, J., Barnhill, S., Vapnik, V.: Gene selection for cancer classification using support vector machines. IEEE Trans. Knowl. Data Eng. **25**(1), 1–14 (2002)
4. Lai, C., Reinders, M.J.T., van't Veer, L.J., Wessels, L.F.A.: A comparison of univariate and multivariate gene selection techniques for classification of cancer datasets. BMC Bioinf. **7**(235) (2006)
5. Lê Cao, K., Boitard, S., Besse, P.: Sparse PLS discriminant analysis: biologically relevant feature selection and graphical displays for multiclass problems. BMC Bioinf. **12**(253) (2011). https://doi.org/10.1186/1471-2105-12-253
6. Rana, M.M., Ahmed, K.: Feature selection and biomedical signal classification using minimum redundancy maximum relevance and artificial neural network. In: Proceedings of International

Joint Conference on Computational Intelligence. Algorithms for Intelligent Systems, Springer, Singapore (2020)
7. Nancy, S.G., Saranya, K., Rajasekar, S.: Neuro-Fuzzy ant bee colony based feature selection for cancer classification. Springer Innovations in Communication and Computing, Springer, Cham (2020)
8. Shukla, A.K., Tripathi, D.: Detecting biomarkers from microarray data using distributed correlation based gene selection. Genes Genomic. (2020). https://doi.org/10.1007/s13258-020-00916-w
9. Almugren, Nada, Alshamlana, Hala: Survey on hybrid feature selection methods in microarray gene expression data for cancer classification. IEEE Access **7**, 75833–75844 (2019). https://doi.org/10.1109/ACCESS.2019.2922987
10. Algamal, Z.Y., Lee, M.H.: A two-stage sparse logistic regression for optimal gene selection in highdimensional microarray data classification. Adv. Data Anal. Classif. **13**, 753–771 (2019). https://doi.org/10.1007/s11634-018-0334-1
11. Jansi Rani, M., Devaraj, D.: Two-stage hybrid gene selection using mutual information and genetic algorithm for cancer data classification. J. Med. Syst. **43**(235) (2019). https://doi.org/10.1007/s10916-019-1372-8
12. Medjahed, S.A., Saadi, T.A., Benyettou, A., Ouali, M.: Kernel-based learning and feature selection analysis for cancer diagnosis. Appl. Soft Comput. **51**, 39–48 (2017)
13. Sharbaf, F.V., Mosafer, S., Moattar, M.H.: A hybrid gene selection approach for microarray data classification using cellular learning automata and ant colony optimization. Genomics **107**, 231–238 (2016)
14. Miyano, S., Imoto, S., Sharma, A.: A top-r feature selection algorithm for microarray gene expression data. IEEE/ACM Trans. Comp. Biol. Bioinf. **9**(3), 754–764 (2012)
15. Mundra, P., Rajapakse, J.: SVM-RFE with mRMR filter for gene selection. IEEE Trans. Nano. Biosci. **9**(2) (2016)

Author Index

C. K. Mai et al. (eds.), *Machine Learning Technologies and Applications*, Algorithms for Intelligent Systems,
https://doi.org/10.1007/978-981-33-4046-6